World Sustainability Series

Series Editor

Walter Leal Filho, European School of Sustainability Science and Research, Research and Transfer Centre "Sustainable Development and Climate Change Management", Hamburg University of Applied Sciences, Hamburg, Germany

Due to its scope and nature, sustainable development is a matter which is very interdisciplinary, and draws from knowledge and inputs from the social sciences and environmental sciences on the one hand, but also from physical sciences and arts on the other. As such, there is a perceived need to foster integrative approaches, whereby the combination of inputs from various fields may contribute to a better understanding of what sustainability is, and means to people. But despite the need for and the relevance of integrative approaches towards sustainable development, there is a paucity of literature which address matters related to sustainability in an integrated way.

Notes on the quality assurance and peer review of this publication

Prior to publication, the works published in this book are initially assessed and reviewed by an in-house editor. If suitable for publication, manuscripts are sent for further review, which includes a combined effort by the editorial board and appointed subject experts, who provide independent peer-review. The feedback obtained in this way was communicated to authors, and with manuscripts checked upon return before finally accepted. The peer-reviewed nature of the books in the "World Sustainability Series" means that contributions to them have, over many years, been officially accepted for tenure and promotion purposes.

Mona-Maria Narra · Satyanarayana Narra
Editors

Innovations in Circular Economy and Renewable Energy in Africa

Springer

Editors
Mona-Maria Narra
Professorship of Material and Energy Valorisation of Biogenous Residues
Professorship of Waste and Resource Management
University of Rostock
Rostock, Mecklenburg-Vorpommern, Germany

Satyanarayana Narra
Professorship of Material and Energy Valorisation of Biogenous Residues
Professorship of Waste and Resource Management
University of Rostock
Rostock, Mecklenburg-Vorpommern, Germany

ISSN 2199-7373 ISSN 2199-7381 (electronic)
World Sustainability Series
ISBN 978-3-031-68329-9 ISBN 978-3-031-68330-5 (eBook)
https://doi.org/10.1007/978-3-031-68330-5

This Springer imprint is published by the registered company Springer Nature Switzerland AG
The registered company address is: Gewerbestrasse 11, 6330 Cham, Switzerland

Preface

In most of the African countries, access to electricity and modern energy fuels is still a big concern for the population. More than half of the population in Sub-Saharan Africa (SSA), representing about 570 million people, is still lacking access to electricity, while only about 18% of the population has access to modern fuels for cooking.

What is more worrying is the fact that many of the countries with even high access to electricity are overly dependent on fossil fuels as a primary energy source. For instance, Ghana's electric generation capacity is heavily dependent on light crude oil and natural gas, meaning price fluctuations in the international market have a direct effect on consumers.

Similarly, fossil fuels account for about 86% of South Africa's primary energy demand. These conditions as examples are alarming because many countries in Africa are facing major challenges due to massive debts, lack of maintenance and system failures raising the need for new investment in the transition of the sector. The need for diversification of energy and fuel source, investment in technologies and a new paradigm shift in energy and fuel access is more urgent than ever before.

In addition, even though Africa is not a heavy emitter of greenhouse gases, the current situation of overdependence on fossil fuel coupled with the growing demand for energy and fuel as the economy expands and a strong positive population growth rate means the business-as-usual case will not persist for long. Rather sharp increases in greenhouse gas emissions are expected over the near to long term future.

The use of biomass as an energy source is predominant in Sub-Saharan Africa but is mostly used in its raw state thus posing health hazards as a result of incomplete combustion. On the other hand, the amount of waste produced is also increasing due to various factors such as urbanization, increased economic activities and rise in living standards. Based on which waste management is also becoming one of the most pressing challenges faced by most of the countries in the continent.

Although the per capita waste generation of Africa is lower comparatively to other, more developed continents, the lack of treatment technologies, finance, legal and regulatory mechanisms makes the management of waste a very challenging aspect.

The lack of treatment has resulted in many calamities in several African countries including flooding, cholera outbreaks, pollution of surface water bodies and the ocean and the pollution of the marine environment. Lastly, the generation of agricultural residues as a result of agriculture and agro-processing activities across the continent presents a great opportunity for energy transition.

The diverse and large quantities of agricultural residues generation, amounts of energetic valuable wastes and utilization of non-conventional fuels mean various energy carriers can be developed and generated using these residues to meet or supplement Africa's energy and fuel demand.

The current situation offers a window of opportunity for Africa to transit from a more dependent fossil-based economy to a more sustainable green economy pathway through technology, knowledge and skills leveraging to support rapid transition and economic growth.

"Innovations in circular economy and renewable energy" highlights current developments in the field of waste management, the utilization of biomass residues for energetic purposes as well as innovations and potentials in green hydrogen production in Africa, from and with predominantly African research results. The articles in this book were first presented at the "First International Conference on Circular Economy, Renewable Energy and Green Hydrogen", in Kumasi (Ghana) in September 2023.

Rostock, Germany

Mona-Maria Narra
Satyanarayana Narra

Acknowledgements

The "First International Conference on Circular Economy, Renewable Energy and Green Hydrogen" and the resulting book *Innovations in Circular Economy and Renewable Energy in Africa* have been funded by the German Federal Ministry of Education and Research under the project "Verbundvorhaben Waste2Energy: Hybrid Energie aus Abfall als nachhaltige Lösung für Ghana".

SPONSORED BY THE

Federal Ministry
of Education
and Research

Contents

Waste to Energy

Bibliometric Study of Trends in Waste to Energy Research in Africa

Ekua Afrakoma Armoo, Belinda Boahemaa, Satyanarayana Narra, and Essossinam Beguedou

Abstract "Waste-to-energy" (WtE) has been historically associated with incineration. However, this term has taken on a broader connotation and encompasses currently any processing method that generates energy or alternative fuels from waste. In this chapter, two articles conduct systematic reviews on WtE, one aiming at the research in the management of municipal solid waste in Africa, while the other aims to promote the potential of emerging WtE generation technologies and innovations. The other articles research the background narrative for the "Hybrid Waste-to-Energy as a sustainable solution for Ghana" project, funded by the German Federal Ministry of Education and Research. The analyses provide a true reflection of the waste generation vis-a-vis collection models and their potential for energy generation in Ghana. These analyses could be used to replicate the model for similar communities in other countries. The research also focuses on effective WtE plant operationalization and acceptance. Capacity building and basic education play a vital role in enhancing the waste segregation and treatment, reducing the burden on waste management. Publicizing/disseminating the net social and economic benefits to the target communities should create additional benefits. The chapter is concluded with case specific business model based on a newly built WtE plant in Ghana.

E. A. Armoo (✉) · S. Narra · E. Beguedou
Professorship of Material and Energy Valorisation of Biogenous Residues, Professorship of Waste and Resource Management, University of Rostock, Rostock, Germany
e-mail: ekua.afra2@uni-rostock.de

S. Narra
e-mail: satyanarayana.narra@uni-rostock.de

E. Beguedou
e-mail: essossinam.beguedou@uni-rostock.de

E. A. Armoo
Council for Scientific and Industrial Research – Institute of Industrial Research, Accra, Ghana

B. Boahemaa
The Brew-Hammond Energy Centre, Kwame Nkrumah University of Science and Technology, Kumasi, Ghana
e-mail: boahemaabelinda30@gmail.com

M. Narra and S. Narra (eds.), *Innovations in Circular Economy and Renewable Energy in Africa*, World Sustainability Series, https://doi.org/10.1007/978-3-031-68330-5_1

1 Introduction

The World Bank estimates that global waste production has increased by nearly 70% and is expected to rise to an estimated 3.4 billion metric tons per annum by 2050 if the current average daily generation rate of 0.74kg per capita continues, mainly due to increasing population growth, urbanization, economic expansion and changing consumer purchasing patterns (Silpa et al. 2018). The waste sector is the third largest contributor (20%) to global anthropogenic methane emissions largely originating from landfills and thus critical to the climate crises (United Nations Environment Program and Climate and Clean Air Coalition 2021). To mitigate and adapt to the impacts of increasing global emissions, many countries have developed action plans known as the Nationally Determined Contributions (NDCs) under the Paris Agreement to meet the 1.5 °C global warming target by the end of this century (United Nations Framework Convention on Climate Change 2022). The European Union has taken bold steps towards mitigation through the European Green Deal; with a target to become the first climate-neutral continent by 2050 and slash greenhouse gas emissions by 55% compared to the 1990 levels. Sustainable resource management will play a vital role in achieving this target (European Commission 2020).

Although most developing countries are signatories to the Paris Agreement and have gone ahead to develop their NDCs, countries in Africa are lagging in both research and the implementation of sustainable solutions to its complex waste management challenges (Ndou and Rampedi 2022). One of the major challenges with waste management practices in developing countries is that waste is treated as an item to be disposed of in a linear consumption-to-disposal model rather than a circular resource recovery model (Sillanpää and Ncibi 2019).

Municipal solid waste management (MSWM) in developing countries is characterized by landfilling or open dumping (Silpa et al. 2018). Although waste is generally abundant, it is not readily accessible for proper collection and treatment due to challenges such as inadequate waste segregation and collection services, poor infrastructure, few treatment facilities, weak enforcement of laws, etc. (Salami et al. 2019; Thabit et al. 2022). The unsegregated waste ends up contaminating the environment by clogging drains, littering streets, and causing severe damage at open dumps. Such waste is difficult to handle and treat, even if collected. For instance, in Ghana, waste collection to landfills is at 21% while the remaining is burned, and disposed into open dumps or indiscriminate spaces (Mudu et al. 2021). Waste management is the single largest budget item for city managers in developing countries. However, the negative impacts of poor MSWM are enormous and cut across all aspects of human life including social, economic, environmental, health, infrastructural, etc., and are more prevalent in developing countries (Hoornweg and Perinaz 2012). This includes greenhouse gas emissions (Verma and Borongan 2022) and contamination of groundwater from landfill leachate (El Maguiri et al. 2017), health risks including air pollution and contamination from open burning of both domestic and industrial waste, disease transmission due to vectors and informal waste collectors and scavengers, (Ferronato and Torretta 2019), flooding due to choked drains and water ways (IBRD/The World Bank 2010), and transmission of diseases such as COVID-19 (Kalina et al. 2022).

Recovering energy from the waste is a well-established strategy for managing waste. However, the development of such systems is subject to extensive research based on major areas such as the technical, economic, environmental and social aspects.

1.1 Waste to Energy Research

In recent years, there has been significant global research and development focused on harnessing energy from various types of waste, employing a diverse range of waste-to-energy technologies. These innovations are crucial components of sustainable integrated waste management systems, aligning with both economic and environmental needs. (Loizidou et al. 2021). This valuable resource holds immense potential for bolstering the socio-economic development of numerous developing nations, particularly in Africa.

Global trends in energy recovery from waste (WtE) indicate that the most important technologies include gasification, incineration, anaerobic digestion, pyrolysis, and landfill gas recovery (Ndou and Rampedi 2022; Tsai et al. 2020). Research into alternative energy sources such as hydrogen is growing at a very fast pace since the onset of the Russia–Ukraine conflict. With natural gas prices increasing at an exponential rate, hydrogen from renewable sources is gaining increasing attention (Mbuk 2022). Hydrogen is not a fuel itself but an energy carrier. It has the good potential to store energy in the form of molecules and can be stored and transported as compared to other renewable energy sources such as solar and wind. It also has major applications in the energy, transport, and chemical manufacturing sectors (Reith et al. 2003; Yang and Wang 2020). In the UK, there is a potential for over 22,000 tons of hydrogen from biomass to replace fossils in public transport in the West Midlands region per year. Economic analysis indicated that a cost-competitive business model must incorporate the trade of CO_2, since the selling price is sensitive to the price of raw materials and the internal rate of return (Nouwe Edou and Onwudili 2022).

However, it is crucial to note that research efforts in Africa have primarily focused on challenges related to waste management, especially waste collection and recycling with limited attention given to understanding WTE technologies (Alao et al. 2022; Simatele et al. 2017). There are limited successful cases or waste to energy systems in developing countries which requires enhanced efforts (Mutz et al. 2017). Unlike developed nations such as Japan, China, the USA, and Germany, where WTE technologies have been successfully adopted for commercial use, many African countries grapple with logistical, technical, financial, socio-environmental, and policy-related hurdles. The absence of comprehensive technical data on waste quantity and quality in developing countries poses a significant obstacle. Insufficient knowledge about waste composition and characterization can lead to misguided choices in equipment and technology, resulting in resource wastage, inefficiencies, and undermining the concept of zero waste (Dlamini et al. 2019; Oelofse et al. 2016).

For the economic aspects, WTE technologies require substantial capital investments in equipment. Many developing nations face financial limitations that hinder their ability to invest in waste-to-energy projects, financial incentives like feed-in tariffs, Carbon Reduction Credits, Tax exemptions, and Renewable Energy Credits have the potential to stimulate private sector investments in the waste-to-energy sector (Alao et al. 2022). Regulatory frameworks and policies need to be established through legislative actions, fostering public–private partnerships in the waste-to-energy market. Furthermore, African local authorities, agencies, and stakeholders should collaboratively identify opportunities to enhance the adoption of modern waste management technologies, and unlocking additional economic potential from solid waste resources (Alao et al. 2022).

Given the diversity of WTE, research has seen many changes over several decades. One of the methods of assessing field-specific trends is bibliometric studies. Several authors have investigated various aspects of WTE trends on a global scale. However, there is a research gap in specific bibliometric studies on WTE research in Africa (continental scale), although some studies have been conducted at the country level such as the Republic of South Africa (RSA), and in areas such as RDF, landfill gas, and renewable energy. For instance Sarquah et al. (2022) carried out a global review of RDF research within the Scopus database and identified that publications from Africa between 2015 and 2022 were less than 1%. RDF application is predominantly in the cement processes and electricity generation and Circular Economy and sustainability are recent themes. Afrane et al. (2022) also carried out review of Renewable energy research in Africa within the Web of Science database and concluded that the most recent emerging areas included green hydrogen and economic growth.

This article therefore seeks to assess the trends in WTE research in Africa and to provide a future outlook based on the identification of relevant themes, collaborations and institutions. Africa is noted to be lagging behind in WTE research. The study further aims to provide evidence to stimulate increased relevant research into WTE in Africa by identifying gaps and suggesting future direction.

2 Methodology

Bibliometrics is an important tool for reviewing trends in research in various fields including MSW research. The PRISMA Checklist and Flow diagram guided the methodology for this study (Liberati et al. 2009). The Elsevier Scopus database was used because it is the largest database of scientific peer-reviewed literature and is popular with similar bibliometric studies. The dataset was downloaded in June 2023. The study focused on publications from all 54 countries in Africa and considered the period from 1975 to 2022. The chosen timeframe allows an exploration of historical trends, developments, and recent advancements in the field. These included articles resulting from collaborations between African authors and authors from other countries. To define the field of interest, an advanced search query was utilized to include documents in the English language only and within the period: 1975 to 2022. The analysis was conducted in three stages as depicted in Fig. 1. The search string below was used in the SCOPUS database:

"Municipal Solid Waste" OR "MSW" AND "energy" OR "Waste to Energy" OR "Waste-to-Energy" OR "Energy from Waste" AND AFFCOUNTRY(Africa) OR AFFILCOUNTRY(Nigeria OR Kenya OR Sudan OR Ghana OR Morocco OR Senegal OR Tanzania OR Ethiopia OR Mali OR Uganda OR Algeria OR Somalia OR Madagascar OR Niger OR Tunisia OR Cameroon OR Angola OR Zimbabwe OR Rwanda OR Namibia OR Gabon OR Zambia OR Mozambique OR Chad OR Liberia OR Togo OR Libya OR Djibouti OR Botswana OR Burundi OR Malawi OR Lesotho OR Benin OR Mauritius OR Eritrea OR Mauritania OR Guinea OR Eswatini OR Seychelles OR Comoros OR Egypt) OR AFFILCOUNTRY ("South Africa" OR "Democratic Republic of the Congo" OR "Cote D'Ivoire" OR "South Sudan" OR "Burkina Faso" OR "Central African Republic" OR "Guinea-Bissau" OR "Cabo Verde" OR "The Gambia" OR "Equatorial Guinea" OR "Sierra Leone" OR "Republic of the Congo" OR "Sao Tome and Principe").

The search query identified an initial hit of 385 documents. The dataset was then downloaded as a CSV file for screening with EXCEL. Repeated and irrelevant entries were excluded, and additional searches were performed for incomplete information to standardize the formats. 378 entries remained as the main dataset for the bibliometric analysis which was performed using VOSviewer® software in combination with EXCEL. VOSviewer is a popularly used software for creating network maps of large numbers of scientific publications for visualization and analysis of network data (Van Eck and Waltman 2010). The bibliographic study considers a range of evaluation factors, encompassing quantitative aspects such as research areas, publication

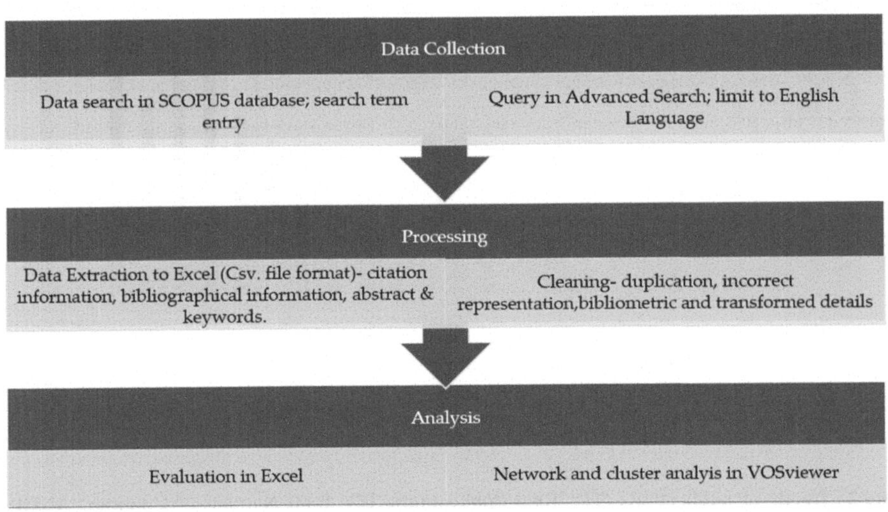

Fig. 1 Representation of research methodology for the study

years, publication count, study locations, language, journal distribution, authors, and institutions. Additionally, the qualitative aspect concentrates on subject-based areas and term analysis. Both the quantitative and qualitative analyses were performed using Microsoft Excel and VOSviewer.

3 Results and Discussion

The dataset of 378 documents, comprised diverse types of literature including, conference papers, book chapters, books, data papers, journal articles (including review papers), and editorials. Among these, journal articles (including review papers) and conference papers dominated the research landscape, accounting for 75.70% (286 documents) and 17.2% (65 documents), respectively, followed by book chapters at 6.1% (23 documents). Books constituted 0.5% (2 documents), data papers comprised 0.3% (1 document), and editorials made up 0.3% (1 document). Figure 2 provides a concise overview of the trends in waste to energy research over the past 47 years.

Although the first paper was published in 1975 on "The energy potential of MSW in South Africa", it took another 20 years before consistent research began within the area (Connor 1975; Letcher and Kolbe 1994) (Fig. 2). However, the number of articles began to shoot up from 2005 which is consistent with global trends (Ndou & Rampedi 2022). Both the 1995 and 2005 spikes could be attributed to the impact of two international treaties–the 1992 United Nations Framework Convention on Climate Change (the Earth Summit) and the Kyoto Protocol. They sparked the environmental movement in most parts of the world (United Nations 2018). The Kyoto

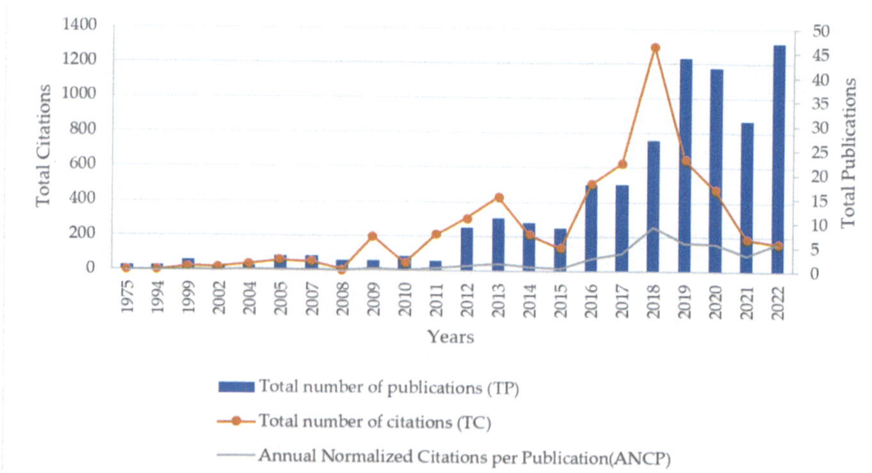

Fig. 2 Trends of publications (TP: Total Publications, TC: Total Number of Citations, ANCP: Annual Normalized Citations Per Publication

Protocol and the subsequent Kyoto Mechanisms particularly kick-started the Clean Development Mechanism (CDM) and joined Implementation projects across many countries (Bogner et al., 2007; Rowlands, 2001). Although the CDM itself is deemed to have failed, several projects were implemented and its principles of sustainability and carbon trade continue to be refined and applied in various contexts (Couth & Trois, 2012; Mutz et al., 2017). However, it must be noted that, although CDM projects might have ignited a lot of interest, there were tremendous challenges with implementation even in advanced African countries like RSA (Mbazima et al., 2022).

4 Thematic Areas of Research

Publication by subject area

The research on WTE in Africa spanned six main subject areas: energy, chemistry, engineering, chemical engineering, mathematics, and physics (see Fig. 3). Among these, engineering emerged as the dominant subject area, with 58 publications, followed by chemical engineering with 23 publications. Energy, chemistry and mathematics and physics accounted for 6, 5, and 2 publications, respectively.

This is an indication that researchers are currently focusing much on the technological aspect, designing various systems for waste conversion and energy recovery of which the engineering community plays a central role in advancing these technologies and optimizing their performance, at the detriment of the type and composition of energy produced. Additionally, the prominence of these subject areas may also be influenced by funding priorities, government policies, and the specific research interests of institutions and researchers (Chand Malav et al. 2020; Williams et al.

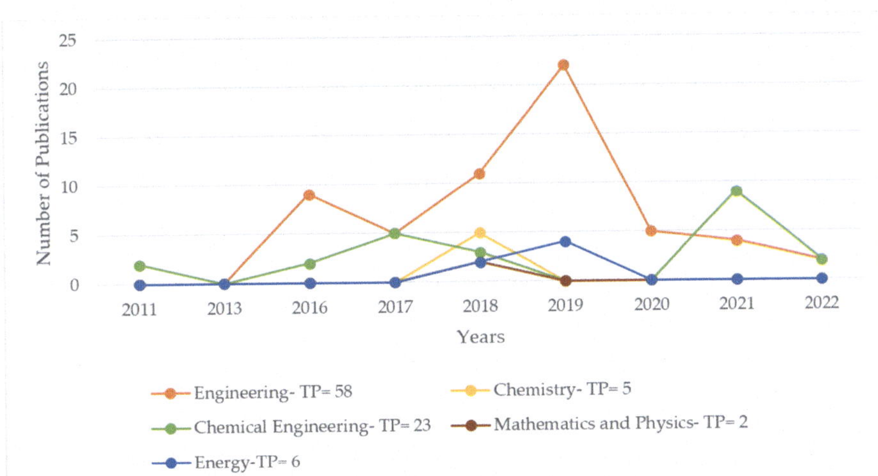

Fig. 3 Number of publications within the different subject areas (2011–2022)

2023). For example, if a country or region places a strong emphasis on sustainable waste management and renewable energy generation, it may lead to more research publications in these fields.

Major thematic areas based on keywords

The study investigated the key concepts and hotspots explored within the field of interest. To achieve this, keyword analysis was employed to explore the knowledge structure, identify emerging trends, and highlight main topics within the research areas. VOSviewer, utilizing the co-occurrence of keywords tool, was used for this analysis. Through cluster analysis, the main research streams and trends were identified. A total of 2,409 keywords were extracted from the publications using both author and index keywords. 405 keywords met the criterion of occurring at least three times. From Table 1 and Fig. 4, there are eight distinct clusters from which the authors distinguished into the following themes within waste-to-energy research based on the various keywords in each cluster. The technical and economic aspects have been thoroughly researched as compared to areas assessing impact of MSW such as health and environment (Medina-Mijangos and Seguí-Amórtegui 2020).

Emerging trends in subject areas (last 10 years)

Over the last ten years, the period from 2016 to 2020 saw a major diversification in the research topics in waste to energy including technologies, valorization, economics and investments as well as environmental aspects. In the past 5 years, the focus has emerged on sustainable management such as hybrid technologies and circular economy. This is consistent with the findings of (Negrete-Cardoso et al. 2022; Tsai et al. 2020) showing that circular economy is an important emerging field in the management of MSW and energy plays an important role in achieving the circular economy, especially in the post COVID-19 era, as can be observed in Fig. 5.

Country analysis

The contribution of different territories to the body of knowledge in WTE research in Africa was analyzed in VOSviewer. Although the research focused on Africa,

Table 1 The subject area clusters and thematic areas

Major subject area (cluster)	Number of keywords within the cluster
1-Thermochemical conversion methods	89
2-Economic value and circular economy	66
3-Bio-conversion methods	62
4-Environmental aspect and emission control	56
5-Renewable energy from waste	54
6-Research methods	39
7-Sanitation and waste recovery	28
8-Human health	11

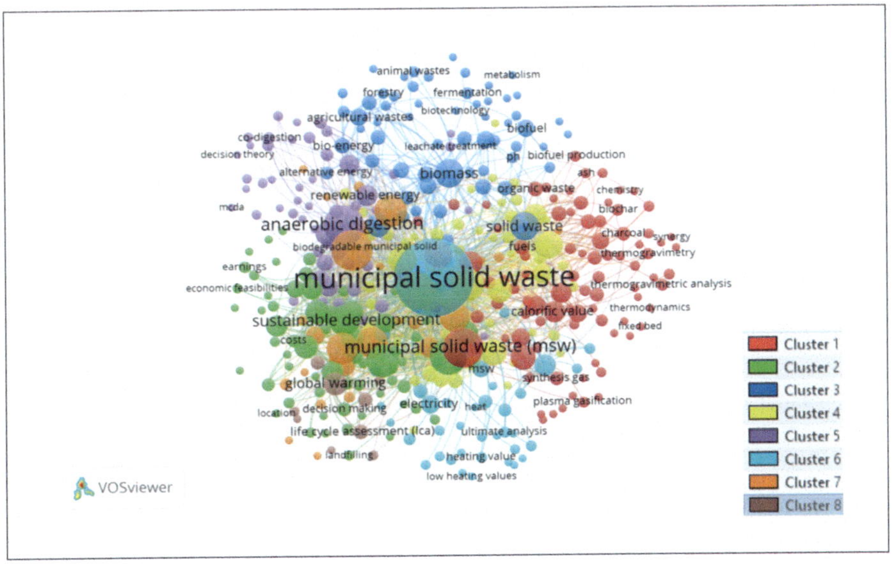

Fig. 4 Clusters representing the major subject areas of research in Africa

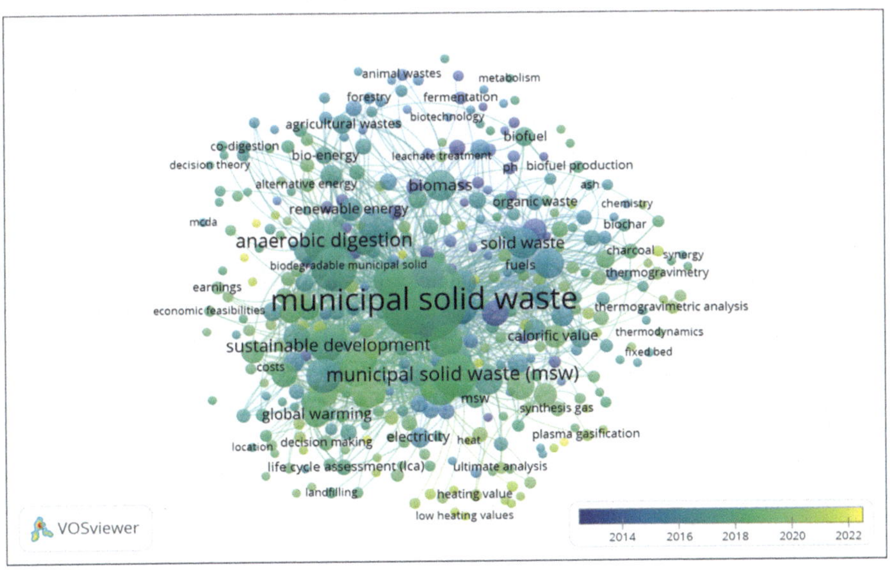

Fig. 5 Keywords incidence in years

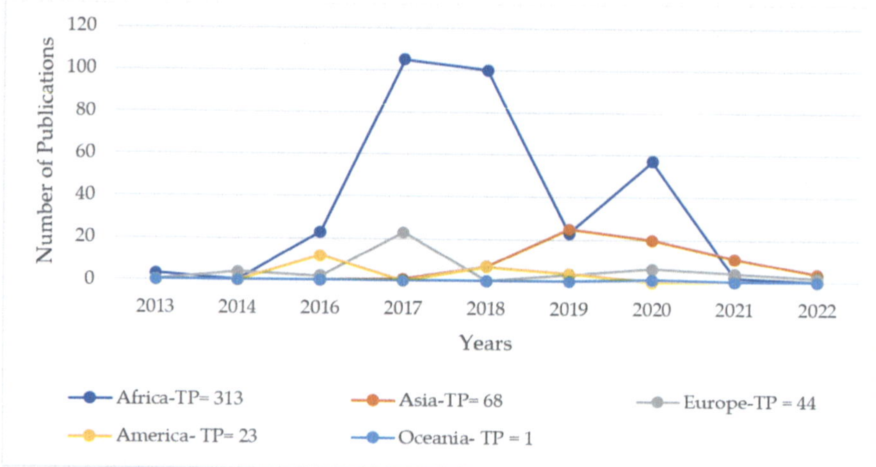

Fig. 6 The contributions of various continents to research based on the number of documents within the dataset

collaborations between authors and institutions outside Africa was significant (see Fig. 6). A total of 57 countries were represented. Although the research was based on Africa, African countries contributing only 33.33% of total countries represented indicating that there was significant co-authoring with institutions outside Africa.

Among African countries, South Africa (86), Nigeria (68), Egypt (32), Morocco (26), and Mauritius (16) were the top five contributors. A positive growth trend in waste-to-energy research in Africa is observed, backed by a significant number of works recorded in recent years (2019–2022), comprising 18.26% of the total publications as can be seen in Fig. 7.

Geographically, all regions (North, East, West, South, and Central) are represented in the top 10 countries. However, most studies are still country-specific. In the entire data, countries such as Eswatini, Comoros, Seychelles, Madagascar, Chad, South Sudan, Namibia, Djibouti are not represented at all. This could be attributed to either a language barrier since Africa is mainly divided along Anglophone or Francophone language lines (Ebong 2022). This research only focused on publications in English thus most Francophone countries may rather publish in French. Also, political and economic instability may have contributed to low availability of funding for research.

Collaboration analysis

The study investigated country collaboration using VOSviewer's co-authorship tool, which revealed 42 documents meeting the criteria of a minimum of 2 documents and 1 citation per country. Figure 8 shows collaborative efforts among countries across the globe.

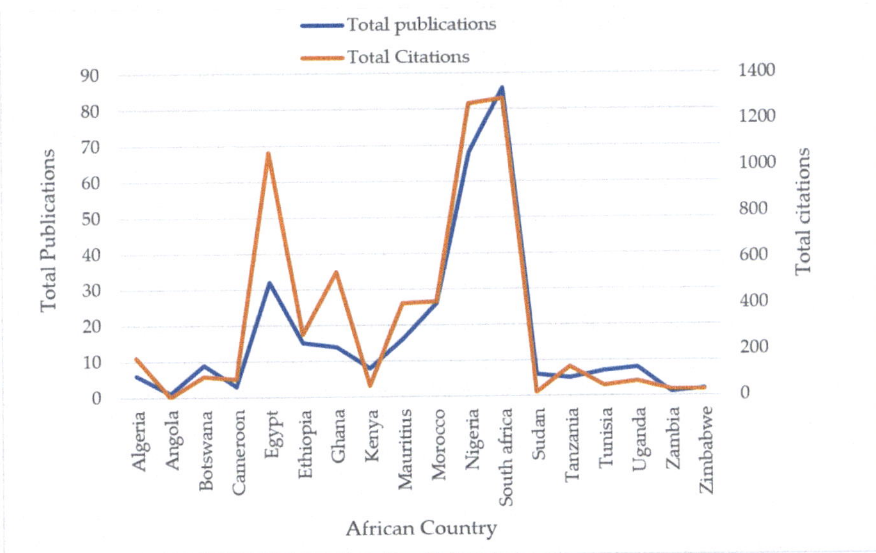

Fig. 7 The contribution of African countries to waste-to-energy research within the continent based on the number of citations and related documents

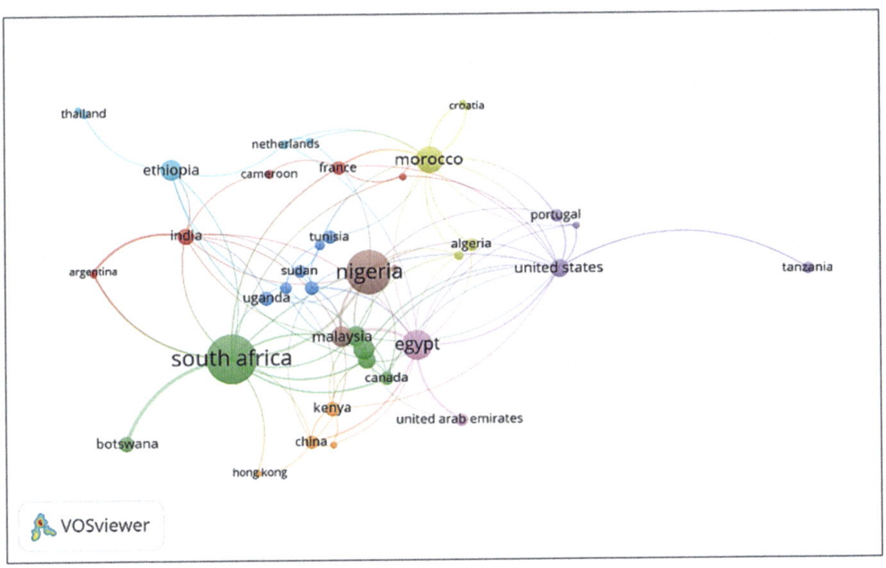

Fig. 8 Country collaborations including those outside of Africa

Analysis of Sources, Authors, and Institutional Contributions

The publications were categorized into two source types: journals and conference proceedings. A total of 174 distinct sources were identified, and the top seven journals include Renewable and Sustainable Energy Reviews, Waste Management, Waste Management and Research, Applied Energy, Biomass and Bioenergy, Energy Sources, and Solid Waste Technology and Management. In terms of affiliation, 52 institutions contributed to the field of Waste -to-Energy research. Among 783 authors, 153 authors had a minimum of 2 publications during the analysis period. A similar work by (Afrane et al. 2022) on renewable energy in Africa using the Web of Science database also revealed that Renewable And Sustainable Energy Reviews has journaled the most number of publications on energy in Africa (see Table 2). Since this is a review publication source, it is inferred that most of the publications on energy and renewable energy in Africa focus on reviews.

An analysis of affiliation based on co-author information revealed a total of 52 institutions, including sub-institutions such as departments, faculties, and centers. These institutions have made contributions to the field of MSW research, both through collaborations and as single institutions. The distribution of these institutions across various countries can be found in Table 3. Only institutions with a minimum of 2 publications are represented in the table.

In the field of study, a total of 783 authors have made contributions. Among them, 153 authors had a minimum of 2 publications during the analysis period. The h-index and affiliation information of these authors were obtained from the Scopus website. Table 4 presents the top six authors who have achieved at least 10 publications in this domain.

Table 2 Characteristics of the top journals of the publications analyzed

Journal name	TP	TC	CPP	Cite Score	SJR	SNIP 2022
Renewable and sustainable energy reviews	16	1072	67	26.3	3.232	3.631
Waste management	9	498	55.33	15.1	1.745	1.912
Waste management and research	8	187	23.38	7.4	0.866	1.328
Applied energy	5	259	51.8	21.1	2.907	2.758
Biomass and bioenergy	5	174	34.8	10.8	1.045	1.305
Energy sources	5	30	6	4.7	0.471	0.807
Solid waste technology and management	5	3	0.6	0.6	0.146	0.187

Note TP = Total Publication, TC = Total Citations, CPP = Citation Per Publication, SJR = Scimago Journal Ranking, 2022 figures from Scopus database, SNIP = Source Normalized Impact per Paper

Table 3 Countries and institutional contributions; TP = Total Publications

Country	Institutions	TP
Nigeria	University of Ibadan	6
	Landmark University	13
	Federal University of Technology	6
	University of Lagos	2
Egypt	Cairo University	4
	Ain Shams University	4
	Suez Canal University	5
South Africa	Council for Scientific and Industrial Research	3
	University of Johannesburg	25
	Cape Peninsula University of Technology	7
	University of Stellenbosch	6
	Tshwane University of Technology	5
	Durban University Technology	2
	Vaal University of Technology	2
	Regional Office for Southern Africa	2
	University of South Africa Gauteng	2
	University of Cape Town	2
Botswana	The Botswana International University of Science and Technology	2
Tanzania	University of Dar es Salaam	2
	The Nelson Mandela African Institution of Science and Technology	2
Malaysia	University of Malaya	2
Ghana	Kwame Nkrumah University of Science and Technology	2
Mauritius	The University of Mauritius	4
Ethiopia	Wollo University	2
Portugal	University of Porto	2
Argentina	Fundación Bariloche	2
Tunisia	Higher Institute of Information and Communication Technologies	2
Morrocco	Chouaïb Doukkali University	4
France	Lille University	2
India	National Institute of Urban Affairs (niua)	2
Brazil	University of Sao Paulo	2
China	Harbin Institute of Technology	2
Tunisia	Higher Institute of Information and Communication Technologies	2

Table 4 Top six authors with 10 or more publications from 2001–2022

Author name	TP	TC	h-index	Current affiliation	Country
Ayodele, Temitope Raphael	12	459	29	The University of Ibadan	Nigeria
Elagroudy, Sherien A	10	174	12	Ain Shams University	Eqypt
Ibikunle, R. A	11	123	8	Landmark University	Nigeria
Mbohwa, Charles	16	86	27	University of Johannesburg	South Africa
Muzenda, Edson	16	221	23	Botswana International University of Science and Technology	Botswana
Titiladunayo, Isaac Femi	10	123	7	Federal University of Technology, Akure	Nigeria

Note TP = Total Publications, TC = Total Citations

Citation analysis

An analysis was performed on the highly cited publications in the field of MSW research from 1975 to 2022. The study examined various factors, including total citations (TC), research conducted in specific territories/ areas, annual averages, sources, and year of publication (YP).

Remarkably, a significant number of these highly cited works were found in the Egyptian Journal of Petroleum, predominantly authored within Egypt's territory. Notably, Table 5 revealed that the most cited research was published in 2018, amassing an impressive 577 recorded citations. This publication, which originated from Egyptian territory, was featured in the Egyptian Journal and credited to the Egyptian Petroleum Research Institute, with an average of 21.37 citations per year.

In terms of geographical scope, east and central Africa recorded no entry. It is expected that the establishment of the first large-scale waste-to-energy plant in Ethiopia will stimulate increased interest and more research into the impacts of such interventions in East Africa (UNEP 2017). Table 6 also confirms Fig. 6; with the top 3 countries also appearing in the top 10 most cited articles.

Table 5 Top 5 sources for publication, total citations, and the average publication year

Year	Title	TC	TC/ Year	Source	CA
2018	Solid waste issue: composition, disposal, recycling, and valorization	577	21.37	Egyptian Journal of Petroleum	Egypt
2011	A comprehensive review of biomass resources and biofuels potential in Ghana	206	103	Renewable and Sustainable Energy Reviews	Ghana, and the United Kingdom
2016	Liquid biofuels from food waste: current trends, prospect and limitation	160	8.89	Renewable and Sustainable Energy Reviews	India
2018	Current updates on waste to energy (WtE) technologies: a review	145	5.37	Renewable Energy Focus	Ethiopia, and Italy
2012	Enzyme research and applications in biotechnological intensification of biogas production	140	15.56	Critical Review in Biotechnology	Zimbabwe
2013	An Overview of Energy Recovery from Municipal Solid Wastes (MSW) in Malaysia Scenario	135	12.27	Renewable and Sustainable Energy Reviews	Malaysia, and Nigeria
2009	Comparison of Municipal solid waste management systems in Canada and Ghana: A case study of the Cities of London, Ontario, and Kumasi, Ghana	129	64.5	Waste Management	Ghana, and the United Kingdom
2017	Life cycle assessment of waste-to-energy (WtE) technologies for electricity generation using municipal solid waste in Nigeria	121	6.72	Applied Energy	Nigeria
2017	Electricity generation from municipal solid waste in some selected cities in Nigeria: An assessment of feasibility, potential, and technologies	107	5.94	Renewable and Sustainable Energy Review	Nigeria

Note TP = Total Publications, TC/year = Total Citations per year, CA = Country of Authors

Table 6 The top 10 most cited articles, the areas of research, and the total citations per year

References	Area of research	Year	Source	Country	Total citations	TC/ year
Tizaoui et al. (2007)	Landfill leachate treatment	2007	Journal of Hazardous Materials	RSA	278	19
Scarlat et al. (2015)	Energy potential of municipal solid waste	2015	Renewable and Sustainable Energy Reviews	RSA	215	31
Osibanjo and Nnorom (2007)	Electronic waste (e-waste) management	2007	Waste Management and Research	Nigeria	201	13
Salem et al. (2008)	Landfill leachate treatment	2008	Desalination	Algeria	158	11
Achiba et al. (2009)	Compost	2009	Agriculture, Ecosystems and Environment	Tunisia	156	12
Asase et al. (2009)	MSWM systems	2009	Waste Management	Ghana	122	9
Mohammed et al. (2013)	Renewable energy resources	2013	Renewable and Sustainable Energy Reviews	Sub-Sahara Africa	117	13
Kofoworola, (2007)	Recovery and recycling practices	2007	Waste Management	Nigeria	100	7
Giwa et al., (2017)	Sustainable energy	2017	Renewable and Sustainable Energy Reviews	Nigeria	99	20
Idowu et al., (2019)	Landfill classification systems	2019	Waste Management	Sub-Sahara Africa	91	30

5 Conclusion

The study provides an insightful and comprehensive analysis of research on municipal solid waste-to-energy in Africa over the past 47 years. It sheds light on key aspects such as document types, subject areas, research concepts, country/ territory contributions, collaboration patterns, sources, authors, and citation trends. The findings highlight Africa's significant role in WTE research and reveal the growing interest and collaboration within the field. The major challenges identified is the low contribution of most countries on the continent to this research area indicating a need for increased research funding and support. This study serves as a valuable resource for researchers and policymakers seeking to advance sustainable waste management and energy generation in the African context. Also, the growing interest in circular economy indicates that this may be a sustainable solution for countries seeking to solve Municipal Solid Waste management challenges.

References

Achiba, W. Ben, Gabteni, N., Lakhdar, A., Laing, G. Du, Verloo, M., Jedidi, N., & Gallali, T. (2009). Effects of 5-year application of municipal solid waste compost on the distribution and mobility of heavy metals in a Tunisian calcareous soil. Agriculture, Ecosystems & Environment, 130(3), 156–163. https://doi.org/10.1016/j.agee.2009.01.001

Afrane, S.; Ampah, J.D.; Aboagye, E.M. (2022). Investigating Evolutionary Trends and Characteristics of Renewable Energy Research in Africa: A Bibliometric Analysis from 1999 to 2021. Environ. Sci. Pollut. Res., 29, 59328–59362, https://doi.org/10.1007/s11356-022-20125-0

Alao, M. A., Popoola, O. M., & Ayodele, T. R. (2022). Waste-to-energy nexus: An overview of technologies and implementation for sustainable development. Cleaner Energy Systems, 3(January), 100034. https://doi.org/10.1016/j.cles.2022.100034

Asase, M., Yanful, E. K., Mensah, M., Stanford, J., & Amponsah, S. (2009). Comparison of municipal solid waste management systems in Canada and Ghana: A case study of the cities of London, Ontario, and Kumasi, Ghana. Waste Management, 29(10), 2779–2786. https://doi.org/10.1016/j.wasman.2009.06.019

Bogner, J., Abdelrafie Ahmed, M., Diaz, C., Faaij, A., Gao, Q., Hashimoto, S., Mareckova, K., Pipatti, R., & Zhang, T. (2007). Waste Management. In L. A. M. B. Metz, O.R. Davidson, P.R. Bosch, R. Dave (Ed.), Contribution of Working Group III to the Fourth Assessment Report of the Intergovernmental Panel on Climate Change. Cambridge University Press. https://archive.ipcc.ch/publications_and_data/ar4/wg3/en/contents.html

Chand Malav, L., Yadav, K. K., Gupta, N., Kumar, S., Sharma, G. K., Krishnan, S., Rezania, S., Kamyab, H., Pham, Q. B., Yadav, S., Bhattacharyya, S., Yadav, V. K., & Bach, Q. V. (2020). A review on municipal solid waste as a renewable source for waste-to-energy project in India: Current practices, challenges, and future opportunities. Journal of Cleaner Production, 277, 123227. https://doi.org/10.1016/j.jclepro.2020.123227

European Commission. (2020). Stepping up Europe's 2030 climate ambition (No. 98; COMMUNICATION FROM THE COMMISSION TO THE EUROPEAN PARLIAMENT, THE COUNCIL, THE EUROPEAN ECONOMIC AND SOCIAL COMMITTEE AND THE COMMITTEE OF THE REGIONS). https://commission.europa.eu/strategy-and-policy/priorities-2019-2024/european-green-deal_en

Connor, M. A. (1975). The energy potential of municipal solid wastes in South Africa. THE CIVIL ENGINEER in South Africa -, 17(11), 295–301. https://journals.co.za/doi/pdf/https://doi.org/10.10520/AJA10212019_13639

Couth, R., & Trois, C. (2012). Sustainable waste management in Africa through CDM projects. Waste Management, 32(11), 2115–2125. https://doi.org/10.1016/j.wasman.2012.02.022

Dlamini, S., Simatele, M. D., & Serge Kubanza, N. (2019). Municipal solid waste management in South Africa: from waste to energy recovery through waste-to-energy technologies in Johannesburg. Local Environment, 24(3), 249–257. https://doi.org/10.1080/13549839.2018.1561656

Ebong, O. E. (2022). Overcoming the Francophone and Anglophone Dichotomy in the 21st Century to Promote African Unity. PINISI JOURNAL OF ART, HUMANITY AND SOCIAL STUDIES, 2(2), 107–114

Van Eck, N. J., & Waltman, L. (2010). Software survey: VOSviewer, a computer program for bibliometric mapping. Scientometrics, 84(2), 523–538

Ferronato, N., & Torretta, V. (2019). Waste mismanagement in developing countries: A review of global issues. International Journal of Environmental Research and Public Health, 16(6). https://doi.org/10.3390/ijerph16061060

Giwa, A., Alabi, A., Yusuf, A., & Olukan, T. (2017). A comprehensive review on biomass and solar energy for sustainable energy generation in Nigeria. Renewable and Sustainable Energy Reviews, 69, 620–641. https://doi.org/10.1016/j.rser.2016.11.160

United Nations Environment Program and Climate and Clean Air Coalition. (2021). Global Methane Assessment: Benefits and Costs of Mitigating Methane Emissions. https://wedocs.unep.org/bitstream/handle/20.500.11822/35917/GMA_ES.pdf

IBRD/The World Bank. (2010). City of Accra, Ghana: Consultative Citizens' Report Card (Issues 55117-GH). http://www-wds.worldbank.org/external/default/WDSContentServer/WDSP/IB/2010/08/01/000333037_20100801234448/Rendered/PDF/551170ESW0P1131Citizens0Report0Card.pdf

Idowu, I. A., Atherton, W., Hashim, K., Kot, P., Alkhaddar, R., Alo, B. I., & Shaw, A. (2019). An analyses of the status of landfill classification systems in developing countries: Sub Saharan Africa landfill experiences. Waste Management, 87, 761–771. https://doi.org/10.1016/j.wasman.2019.03.011

Kalina, M., Kwangulero, J., Ali, F., & Tilley, E. (2022). "You need to dispose of them somewhere safe": Covid-19, masks, and the pit latrine in Malawi and South Africa. PLoS ONE, 17(2 February), 1–13. https://doi.org/10.1371/journal.pone.0262741

Kofoworola, O. F. (2007). Recovery and recycling practices in municipal solid waste management in Lagos, Nigeria. Waste Management, 27(9), 1139–1143. https://doi.org/10.1016/j.wasman.2006.05.006

Letcher, T. M., & Kolbe, F. F. (1994). Energy from landfill and sewage gas in South Africa. Journal of Energy in Southern Africa, 5(3), 87–91.

A Liberati DG Altman J Tetzlaff C Mulrow PC Gøtzsche JPA Ioannidis M Clarke PJ Devereaux J Kleijnen D Moher 2009 The PRISMA statement for reporting systematic reviews and meta-analyses of studies that evaluate healthcare interventions: explanation and elaboration BMJ 339 https://doi.org/10.1136/bmj.b2700

Loizidou, Maria, Konstantinos Moustakas, Mohammed Rehan, Abdul-Sattar Nizami, M. T. (2021). New developments in sustainable waste-to-energy Systems. 151(August). https://doi.org/10.1016/j.rser.2021.111581

El Maguiri, A., Zegzouti, Y., Souabi, S., Idrissi, L., El Hadek, M., Rene, E. R., Pala, A. I., & Hafidi, M. (2017). Landfills : Environmental Impacts, Assessment and Management Complimentary Contributor Copy. In C. Norma (Ed.), Landfills: Environmental impacts, assessment and management (Issue October, p. 298). Nova Science Publishers, Inc. https://www.researchgate.net/profile/Abdelhakim-El-Maguiri/publication/322053049_Landfills_Environmental_Impacts_Assessment_and_Management/links/5aae7990aca2721710fc4629/Landfills-Environmental-Impacts-Assessment-and-Management.pdf#page=17

Mbazima, S. J., Masekameni, M. D., & Mmereki, D. (2022). Waste-to-energy in a developing country: The state of landfill gas to energy in the Republic of South Africa. Energy Exploration and Exploitation, 40(4), 1287–1312. https://doi.org/10.1177/01445987221084376

Mbuk, K. (2022). Clean Hydrogen ' s Place in the Energy Transition (Issue October). https://carbontracker.org/wp-content/uploads/2022/10/Clean-Hydrogens-Place-in-the-Energy-Transition-1.pdf?lang=ja

Medina-Mijangos, R., & Seguí-Amórtegui, L. (2020). Research trends in the economic analysis of municipal solid waste management systems: A bibliometric analysis from 1980 to 2019. Sustainability (Switzerland), 12(20), 1–20. https://doi.org/10.3390/su12208509

Mohammed, Y. S., Mustafa, M. W., Bashir, N., & Mokhtar, A. S. (2013). Renewable energy resources for distributed power generation in Nigeria: A review of the potential. Renewable and Sustainable Energy Reviews, 22, 257–268. https://doi.org/10.1016/j.rser.2013.01.020

Mudu, P., Nartey, B. A., Kanhai, G., Spadaro, J. V, & Fobil, J. (2021). Solid waste management and health in Accra, Ghana. In Geneva:World Health Oragnization. Geneva:World Health Oragnization. https://apps.who.int/iris/bitstream/handle/10665/340952/9789240024250-eng.pdf?sequence=1&isAllowed=y

Mutz, D., Hengevoss, D., Hugi, C., Gross, T. (2017). Waste-to-energy options in municipal solid waste management: A guide for decision makers in developing and emerging countries, Deutsche Gesellschaft für Internationale Zusammenarbeit (GIZ) GmbH. Eschborn, Germany. https://www.giz.de/en/downloads/GIZ_WasteToEnergy_Guidelines_2017.pdf. Accessed 15 March 2021

Ndou, V., & Rampedi, I. T. (2022). Bibliometric Analysis of Municipal Solid Waste Management Research: Global and South African Trends. Sustainability (Switzerland), 14(16). https://doi.org/10.3390/su141610229

Negrete-Cardoso, M.; Rosano-Ortega, G.; Álvarez-Aros, E.L.; Tavera-Cortés, M.E.; Vega-Lebrún, C.A.; Sánchez-Ruíz, F.J. (2022). Circular Economy Strategy and Waste Management: A Bibliometric Analysis in Its Contribution to Sustainable Development, toward a Post-COVID-19 Era. Environ. Sci. Pollut. Res., 29, 61729–61746, https://doi.org/10.1007/s11356-022-18703-3

Nouwe Edou, D. J., & Onwudili, J. A. (2022). Comparative techno-economic modelling of large-scale thermochemical biohydrogen production technologies to fuel public buses: A case study of West Midlands region of England. Renewable Energy, 189, 704–716. https://doi.org/10.1016/j.renene.2022.02.074

Oelofse, S. H. H., Muswema, A. P., & Koen, R. (2016). The changing face of waste management – considerations when conducting a waste characterisation study. Proceedings of the 23rd WasteCon Conference 17–21 October 2016, Emperors Palace, Johannesburg, South Africa, October, 345–349.

Osibanjo, O., & Nnorom, I. C. (2007). The challenge of electronic waste (e-waste) management in developing countries. Waste Management and Research, 25(6), 489–501. https://doi.org/10.1177/0734242X07082028

Reith, R. H., Wijffels, R. H., & Barten, H. (2003). Bio_methane & Bio-hydrogen: Status and perspectives of biological methane and hydrogen production (J. H. Reith, R. H. Wijffels, & H. Barten (eds.)). Dutch Biological Hydrogen Foundation. https://www.ircwash.org/sites/default/files/Reith-2003-Bio.pdf

Rowlands, I. H. (2001). The Kyoto Protocol's 'Clean Development Mechanism': a sustainability assessment. Third World Quaterly, 22(5), 795–811. https://library.fes.de/libalt/journals/swetsfulltext/11833477.pdf

Salami, H. A., Adegite, J. O., Bademosi, T. T., Lawal, S. O., Olutayo, O. O., & Olowosokedile, O. (2019). A Review on the Current Status of Municipal Solid Waste Management in Nigeria: Problems and Solutions. Journal of Engineering Research and Reports, February, 1–16. https://doi.org/10.9734/jerr/2018/v3i416884

Salem, Z., Hamouri, K., Djemaa, R., & Allia, K. (2008). Evaluation of landfill leachate pollution and treatment. Desalination, 220(1), 108–114. https://doi.org/10.1016/j.desal.2007.01.026

Sarquah, K., Narra, S., Beck, G., Awafo, E. A., & Antwi, E. (2022). Bibliometric Analysis; Characteristics and Trends of Refuse Derived Fuel Research. Sustainability (Switzerland), 14(4). https://doi.org/10.3390/su14041994

Scarlat, N., Motola, V., Dallemand, J.F., Monforti-Ferrario, F., Mofor, L.(2015). Evaluation of energy potential of Municipal Solid Waste from African urban areas. Renew. Sust. Energ. Rev. 50, 1269–1286. https://doi.org/10.1016/j.rser.2015.05.067

Sillanpää, M., & Ncibi, C. (2019). The circular economy: Case studies about the transition from the linear economy. In The Circular Economy: Case Studies about the Transition from the Linear Economy (Issue August). Academic Press. https://doi.org/10.1016/C2017-0-02916-6

Silpa, K.; Lisa, Y.; Perinaz, B.-T.; Woerden, F. Van (2018). What a Waste 2.0: A Global Snapshot of Solid Waste Management to 2050; Urban Deve.; World Bank: Washington, DC: World Bank.

Simatele, D. M., Dlamini, S., & Kubanza, N. S. (2017). From informality to formality: Perspectives on the challenges of integrating solid waste management into the urban development and planning policy in Johannesburg, South Africa. Habitat International, 63, 122–130. https://doi.org/10.1016/j.habitatint.2017.03.018

Thabit, Q., Nassour, A., & Nelles, M. (2022). Facts and Figures on Aspects of Waste Management in Middle East and North Africa Region. Waste, 1(1), 52–80. https://doi.org/10.3390/waste1010005

Tizaoui, C., Bouselmi, L., Mansouri, L., & Ghrabi, A. (2007). Landfill leachate treatment with ozone and ozone/hydrogen peroxide systems. Journal of Hazardous Materials, 140(1–2), 316–324. https://doi.org/10.1016/j.jhazmat.2006.09.023

Tsai, F.M.; Bui, T.D.; Tseng, M.L.; Lim, M.K.; Hu, J.(2020). Municipal Solid Waste Management in a Circular Economy: A Data-Driven Bibliometric Analysis. J. Clean. Prod., 275, 124132, https://doi.org/10.1016/j.jclepro.2020.124132.

UNEP. (2017). Ethiopia's waste-to-energy plant is a first in Africa. CHEMICALS & POLLUTION ACTION. https://www.unep.org/news-and-stories/story/ethiopias-waste-energy-plant-first-africa

United Nations Framework Convention on Climate Change. (2022). Nationally determined contributions under the Paris Agreement. https://unfccc.int/documents/619180?gclid=Cj0KCQjw2v-gBhC1ARIsAOQdKY0m0PiVBu-FvQbpu90jVwkAcALf92RXKTWFBaMLComxzrdp0zq0a7UaAiOkEALw_wcB

Verma, R. L., & Borongan, G. (2022). Emissions of Greenhouse Gases from Municipal Solid Waste Management System in Ho Chi Minh City of Viet Nam. Urban Science, 6(4), 78. https://doi.org/10.3390/urbansci6040078

Williams, P. A., Narra, S., Antwi, E., Quaye, W., Hagan, E., Asare, R., Owusu-Arthur, J., & Ekanthalu, V. S. (2023). Review of Barriers to Effective Implementation of Waste and Energy Management Policies in Ghana: Implications for the Promotion of Waste-to-Energy Technologies. Waste, 1(2), 313–332. https://doi.org/10.3390/waste1020021

Yang, G., & Wang, J. (2020). Biohydrogen production from waste activated sludge pretreated by combining sodium citrate with ultrasonic: Energy conversion and microbial community. Energy Conversion and Management, 225(September), 113436. https://doi.org/10.1016/j.enconman.2020.113436

Perspective Review on Emerging Waste to Energy (WtE) Technologies for Effective Waste Management

Abdul-Wahab Tahiru, Samuel Jerry Cobbina, Mutala Mohammed, and Wilhemina Asare

1 Introduction

As the globe rushes headlong towards its urban future, the volume of municipal solid waste (MSW), one of the critical by-products of an urban lifestyle, is increasing at a rate that is outpacing the rate of urbanization (Tahir et al. 2015). The quantity of municipal solid waste (MSW) generated by cities throughout the world is expected to rise from 1.3 billion tons per year to 2.2 billion tons per year by 2025, with waste output rates in emerging economies expected to escalate over the next two decades (Kaur et al. 2021). Of this, only 30–70% of waste can be collected in most developing countries whereas developed countries have a 76–90% collection efficiency (Dhokhikah and Trihadiningrum 2012; Sharma and Jain 2020). Furthermore, the world population is expected to reach 8.6 billion by 2030, 9.8 billion by 2050, and 11.2 billion by 2100, accompanied by the aforementioned urbanization, economic expansion, and changing lifestyles (Shivanna 2020), all of which directly influence the growth of power consumption and will force countries to more than quadruple their generation capacity by 2040 (Sönmez et al. 2019). As a result of growing concerns about climate change and the imminent depletion of fossil fuels, governments have begun to turn to renewable and alternative energy sources (Tun et al. 2019). This way, they will be able to equalize their respective energy demands while simultaneously reducing greenhouse gas emissions and other environmental concerns connected

A.-W. Tahiru (✉) · S. J. Cobbina · W. Asare
Department of Environment and Sustainability Sciences, Faculty of Natural Resources and Environment, University for Development Studies, P.O. Box TL, 1350 Tamale, Ghana
e-mail: tahiru.abdulwahab21@uds.edu.gh

M. Mohammed
CSIR-Institute of Industrial Research, Box LG 576, Accra, Ghana

S. J. Cobbina
West African Centre for Water, Irrigation and Sustainable Agriculture (WACWISA), Tamale, Ghana

M. Narra and S. Narra (eds.), *Innovations in Circular Economy and Renewable Energy in Africa*, World Sustainability Series, https://doi.org/10.1007/978-3-031-68330-5_2

with the prevailing fossil-fuel-dependent energy industry. Among the apparent and immediate solutions to these demands is waste-to-energy, which is an exhaustive term for several technologies that allow countries to dispose waste while also generating green power at the same time (Tun et al. 2020). Thus, the technologies serve as a point of convergence for the waste management and energy sectors, allowing them to interact and benefit from one another in the most effective way practicable. To be precise, waste-to-energy (WTE) is a sort of biomass energy that offers advantages not only in terms of lowering the issue of waste, but also in terms of serving as a viable alternative and substitute for fossil fuels, therefore minimizing the consequences of global warming (Sovacool et al. 2021).

It is anticipated that by 2025, more than 600 new waste-to-energy facilities will be developed globally, with a combined capacity of 170 million tons per year (Thiel et al. 2018). Currently, there are more than 2200 thermal waste treatment plants across the world, with a combined capacity of 300 million tons per year (Thiel et al. 2018). In 2017, China produced 7.3 gigatons of energy from 339 power plants, and the country expects to increase that figure to 10 gigatons and 600 plants by 2020 (Capizzi et al. 2019). Environmental strategies and social, economic, and technological variables are considered strongly in the selection of the most appropriate waste-to-energy technology in order to achieve the greatest possible outcomes (Klavenieks et al. 2017). Regional preferences exist when it comes to waste-to-energy technologies, for example, gasification is popular in South Asia, whereas grate incineration is popular in America and Europe, among other places (AlQattan et al. 2018; Yan and Waluyo 2020). For waste-to-energy conversion in Europe, the most common solutions are the incineration of grey waste and anaerobic digestion (AD), which are commonly coupled with the composting of the sorted part of biowaste (composting) (Bolzonella et al. 2006; Gurgul et al. 2018). According to Kumar and Samadder (2017), even though on a small scale, the following waste-to-energy technologies are often encountered in the developing world: thermal conversion (incineration and mass combustion), biochemical transformation (anaerobic digestion, landfills with gas, and composting) and chemical transformation (esterification). Despite shortcomings such as greenhouse gas (GHG) emissions, high upfront costs, low calorific value, and other factors, prevalent combustion and thermal technologies such as incineration, mass burn, pyrolysis, and gasification have proven beneficial not only for reducing the bulk and volume of initial waste, but also for energy recovery and reducing landfill land usage, making it one of the most common waste management solutions (Kumar and Samadder 2017). Even in the face of a variety of obstacles such as public perception, public–private partnerships, stakeholder indifference, discriminatory policies, financial constraints and climatic conditions, the adoption of waste–to–energy technologies have progressed rather well in several developed and developing countries (Abdul-wahab and Mohammed 2019). To address the issues associated with prevalent waste-to-energy technologies, such as surface and ground water pollution, greenhouse gas emissions, and insect breeding (Goud and Mohan 2011), and also in light of the fact that organic waste constitutes a significant proportion of municipal solid waste, which contributes around 9.6% of global CO_2 emissions (Kiran et al. 2014), emerging waste-to-energy technologies, including but not limited to biological hydrogen generation (photo-fermentation and dark fermentation), bio-electrochemical processes (microbial fuel cells (MFCs), microbial electrolysis cells (MECs), torrefaction, and hydrothermal carbonization (HTC) are at various stages of development and deployment across the world (Rasheed et al. 2021). Several studies have, to a considerable extent, examined waste generation and management as one entity and renewable energy development as another (Young 2010; Zimon et al. 2020). However, a comprehensive review of emerging waste-to-energy technologies is limited in literature. It was for this reason that this study set out to review some emerging waste-to-energy technologies, specifically current advancements and their respective limitations.

2 Methodology

In this study, a systematic desk review with much reliance on secondary data was employed in reviewing emerging waste to energy technologies and their effect on environmental sustainability.

Systemic review

Several review types have been identified in literature including literature reviews, scoping reviews, overviews, critical reviews, systematic reviews, and meta-analyses with some of the assessments more comprehensive than others (Peters et al. 2020). A systematic review is a method of searching for, selecting, analyzing, and synthesizing evidence from prior research. These kinds of studies give a comprehensive response to the questions of what is known, what is unknown, and what recommendations for further research exist (Peters et al. 2020). The Systematic Review framework is based on internationally recognized criteria that are dependent on the works of other researchers. Systematic review is the most effective way of developing a framework since it allows the collection of unbiased secondary data. It aids in the analysis of emerging waste-to-energy technologies and their respective potential in this research. The study employed the widely established Preferred Reporting Items for Systematic Review.

Search strategy

For each stage of the systematic review, such as selecting search terms, keywords, criteria, and data extraction, a method was followed. As shown in Table 1, a systematic database search was conducted followed by a thorough search for several review articles on WtE technology and environmental waste management. The first keywords were found by trial and error, and the results were monitored each time to identify the terms, enabling the selection of the highest number of relevant publications connected to the systematic review topic. As a direct result, several synonyms and abbreviations were ultimately chosen.

Table 1 Literature search strings in Google scholar and ProQuest (Authors construct, 2023)

Search string	Search by:	Records identified	Custom range (by date)
"Waste to energy" OR "energy from waste")	Topic	701,000	Since 2010
"Emerging wte"	Topic	3,970	Since 2010
"Waste" AND "Energy	Topic	657,000	Since 2010
"Emerging" AND "Waste to energy technologies"	Topic	8,700	Since 2010
"Municipal solid waste" AND "Environment"	Topic	173,384	Since 2010

Inclusion and exclusion criteria

To begin, non-English publications were removed from the research articles and review studies. Only original, peer-reviewed research articles and high-quality reviews of emerging WtE technologies published in scholarly journals were used to ensure the review's quality. As a result, the titles, abstracts, and keywords of publications published between 2016 and 2022 were reviewed, and those with search strings were added to the database. As shown in Table 1, Google Scholar and ProQuest databases were searched for reviews of publications in English using the search terms ("waste to energy" OR "energy from waste"), ("environment"*), and ("municipal solid waste"), as well as their combinations in 'Topic'. The abstracts of the publications obtained were then examined to have a preliminary look at the relevant literature.

A flowchart was developed to represent the process of choosing and adding the relevant publications to the database (see Fig. 1). 59 research articles were considered irrelevant in the first phase because their topics did not incorporate the strings ("Waste") and ("Energy"). Then, five articles were eliminated based on their language and document format. As a result, 49 English articles and reviews with the topic strings "waste to energy" or "energy from waste" and "environment" were considered. However, eight studies were judged irrelevant and were removed from consideration due to a lack of extractable data. The final sample (41) for the review was synthesized and compiled. A comprehensive review of the compiled literature was performed to achieve the study objectives. This technique was selected because it decreases the possibility of errors and replication (Mulrow, 1994). The review was completed in five steps: (I) establishing the study aim, (ii) selecting the database, (iii) locating the keywords, (iv) selecting the articles, and (v) obtaining the data (Pietsch et al. 2017; Tahiru et al. 2024).

Content analysis

In this review, the technologies for converting waste into energy were broken down into the following categories: combined treatment (sequential, Combined systems,

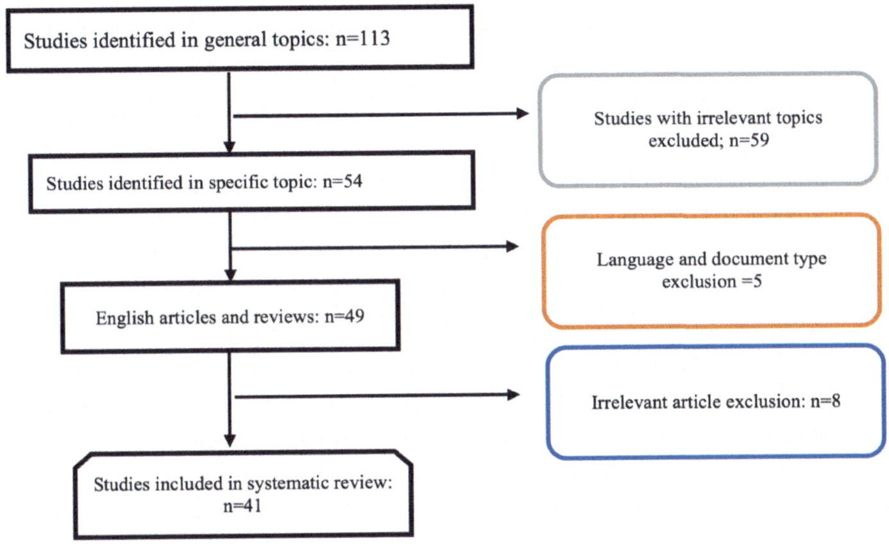

Fig. 1 Flow chart of search and selection of waste to energy technologies (author's construct)

Three-component system of bio-photolysis, dark and photo-fermentation), biological hydrogen production (photo-biological process & dark fermentation), bio-electrochemical processes (microbial fuel cells (MFCs and microbial electrolysis cells (MECs), biochemical (Ethanol fermentation), torrefaction, and hydrothermal.

3 Results and Discussion

The different WtE technologies identified during the literature search relevant to the objectives of the study are thoroughly discussed in the sections below.

<u>Biological hydrogen production (photo-biological process and dark fermentation)</u>

A promising green fuel and renewable energy among a variety of alternative clean fuels, hydrogen (H_2) has high energy content (122 kJ/g) (Hosseini and Wahid 2016; Pandey et al. 2019), and emits water as the final by-product of its combustion (Akhlaghi and Najafpour-Darzi 2020). Currently, steam reformation of CH_4, coal gasification, and electrolysis of water are the primary sources of commercially produced hydrogen, accounting for more than 90 percent of total production (Balat 2008; Mizeraczyk et al. 2014). However, because these processes are unsustainable, highly energy-intensive, and reliant on the use of fossil fuel reserves, they emit greenhouse gases into the atmosphere (Wang et al. 2020). Under ambient temperature and pressure, biological hydrogen generation (Savla et al. 2020) is an efficacious alternative approach for producing fuel from low-cost, renewable, and environmentally friendly resources. Different techniques, such as photo-biological processes and dark fermentation, can be used to generate bio-hydrogen from bio waste.

Photo-biological process

Photo fermentation is a biological process in which photosynthetic bacteria, primarily purple non-sulfur bacteria (PNSB), produce hydrogen in the absence of oxygen and nitrogen but in the presence of light and a carbon source (Ghosh et al. 2017).

The light source can be solar or various artificial sources, and the carbon source can be simple sugar or various wastes containing volatile fatty acids. One mol of glucose can theoretically produce 12 mol of hydrogen, but other volatile fatty acids (VFAs) can produce varying amounts of hydrogen depending on their chemical properties. However, when compared to dark fermentation, photosynthetic bacteria degrade sugar and various VFAs at a slower pace (Mizeraczyk et al. 2014). Furthermore, PNSB appears to prefer VFAs over sugars as a substrate, as indicated by its quicker hydrogen generation rates with VFAs compared to pure sugars (Tian et al. 2019). As a result, combining dark and photo fermentation has been proposed and developed in recent years, and an extensive study has been conducted (Ventura et al. 2021).

$$N_2 + 8H + +8e - +16ATP \rightarrow 2NH_3 + H2 + 16ADP + 16Pi \tag{1}$$

$$8H + +8e - +16ATP \rightarrow 4H_2 + 16ADP + 16Pi \tag{2}$$

Unlike dark fermentation, which employs hydrogenase to produce hydrogen, photo fermentation employs a nitrogenase vehicle to drive hydrogen production (Ma et al. 2021). When one mol of nitrogen is fixed to ammonia by nitrogenase, one mol of hydrogen is generated, as described in Eq. (1). In the absence of nitrogen, hydrogen can be synthesized more effectively (Chernova and Kiseleva 2017), as indicated in Eq. (2). As a result, the nitrogen source and the C/N ratio in the substrate are crucial in photo fermentation hydrogen generation. Finally, because light is required for photo fermentation, light efficiency is an important indicator for determining the system's energy conversion rate (Tian et al. 2019).

Dark fermentation

Toledo-Alarcón et al. (2018), identified dark fermentation as one of the most investigated methodologies for producing hydrogen using biological methods. In dark fermentation, through the use of anaerobic microbes, hydrogen was produced from a variety of waste products (e.g., food waste or agricultural waste) or wastewater (Kadier et al. 2016).

The acetate-mediated route is used to produce hydrogen through dark fermentation. Equation 3 was the overall reaction:

$$C_6H_{12}O_6 + \frac{2H_2O}{4H_2} + 2CO_2 + 2CH_3COOH \ DGO^1/_4 \rightarrow 206\ kj \tag{3}$$

As shown in Eq. (3), the maximum amount of hydrogen produced per 1 mol of glucose used as a carbon substrate is 4 mol. However, because microorganisms consume a portion of glucose for development and maintenance (Akinbomi & Taherzadeh, 2015) and produce volatile fatty acids (VFAs), the actual yield of hydrogen production is lower than the theoretical yield. Reactor parameters, MSW components, and substrate types all influence biohydrogen yield. Apart from that, the most prevalent pathway for dark fermentative hydrogen production from diverse types of bio waste materials involves the breakdown of monosaccharides and glycerol. According to Beyene et al. (2020), nutritional composition and feedstock from a carbohydrate-rich group produce higher H_2 yields than feedstock from the other groups. Carbohydrates outperform proteins, lipids, and lignocelluloses as a dark fermentative H_2 substrate in MSW (Dong et al. 2009). According to De Luca et al. (2012), waste variability in MSW has a significant impact on the production of H_2 potential. Variations in waste composition can cause significant changes in H_2 output but have less impact on methane generation. One of the main problems with dark fermentation for producing biohydrogen out of waste is that the yield is low (Sekoai et al. 2018).

Two-component systems of dark and photo-fermentation

In comparison to single-stage dark fermentation or photo-fermentation processes, the hydrogen output in these systems is higher. Additionally, photosynthetic bacteria have a reduced overall energy need (Bundhoo, 2017). Hydrogen is generated as a by-product of acidogenic bacteria's anaerobic breakdown of organic wastes into VFAs and alcohols during the dark fermentation process (Annamalai et al. 2020). However, the yield and rate of hydrogen production are limited due to the accumulation of VFAs in the medium (Annamalai et al. 2020). PNS bacteria can overcome the free energy barrier in the catabolism of VFAs and transform them into hydrogen during the nitrogen fixation process, utilizing light energy during the photo-fermentation process (Chandrasekhar et al. 2015). In general, three types of integrated dark and photo-fermentation hydrogen production processes are available: sequential (consecutive operation), combined (simultaneous operation), and three-component systems of bio-photolysis, dark, and photo-fermentation bio-process approaches (Akhlaghi & Najafpour-Darzi, 2020).

Sequential

The waste effluents (organic acids) from dark fermentation are employed as a substrate for photosynthetic bacteria in the photo-fermentation stage of the sequential dark-and-photo-fermentation processes (Seifert et al. 2018). Using *C. butyricum* (dark fermentative) and *R. palustris* (photo fermentative bacteria), Mishra et al. (2019) showed significant increases in overall hydrogen production. In addition to increased hydrogen production, 93% of total COD was removed from palm oil mill effluent (Mishra et al. 2016; Krishnan et al. 2016). In a similar study, using a sequential approach enhanced hydrogen generation from corn cob while also removing the brown color of dark fermentation effluent by color component digestion during the photo-fermentation stage (Akhlaghi and Najafpour-Darzi 2020; Rodrigues et al. 2020).

Combined systems

While dark-fermentative bacteria produce VFAs, in-situ photo-fermentative bacteria convert the VFAs to hydrogen in a single bioreactor (Ghosh et al. 2018). Because PNS bacteria use the short-chain organic acids generated by dark fermentation as carbon sources, a larger hydrogen output can be achieved when the two systems are combined than when dark or photo-fermentation is used alone (Mahidhara et al. 2019). However, the ideal environmental parameters for dark-and photo-fermentative bacteria differ, limiting their ability to operate in a stable bioreactor (Argun & Kargi 2011).

Three-component system of bio-photolysis, dark and photo-fermentation

This system is built on a commensalism relationship between strains of organisms, feedstock, and products, with three stages: biomass is collected by green algae or cyanobacteria by photosynthetic CO_2 fixation in the first stage. The accumulated biomass can be processed to produce hydrogen and organic acids in the second step by using dark-fermentative bacteria in a dark fermenter. Finally, the organic acids are employed as a substrate source for photo-fermentative hydrogen generation using PNS bacteria (Nabati 2017).

Bio-electrochemical processes

Microbial fuel cells (MFC) and microbial electrolysis cells (MEC) are emerging as sustainable power source technologies that can be used to process waste and produce hydrogen. Hydrogen gas has enormous promise as an environmentally friendly automotive fuel. MFC is a bio-electrochemical process that uses microorganisms as an impetus to oxidize organic and inorganic substances and generate energy, whereas MEC is a reactor that combines MFC and electrolysis to produce biohydrogen. The cleanest, most ecologically friendly, and most efficient ways for biohydrogen production have been found in MEC and MFC (Srikanth et al. 2018; Tang et al. 2019).

Microbial fuel cells (MFCs)

Microbial fuel cells rely on microorganisms like bacteria as catalysts to produce electricity from organic waste. These microorganisms facilitate electrochemical oxidations and reductions at an anode and a cathode, respectively (Logroño et al. 2015). However, excessively concentrated substrates might cause fouling of the separator and/or exchange membrane, resulting in poor substrate degradation and poorer total energy generation (Kumar et al. 2018). Furthermore, certain bacteria can only digest specific fermentation products to generate energy, restricting their application (Caccavo et al. 1994; Xu et al. 2012). While microbial fuel cells (MFCs) are a promising strategy for biohydrogen synthesis and energy production from a variety of organic substrates such as household waste, animal waste, and waste sludge (Beyene et al. 2018), MFCs perform less efficiently for industrial wastewater treatment than simple and specified substrates (Gude 2016). This can be linked to variables such as lower complex organic breakdown rates and substrate competitive processes in the anodic community, such as the competition process for methane production, which reduces the amount of available substrate for electricity generation (Lee et al. 2009). As a result, pre-treatment methods such as dilution, autoclave, and sonication have been developed to increase the biological properties of wastewater and improve MFC performance (Arun et al. 2020).

Microbial electrolysis cells (MECs)

This process is akin to that of MFCs, with the exception that the cathode of MECs is not exposed to the atmosphere (Shehab et al. 2013). The use of microbial electrolysis cells (MECs) has attracted increasing attention in recent years as a viable option for getting clean and acceptable energy from waste (Rabaey et al. 2009; Dutta 2014). Furthermore, as compared to dark, photo fermentation, and MFC, MEC has a better H_2 recovery and a greater variety of substrates (Kadier et al. 2016). The MEC method takes more than 90% of H_2 retrieval, compared to just 33% with the dark fermentation process (Beyene et al. 2018). In this system, a low voltage is generated at the anode, and an external voltage is applied to the circuit, resulting in the production of hydrogen at the cathode (Hallenbeck et al. 2012; Kadier et al. 2016). In MECs, bacteria such as geobacter sulfurreducens and Shewanella oneidensis grow on and adhere to the anode. The properties of these microorganisms affect the products produced and the efficiency with which the MECs function (Kumar et al. 2018). The anode material in a MEC can be the same as the anode material in an MFC, such as carbon fabric, carbon paper, graphite felt, graphite granules, or graphite brushes, or it can be a completely different material (Mier et al. 2021). To lower the overpotential necessary for hydrogen synthesis, platinum can be utilized as a catalyst. However, biocathodes are being investigated as an alternative to platinum because of the high cost of platinum (Kundu et al. 2013).

Torrefaction

When heat is sufficiently applied to biomass under controlled thermodynamic conditions, torrefying occurs, causing chemical, physical, and mechanical changes in the biomass. The torrefied biomass can be cofired with coal to reduce the reliance on fossil fuels while curbing emissions of CO_2, NOx, and SOx (Huang et al. 2013; Arteaga-Pérez et al. 2015). Due to the lower energy density of biomass compared to fossil fuels such as coal or diesel, when biomass is directly consumed, the utilization and efficiency are extremely low (Sharma et al. 2014). Several studies have demonstrated that torrefaction can boost the energy density of biomass to levels that are equivalent to or even greater than those of coal (Pahla et al. 2018; Acharya et al. 2012; Tchapda & Pisupati 2014). When compared to pyrolysis, torrefaction produces char with a greater energy content, lower water adsorption capacity and enhanced stability without additional microbial degradation (Okolie et al. 2022). To effectively utilize MSW as a fuel for combustion, it is necessary to address the limitations of pyrolysis, incineration, mas burning, and gasification, such as high moisture content and low calorific value of the waste, poor grindability, poor combustion properties, low energy density, low H/C and O/C ratios of the waste, as well as low H/C and O/C ratios of the waste (Abdulyekeen et al. 2021). Pretreatment of MSW such as torrefaction is an efficient method of reducing these problems. Torrefaction is therefore an environmentally friendly and promising thermochemical process that is used for the pretreatment of various biomass resources to reduce the aforementioned MSW limitations (Ciolkosz and Wallace 2011; Tumuluru et al. 2011).

Dry and Wet Torrefaction

Torrefaction processes can be categorized into two types depending on their environmental circumstances (Siwal et al. 2021). The first is termed "dry torrefaction." It entails enabling the controlled heating of organic waste in inert atmospheres with limited oxygen availability for partial breakdown of the material without combustion (Sarker et al. 2021). These conditions can be obtained by applying a partial or complete vacuum and raising the concentration of inert gases such as nitrogen gas (N_2) or carbon dioxide (CO_2) (Arauzo et al. 2020). The second type, called "wet torrefaction" or "hydrothermal treatment," is when biomass is partially broken down by heating it in hot water or steam in high-pressure reactors in a controlled manner (Shen 2020). Wet torrefaction requires a smaller temperature range and a shorter heating time than dry torrefaction. While water treatment typically uses a temperature range of 150–260 °C, traditional dry treatment is often performed at 200–300 °C for a few minutes to a few hours (Yek et al. 2021).

According to Lui et al. (2020), wet torrefaction incurs fewer constraints on the moisture and purity of the organic waste to be treated. It is also usual for organic waste, particularly from municipal solid waste, to have soil and other impurities impregnated throughout the collection, transport, and handling stages. There is a natural washing of biomass during wet torrefaction, allowing the virtually complete removal of impurities (Negi et al. 2020). Because of the transport and solubilization capabilities of water, biomasses treated to this kind of torrefaction have low ash levels. The primary drawbacks of wet torrefaction are the technical, financial, and environmental difficulties associated with its application (Chen and Liu 2021). On a commercial scale, industrial manufacture of reactors for hydrothermal treatment of biomass at high temperatures is almost non-existent worldwide. The primary challenges are in developing reactors that are resistant to high internal pressures attained during treatment (Chen and Liu 2021).

Hydrothermal Carbonization

According to Liu et al. (2020), hydrothermal carbonization HTC is a thermochemical process in which waste biomass is converted into carbon-rich products using saturated water and vapor pressure. In most cases, HTC is performed at temperatures ranging from 150 to 350°C and under autogenous pressure (Al Ramahi et al. 2021). Thus, the water content remains in the aqueous phase during the HTC process, but the density and dielectric constant of the water decreases considerably. In a similar vein, the O and H contents of the feedstock decrease as a result of the severe dehydration and decarboxylation events that take place during the process (Wang et al. 2018). Additionally, liquid and gas products are produced during the hydrothermal carbonization process in addition to its solid residues. The liquids obtained from the hydrothermal carbonization process of Human excreta (Spitzer et al. 2018), sewage and dairy sludge (Atallah et al. 2020), agricultural waste (Lucian et al. 2018), and poultry litter (Mau et al. 2016) have all been used to produce liquid products for a variety of applications, including soil amendment and bioenergy generation. Because of its extremely hydrophobic and friable qualities (Wang et al. 2018), HTC-derived hydrochar can be easily recovered from its aqueous phase during the separation process. Furthermore, HTC hydrochar has a high mass and energy density, making it a viable candidate for utilization as a clean energy source (Mau and Gross, 2018). For its promising characteristics, there has been a tremendous increase in interest in HTC hydrochar for use in environmental and renewable energy applications in recent years. For example, Al Ramahi et al. (2021) have emphasized the potential of hydrochar in climate change mitigation, while the use of hydrochar as a soil amendment has also been widely documented in literature (Reza et al. 2014).

Ethanol fermentation

Biochemical methods can be used to convert waste/biomass into ethanol for use in vehicles and generator sets. When compared to petroleum fuels, according to Gavahian et al. (2016), bioethanol is a more environmentally friendly and attractive replacement for gasoline due to its high-octane number and greater flammable temperature. Additionally, mixing bioethanol with gasoline in the concentration range of for example 10–8% (v/v) (Vohra et al. 2014) decreases greenhouse gas emissions and other hazardous gases, further demonstrating the sustainable nature of the fuel (Gavahian et al. 2016). In the conversion of waste into bioethanol, the first step is to ensure that the waste is adequately pretreated before proceeding. Enzymes or acids are utilized to break down the waste, and then microbial fermentation takes place, resulting in the production of ethanol. Finally, the fermented broth is distilled to produce fuel-grade bioethanol suitable for use as a fuel (Gavahian et al. 2016; Koti et al. 2016).

According to Koçar and Civaş (2013), the residues from high-yielding sugar liquid, lignocellulose, and starch crops, such as cassava, sugarcane, maize, and other crops, might be a valuable source of raw materials to produce ethanol fuel. Naresh Kumar et al. (2020) used pre-treated woody stem Prosopis juliflora to produce ethanol, with Kluyveromyces marxianus being used to separate the hydrolysis and fermentation processes. It was at a temperature of 41 °C and a pH of 4.9 that the maximum yield of bioethanol (21.45 g/L) was obtained per liter of water. The substrate and inoculum concentrations were 5% (w/v) and 3% (v/v). Narra et al. (2015) used lignocellulosic biomass for bioethanol synthesis, which was pre-treated with acid as well as joint saccharification and fermentation by Kluyveromyces sp. Maximum yields of ethanol were realized from rice straw, wheat straw, and sugarcane bagasse, with yields of 23.23, 18.29, and 17.91 g/L, respectively, under optimal conditions (60 h; 10% solid loading; enzyme loading of 9 FPU/g substrate). Furthermore, due to its ease of use and higher efficiency, ethanol fuel is an ideal option for transportation fuel. Research has revealed that ethanol production and combustion result in a 12% reduction in greenhouse gas emissions compared to gasoline usage

(Hill et al. 2006). As a result, fermentation-produced ethanol accounts for more than 90% of the global ethanol industry (Zi et al. 2013). However, because a significant percentage of ethanol production has been carried out from food-base materials such as sugar cane and corn, bioethanol fuel is more expensive than fossil fuels in comparison (Karmee 2016).

4 Challenges and Prospects Associated with Emerging WtE Technologies

Various challenges are associated with the application of waste-to-energy technologies, particularly for better performing emerging WtE technologies or various combinations of emerging WtE technologies to gain additional benefits. These problems include higher costs, limited understanding of how emerging WtE technologies work and how they can be used with conventional WtE technologies, and the fact that emerging WtE technologies have not been very successful commercially (Harvey et al. 2018). Additionally, emerging WtE technologies have greater capital, compliance, and operational costs than traditional approaches. For example, employing MEC requires 50 kWh of electricity to produce 1 kg of hydrogen gas (Dange et al. 2021). Furthermore, skilled labor expenses are considerable, raising the operating costs of emerging WtE procedures above those of conventional methods. The understanding of emerging WtE processes, fundamental and thermodynamic facts, and reaction kinetics is limited (Munir et al. 2018).

The commercial limitation of emerging WtE techniques for WtE conversion is another obstacle linked with their deployment. Despite the numerous benefits outlined in the results and discussion, emerging WtE technologies have not proven economically viable. This limited commercial success might be attributable to safety concerns and costly prices. Consequently, emerging WtE technologies are less attractive for investment (Afrane et al. 2021). It is obvious from the results and discussion that society, community, and technical readiness levels for emerging WtE technologies range from low to medium. Since the majority of emerging WtE technologies operate under severe conditions, their social preparedness levels are often poor (Hoang et al. 2022). Moderate consumer readiness exists for emerging WtE technologies due to their effective waste treatment and increased net energy generation capacities. In addition, emerging WtE approaches have a moderate level of technological preparedness since they are relatively new and lack fundamental data. Consequently, lower odds of societal, community, and technological preparedness for emerging WtE technologies hinder investment in same. Waste management strategy uncertainty is another problem. This is likely because WtE facilities need a steady supply of MSW, which goes against the global recycling policy. However, as compared to conventional techniques, they provide a significantly larger net income, contribute to the generation of energy, are significantly safer and more environmentally friendly, and are thus sustainable.

5 Conclusion

The utilization of waste in energy production has been one such technological novelty that has been gaining cognizance globally in recent times. The technology is far more impressive since it not only helps to solve the waste management problem in most countries but also generate revenues and aids in closing the energy deficit that most developing countries are grappling with. The findings of this study overwhelmingly support the position that waste generated by cities, farms, industries, and households has great potential for repurposing as energy. This review provides an insight into current research activities that are centered on techniques for concurrent waste reduction and energy generation. Various WtE technologies have been investigated, including biochemical routes (bioethanol production), bio-electrochemical processes (microbial fuel cells (MFC) and microbial electrolysis cells (MEC)), and biological hydrogen production routes (photo biological hydrogen production, dark fermentation), as well as pre-treatment technologies. This study found from most of the literature that, integrated WtE technologies, such as those that combine dark fermentation with photo fermentation and microbial electrolysis cells with microbial fuel cells, are more promising than other WtE technologies due to their higher H_2 recovery, cost effectiveness, environmental friendliness, technological sustainability, and greater substrate diversity. On a one-to-one basis, the use of MECs ranks first in terms of efficiency, followed closely by ethanol fermentation, even though it requires a grueling treatment step and emits trace gases with odor into the environment. Lastly, the study recommends the advancement of research with particular emphasis on (1) making WtE conversion at a large scale more economically feasible and eco-friendly in order to realize sustainable development goals 7 and 11 (target 11.6) and (2) resolve the by-product accumulation problem.

References

Abdulyekeen, K. A., Umar, A. A., Patah, M. F. A., & Daud, W. M. A. W. (2021). Torrefaction of biomass: Production of enhanced solid biofuel from municipal solid waste and other types of biomass. Renewable and Sustainable Energy Reviews, 150, 111436.

Abdul-Wahab, T., & Mohammed, A. T. (2019). Biodiesel Production from Neem (Azadirachtaindica) Seed Oil. International journal of innovative research & development. ISSN 2278 – 0211. DOI NO: 10.24940/IJIRD/2019/V8/I8/AUG1903.

Acharya, B., Sule, I., & Dutta, A. (2012). A review on advances of torrefaction technologies for biomass processing. Biomass Conversion and Biorefinery, 2(4), 349–369.

Afrane, S., Ampah, J. D., Jin, C., Liu, H., & Aboagye, E. M. (2021). Techno-economic feasibility of waste-to-energy technologies for investment in Ghana: A multicriteria assessment based on fuzzy TOPSIS approach. Journal of Cleaner Production, 318, 128515.

Akhlaghi, N., & Najafpour-Darzi, G. (2020). A comprehensive review on biological hydrogen production. International Journal of Hydrogen Energy, 45(43), 22492-22512.

Akinbomi, J., & Taherzadeh, M. J. (2015). Evaluation of fermentative hydrogen production from single and mixed fruit wastes. Energies, 8(5), 4253–4272.

Al Ramahi, M., Keszthelyi-Szabó, G., & Beszédes, S. (2021). Coupling hydrothermal carbonization with anaerobic digestion: an evaluation based on energy recovery and hydrochar utilization. Biofuel Research Journal, 8(3), 1444–1453

AlQattan, N., Acheampong, M., Jaward, F. M., Ertem, F. C., Vijayakumar, N., & Bello, T. (2018). Reviewing the potential of Waste-to-Energy (WTE) technologies for Sustainable Development Goal (SDG) numbers seven and eleven. Renewable Energy Focus, 27, 97–110.

Annamalai, N., Elayaraja, S., Oleskowicz-Popiel, P., Sivakumar, N., & Al Bahry, S. (2020). Volatile fatty acids production during anaerobic digestion of lignocellulosic biomass. In Recent Developments in Bioenergy Research (pp. 237–251). Elsevier.

Arauzo, P. J., Maziarka, P. A., Olszewski, M. P., Isemin, R. L., Muratova, N. S., Ronsse, F., & Kruse, A. (2020). Valorization of the poultry litter through wet torrefaction and different activation treatments. Science of The Total Environment, 732, 139288.

Argun, H., & Kargi, F. (2011). Bio-hydrogen production by different operational modes of dark and photo-fermentation: an overview. International Journal of Hydrogen Energy, 36(13), 7443–7459.

Arteaga-Pérez, L. E., Vega, M., Rodríguez, L. C., Flores, M., Zaror, C. A., & Ledón, Y. C. (2015). Life-Cycle Assessment of coal–biomass based electricity in Chile: Focus on using raw vs torrefied wood. Energy for sustainable development, 29, 81–90.

Arun, S., Sinharoy, A., Pakshirajan, K., & Lens, P. N. (2020). Algae based microbial fuel cells for wastewater treatment and recovery of value-added products. Renewable and Sustainable Energy Reviews, 132, 110041.

Atallah, E., Zeaiter, J., Ahmad, M. N., Kwapinska, M., Leahy, J. J., & Kwapinski, W. (2020). The effect of temperature, residence time, and water-sludge ratio on hydrothermal carbonization of DAF dairy sludge. Journal of Environmental Chemical Engineering, 8(1), 103599.

Balat, M. (2008). Potential importance of hydrogen as a future solution to environmental and transportation problems. International journal of hydrogen energy, 33(15), 4013-4029.

Beyene, H. D., Werkneh, A. A., & Ambaye, T. G. (2018). Current updates on waste to energy (WtE) technologies: a review. Renewable Energy Focus, 24, 1–11.

Bolzonella, D., Pavan, P., Mace, S., & Cecchi, F. (2006). Dry anaerobic digestion of differently sorted organic municipal solid waste: a full-scale experience. Water Science and Technology, 53(8), 23–32.

Bundhoo, Z. M. (2017). Coupling dark fermentation with biochemical or bioelectrochemical systems for enhanced bio-energy production: a review. International Journal of Hydrogen Energy, 42(43), 26667–26686.

Caccavo Jr, F., Lonergan, D. J., Lovley, D. R., Davis, M., Stolz, J. F., & McInerney, M. J. (1994). Geobacter sulfurreducens sp. nov., a hydrogen-and acetate-oxidizing dissimilatory metal-reducing microorganism. Applied and environmental microbiology, 60(10), 3752–3759.

Capizzi, F., Das, A., Dauwe, T., Moorkens, I., Juhana, R., & Tomescu, M. (2019). Renewable energy in Europe—2019. European Environmental Agency-European Topic Centre on Climate Change Mitigation and Energy: Mol, Belgium.

Chandrasekhar, K., Lee, Y. J., & Lee, D. W. (2015). Biohydrogen production: strategies to improve process efficiency through microbial routes. International journal of molecular sciences, 16(4), 8266–8293.

Chen, Y. C., & Liu, H. M. (2021). Evaluation of greenhouse gas emissions and the feed-in tariff system of wasteto-energy facilities using a system dynamics model. Science of the Total Environment, 792, 148445.

Chernova, N. I., & Kiseleva, S. V. (2017). Microalgae biofuels: Induction of lipid synthesis for biodiesel production and biomass residues into hydrogen conversion. International journal of hydrogen energy, 42(5), 2861–2867.

Ciolkosz, D., & Wallace, R. (2011). A review of torrefaction for bioenergy feedstock production. Biofuels, Bioproducts and Biorefining, 5(3), 317-329.

Dange, P., Pandit, S., Jadhav, D., Shanmugam, P., Gupta, P. K., Kumar, S., ... & Bhatia, S. K. (2021). Recent developments in microbial electrolysis cell-based biohydrogen production utilizing wastewater as a feedstock. Sustainability, 13(16), 8796.

Dhokhikah, Y., & Trihadiningrum, Y. (2012). Solid waste management in Asian developing countries: challenges and opportunities. Journal of Applied Environmental and Biological Sciences, 2(7), 329–335.

Dong, L., Zhenhong, Y., Yongming, S., Xiaoying, K., & Yu, Z. (2009). Hydrogen production characteristics of the organic fraction of municipal solid wastes by anaerobic mixed culture fermentation. International Journal of Hydrogen Energy, 34(2), 812–820.

Dutta, S. (2014). A review on production, storage of hydrogen and its utilization as an energy resource. Journal of Industrial and Engineering Chemistry, 20(4), 1148–1156.

Gavahian, M., Farahnaky, A., Shavezipur, M., Sastry, S., 2016. Ethanol concentration of fermented broth by ohmic-assisted hydrodistillation. Innovat. Food Sci.Emerg. Technol. 35, 45e51. https://doi.org/10.1016/j.ifset.2016.04.001.

Ghosh, S., Dairkee, U. K., Chowdhury, R., & Bhattacharya, P. (2017). Hydrogen from food processing wastes via photofermentation using purple non-sulfur bacteria (PNSB)–a review. Energy Conversion and Management, 141, 299–314.

Ghosh, S., Chowdhury, R., & Bhattacharya, P. (2018). A review on single stage integrated dark-photo fermentative biohydrogen production: insight into salient strategies and scopes. International Journal of Hydrogen Energy, 43(4), 2091–2107.

Goud, R.K. & Mohan, S.V. (2011). Pre-fermentation of waste as a strategy to enhance the performance of single chambered microbial fuel cell (MFC). Int J Hydrog Energy 36:13753–13762

Gude, V. G. (2016). Microbial fuel cells for wastewater treatment and energy generation. In Microbial Electrochemical and Fuel Cells (pp. 247–285). Woodhead Publishing.

Gurgul, A., Szczepaniak, W., & Zabłocka-Malicka, M. (2018). Incineration and pyrolysis vs. steam gasification of electronic waste. Science of the total environment, 624, 1119–1124.

Hallenbeck, P. C., Abo-Hashesh, M., & Ghosh, D. (2012). Strategies for improving biological hydrogen production. Bioresource technology, 110, 1–9.

Harvey, H., Orvis, R., & Rissman, J. (2018). Designing climate solutions: a policy guide for low-carbon energy. Island Press.

Hill, J. Nelson, E. Tilman, D. Polasky, S. & Tiffany, D. (2006). Environmental, economic, and energetic costs and benefits of biodiesel and ethanol biofuels. Proc Natl Acad Sci U S A 103(30):11206–11210.

Hoang, A. T., Varbanov, P. S., Nižetić, S., Sirohi, R., Pandey, A., Luque, R., & Ng, K. H. (2022). Perspective review on Municipal Solid Waste-to-energy route: Characteristics, management strategy, and role in circular economy. Journal of Cleaner Production, 131897.

Hosseini, S. E., & Wahid, M. A. (2016). Hydrogen production from renewable and sustainable energy resources: Promising green energy carrier for clean development. Renewable and Sustainable Energy Reviews, 57, 850–866.

Huang, Y. F., Syu, F. S., Chiueh, P. T., & Lo, S. L. (2013). Life cycle assessment of biochar cofiring with coal. Bioresource technology, 131, 166–171.

Kadier, A., Simayi, Y., Abdeshahian, P., Azman, N. F., Chandrasekhar, K., & Kalil, M. S. (2016). A comprehensive review of microbial electrolysis cells (MEC) reactor designs and configurations for sustainable hydrogen gas production. Alexandria Engineering Journal, 55(1), 427–443.

Karmee, S.K. (2016). Liquid biofuels from food waste: current trends, prospect and limitation. Renew Sustain Energ Rev 53:945–953.

Kaur, A., Bharti, R., & Sharma, R. (2021). Municipal solid waste as a source of energy. Materials Today: Proceedings.

Kiran, E.U. & Trzcinski, A.P. Nig, W,J. & Liu, Y. (2014). Bioconversion of food waste to energy: a review. Fuel 134(9):389–399

Klavenieks, K., Dzene, K. P., & Blumberga, D. (2017). Optimal strategies for municipal solid waste treatment–environmental and socio-economic criteria assessment. Energy Procedia, 128, 512–519.

Koçar, G., & Civaş, N. (2013). An overview of biofuels from energy crops: Current status and future prospects. Renewable and Sustainable Energy Reviews, 28, 900–916.

Koti, S., Govumoni, S.P., Gentela, J., Venkateswar Rao, L., 2016. Enhanced bioethanol production from wheat straw hemicellulose by mutant strains of pentose fermenting organisms Pichia stipitis and Candida shehatae. SpringerPlus 5, 1545. https://doi.org/10.1186/s40064-016-3222-1.

Krishnan, S., Singh, L., Sakinah, M., Thakur, S., Wahid, Z. A., & Alkasrawi, M. (2016). Process enhancement of hydrogen and methane production from palm oil mill effluent using two-stage thermophilic and mesophilic fermentation. international journal of hydrogen energy, 41(30), 12888–12898.

Kumar, A., & Samadder, S. R. (2017). A review on technological options of waste to energy for effective management of municipal solid waste. Waste Management, 69, 407–422.

Kumar, R., Singh, L., Zularisam, A. W., & Hai, F. I. (2018). Microbial fuel cell is emerging as a versatile technology: a review on its possible applications, challenges and strategies to improve the performances. International Journal of Energy Research, 42(2), 369–394.

Kundu, A., Sahu, J. N., Redzwan, G., & Hashim, M. A. (2013). An overview of cathode material and catalysts suitable for generating hydrogen in microbial electrolysis cell. international journal of hydrogen energy, 38(4), 1745–1757.

Lee, D. H., Behera, S. K., Kim, J. W., & Park, H. S. (2009). Methane production potential of leachate generated from Korean food waste recycling facilities: a lab-scale study. Waste Management, 29(2), 876–882.

Logroño, W., Ramírez, G., Recalde, C., Echeverría, M., & Cunachi, A. (2015). Bioelectricity generation from vegetables and fruits wastes by using single chamber microbial fuel cells with high Andean soils. Energy Procedia, 75, 2009–2014.

Liu, X., Hou, Y., Li, Z., Yu, Z., Tang, J., Wang, Y., & Zhou, S. (2020). Hyperthermophilic composting of sewage sludge accelerates humic acid formation: Elemental and spectroscopic evidence. Waste management, 103, 342–351.

De Luca, L., Donato, A., Santangelo, S., Faggio, G., Messina, G., Donato, N., & Neri, G. (2012). Hydrogen sensing characteristics of Pt/TiO2/MWCNTs composites. international journal of hydrogen energy, 37(2), 1842–1851.

Lucian, M., Volpe, M., Gao, L., Piro, G., Goldfarb, J. L., & Fiori, L. (2018). Impact of hydrothermal carbonization conditions on the formation of hydrochars and secondary chars from the organic fraction of municipal solid waste. Fuel, 233, 257-268.

Lui, J., Chen, W. H., Tsang, D. C., & You, S. (2020). A critical review on the principles, applications, and challenges of waste-to-hydrogen technologies. Renewable and Sustainable Energy Reviews, 134, 110365.

Ma, H., Yang, H., Zheng, X., Lie, T., & Yan, W. (2021). Promoting photo-fermentative hydrogen production performance by substituting the rnf promoter in Rhodobacter capsulatus. International Journal of Hydrogen Energy, 46(5), 3742–3752.

Mahidhara, G., Burrow, H., Sasikala, C., & Ramana, C. V. (2019). Biological hydrogen production: Molecular and electrolytic perspectives. World Journal of Microbiology and Biotechnology, 35(8), 1–13.

Mau, V., & Gross, A. (2018). Energy conversion and gas emissions from production and combustion of poultry-litter-derived hydrochar and biochar. Applied Energy, 213, 510–519.

Mau, V., Quance, J., Posmanik, R., & Gross, A. (2016). Phases' characteristics of poultry litter hydrothermal carbonization under a range of process parameters. Bioresource technology, 219, 632–642.

Mier, A. A., Olvera-Vargas, H., Mejía-López, M., Longoria, A., Verea, L., Sebastian, P. J., & Arias, D. M. (2021). A review of recent advances in electrode materials for emerging bioelectrochemical systems: From biofilm-bearing anodes to specialized cathodes. Chemosphere, 283, 131138.

Mishra, P., Thakur, S., Singh, L., Ab Wahid, Z., & Sakinah, M. (2016). Enhanced hydrogen production from palm oil mill effluent using two stage sequential dark and photo fermentation. international journal of hydrogen energy, 41(41), 18431-18440.

Mishra, P., Ameen, F., Zaid, R. M., Singh, L., Ab Wahid, Z., Islam, M. A., ... & Al Nadhari, S. (2019). Relative effectiveness of substrate-inoculum ratio and initial pH on hydrogen production

from palm oil mill effluent: kinetics and statistical optimization. Journal of Cleaner Production, 228, 276–283.
Mizeraczyk, J., Urashima, K., Jasinski, M., & Dors, M. (2014). Hydrogen production from gaseous fuels by plasmas—a review. Int. J. Plasma Environ. Sci. Technol, 8(2), 89–97.
Mulrow, C. D. (1994). Systematic reviews: rationale for systematic reviews. Bmj, 309(6954), 597–599.
Munir, M. T., Mansouri, S. S., Udugama, I. A., Baroutian, S., Gernaey, K. V., & Young, B. R. (2018). Resource recovery from organic solid waste using hydrothermal processing: Opportunities and challenges. Renewable and Sustainable Energy Reviews, 96, 64–75.
Nabati, Z. (2017). Effect of Yarrowia Lipolytica biofilm on corrosion behavior of carbon steel in simulated biodiesel storage tanks (Doctoral dissertation, University of Toledo).
Naresh Kumar, A. Chatterjee, S. Hemalatha, M. Althuri, A. Min, B. Kim, S.-H.,Venkata & Mohan, S. (2020). Deoiled algal biomass derived renewable sugars for bioethanol and biopolymer production in biorefinery framework. Bioresour. Technol. 296, 122315. https://doi.org/10.1016/j.biortech.2019.122315.
Narra, M. James, J.P. & Balasubramanian, V. (2015). Simultaneous saccharification and fermentation of delignified lignocellulosic biomass at high solid loadings by a newly isolated thermotolerant Kluyveromyces sp. for ethanol production. Bioresour. Technol. 179, 331e338. https://doi.org/10.1016/j.biortech.2014.11.116.
Negi, S., Jaswal, G., Dass, K., Mazumder, K., Elumalai, S., & Roy, J. K. (2020). Torrefaction: a sustainable method for transforming of agri-wastes to high energy density solids (biocoal). Reviews in Environmental Science and Bio/Technology, 19(2), 463–488.
Okolie, J. A., Epelle, E. I., Tabat, M. E., Orivri, U., Amenaghawon, A. N., Okoye, P. U., & Gunes, B. (2022). Waste biomass valorization for the production of biofuels and value-added products: A comprehensive review of thermochemical, biological and integrated processes. Process Safety and Environmental Protection, 159, 323–344.
Pahla, G., Mamvura, T. A., & Muzenda, E. (2018). Torrefaction of waste biomass for application in energy production in South Africa. South African Journal of Chemical Engineering, 25(1), 1–12.
Pandey, B., Prajapati, Y. K., & Sheth, P. N. (2019). Recent progress in thermochemical techniques to produce hydrogen gas from biomass: A state of the art review. International Journal of Hydrogen Energy, 44(47), 25384–25415.
Peters, M. D., Marnie, C., Tricco, A. C., Pollock, D., Munn, Z., Alexander, L., ... & Khalil, H. (2020). Updated methodological guidance for the conduct of scoping reviews. JBI evidence synthesis, 18(10), 2119–2126.
Pietsch, S. Heinrich, S., Karpinski, K., Müller, M., Schönherr, M., Jäger, F. K. (2017). CFD-DEM modeling of a threedimensional prismatic spouted bed. Powder Technology, 316, 245–255.
Rabaey, K., Angenent, L., Schroder, U., & Keller, J. (Eds.). (2009). Bioelectrochemical systems. IWA publishing.
Rasheed, T., Anwar, M. T., Ahmad, N., Sher, F., Khan, S. U. D., Ahmad, A., ... & Wazeer, I. (2021). Valorisation and emerging perspective of biomass based waste-to-energy technologies and their socio-environmental impact: A review. Journal of environmental management, 287, 112257.
Reza, M. T., Becker, W., Sachsenheimer, K., & Mumme, J. (2014). Hydrothermal carbonization (HTC): Near infrared spectroscopy and partial least-squares regression for determination of selective components in HTC solid and liquid products derived from maize silage. Bioresource technology, 161, 91–101.
Rodrigues, C. V., Alcaraz, F. A. R., Nespeca, M. G., Rodrigues, A. V., Motteran, F., Adorno, M. A. T., ... & Maintinguer, S. I. (2020). Biohydrogen production in an integrated biosystem using crude glycerol from waste cooking oils. Renewable Energy, 162, 701–711.
Sarker, T. R., Nanda, S., Dalai, A. K., & Meda, V. (2021). A review of torrefaction technology for upgrading lignocellulosic biomass to solid biofuels. BioEnergy Research, 14(2), 645–669.

Savla, N., Shinde, A., Sonawane, K., Mekuto, L., Chowdhary, P., & Pandit, S. (2020). Microbial hydrogen production: Fundamentals to application. In Microorganisms for Sustainable Environment and Health (pp. 343–365). Elsevier.

Seifert, K., Zagrodnik, R., Stodolny, M., & Łaniecki, M. (2018). Biohydrogen production from chewing gum manufacturing residue in a two-step process of dark fermentation and photofermentation. Renewable Energy, 122, 526–532.

Sekoai, P. T., Yoro, K. O., Bodunrin, M. O., Ayeni, A. O., & Daramola, M. O. (2018). Integrated system approach to dark fermentative biohydrogen production for enhanced yield, energy efficiency and substrate recovery. Reviews in Environmental Science and Bio/Technology, 17(3), 501–529.

Sharma, K. D., & Jain, S. (2020). Municipal solid waste generation, composition, and management: the global scenario. Social Responsibility Journal.

Sharma, S., Meena, R., Sharma, A., & Goyal, P. (2014). Biomass conversion technologies for renewable energy and fuels: A review note. IOSR J. Mech. Civ. Eng, 11(2), 28–35.

Shehab, N., Li, D., Amy, G. L., Logan, B. E., & Saikaly, P. E. (2013). Characterization of bacterial and archaeal communities in air-cathode microbial fuel cells, open circuit and sealed-off reactors. Applied microbiology and biotechnology, 97(22), 9885–9895.

Shen, Y. (2020). A review on hydrothermal carbonization of biomass and plastic wastes to energy products. Biomass and Bioenergy, 134, 105479.

Shivanna, K. R. (2020). The sixth mass extinction crisis and its impact on biodiversity and human welfare. Resonance, 25(1), 93–109.

Siwal, S. S., Zhang, Q., Devi, N., Saini, A. K., Saini, V., Pareek, B., ... & Thakur, V. K. (2021). Recovery processes of sustainable energy using different biomass and wastes. Renewable and Sustainable Energy Reviews, 150, 111483.

Sovacool, B. K., Griffiths, S., Kim, J., & Bazilian, M. (2021). Climate change and industrial F-gases: A critical and systematic review of developments, sociotechnical systems and policy options for reducing synthetic greenhouse gas emissions. Renewable and Sustainable Energy Reviews, 141, 110759.

Sönmez, S., Wiitala, J., & Apostolopoulos, Y. (2019). How complex travel, tourism, and transportation networks influence infectious disease movement in a borderless world. In Handbook of Globalisation and Tourism. Edward Elgar Publishing.

Spitzer, R. Y., Mau, V., & Gross, A. (2018). Using hydrothermal carbonization for sustainable treatment and reuse of human excreta. Journal of Cleaner Production, 205, 955–963.

Srikanth, S., Kumar, M., & Puri, S. K. (2018). Bio-electrochemical system (BES) as an innovative approach for sustainable waste management in petroleum industry. Bioresource technology, 265, 506–518.

Tahir, M., Hussain, T., & Behaylu, A. (2015). Scenario of present and future of solid waste generation in India: A case study of Delhi Mega City. J. of Environ and Earth Sci, 5(8), 83–91.

Tahiru, A. W., Cobbina, S. J., Asare, W., & Takal, S. U. (2024). Unlocking Energy from Waste: A Comprehensive Analysis of Municipal Solid Waste Recovery Potential in Ghana. World, 5(2), 192–218.

Tang, J., Zhang, C., Shi, X., Sun, J., & Cunningham, J. A. (2019). Municipal wastewater treatment plants coupled with electrochemical, biological and bio-electrochemical technologies: Opportunities and challenge toward energy self-sufficiency. Journal of environmental management, 234, 396–403.

Tchapda, A. H., & Pisupati, S. V. (2014). A review of thermal co-conversion of coal and biomass/waste. Energies, 7(3), 1098–1148.

Thiel, S. Thomé-Kozmiensky, E. Winter, F. Juchelková, D. (2018). Waste Management, Waste-to-Energy Vol. 8 ISBN 978-3-944310-42-8

Tian, H., Li, J., Yan, M., Tong, Y. W., Wang, C. H., & Wang, X. (2019). Organic waste to biohydrogen: a critical review from technological development and environmental impact analysis perspective. Applied Energy, 256, 113961.

Toledo-Alarcón, J., Capson-Tojo, G., Marone, A., Paillet, F., Júnior, A. D. N. F., Chatellard, L., ... & Trably, E. (2018). Basics of bio-hydrogen production by dark fermentation. In Bioreactors for microbial biomass and energy conversion (pp. 199–220). Springer, Singapore.

Tumuluru, J. S., Sokhansanj, S., Wright, C. T., Hess, J. R., & Boardman, R. D. (2011). A review on biomass torrefaction process and product properties.

Tun, M. M., Juchelkova, D., Win, M. M., Thu, A. M., & Puchor, T. (2019). Biomass energy: An overview of biomass sources, energy potential, and management in Southeast Asian countries. Resources, 8(2), 81.

Tun, M. M., Palacky, P., Juchelkova, D., & Síťař, V. (2020). Renewable waste-to-energy in southeast Asia: Status, challenges, opportunities, and selection of waste-to-energy technologies. Applied Sciences, 10(20), 7312.

Ventura, J. R. S., Rojas, S. M., Ventura, R. L. G., Nayve, F. R. P., & Lantican, N. B. (2021). Potential for biohydrogen production from organic wastes with focus on sequential dark-and photofermentation: The Philippine setting. Biomass Conversion and Biorefinery, 1–14.

Vohra, M. Manwar, J. Manmode, R. Padgilwar, S. & Patil, S. (2014). Bioethanol production: feedstock and current technologies. J Environ Chem Eng 2(1):573–584.

Wang, T., Zhai, Y., Zhu, Y., Li, C., & Zeng, G. (2018). A review of the hydrothermal carbonization of biomass waste for hydrochar formation: Process conditions, fundamentals, and physicochemical properties. Renewable and Sustainable Energy Reviews, 90, 223–247.

Wang, Q., Xue, M., Lin, B. L., Lei, Z., & Zhang, Z. (2020). Well-to-wheel analysis of energy consumption, greenhouse gas and air pollutants emissions of hydrogen fuel cell vehicle in China. Journal of Cleaner Production, 275, 123061.

Xu, J., Sheng, G. P., Luo, H. W., Li, W. W., Wang, L. F., & Yu, H. Q. (2012). Fouling of proton exchange membrane (PEM) deteriorates the performance of microbial fuel cell. Water research, 46(6), 1817–1824.

Yan, M., & Waluyo, J. (2020). Challenges for sustainable development of waste to energy in developing countries. Waste Management & Research, 38(3), 229-231.

Yek, P. N. Y., Cheng, Y. W., Liew, R. K., Mahari, W. A. W., Ong, H. C., Chen, W. H., ... & Lam, S. S. (2021). Progress in the torrefaction technology for upgrading oil palm wastes to energy-dense biochar: A review. Renewable and Sustainable Energy Reviews, 151, 111645.

Young, G. C. (2010). Municipal solid waste to energy conversion processes: economic, technical, and renewable comparisons. John Wiley & Sons.

Zi, L.H., Liu, C.G., Xin, C.B., & Bai, F.W. (2013). Stillage back set and its impact on ethanol fermentation by the flocculating yeast. Process Biochem 48:753–758.

Zimon, G., Sobolewski, M., & Lew, G. (2020). An influence of group purchasing organizations on financial security of SMEs operating in the renewable energy sector—Case for Poland. Energies, 13(11), 2926.

Spatial Distribution and Resource Mobilization of Waste for a Hybrid Waste-to-Energy Plant in Ghana

Mutala Mohammed, Pax D. Dzamboe, Mizpah A. D. Rockson, Susana Ayittah, Satyanarayana Narra, Francis B. Agyenim, Kafui Joshua, Eric B. Boakye-Danquah, Rafia Abdul-Samii, Ekua A. Armoo, Latifatu Mohammed, Rejart Adomah, Juliet Yeboa, and Edward Antwi

Abstract Waste generation is an essential characteristic of human activities as long as human continues to engage in economic activities and has become a major concern. In the context of circular economy, efficiency and energy security, it is fundamental to develop alternative energies that meet the needs of both the energy supply and sustainable development. Municipal solid waste generation is ascending and potentially capable of generating energy, thus they can serve as an alternative fuel for generating energy. This work aims to assess the potential bio-waste in terms of quality and quantity for energy generation using municipal solid waste waste, as well as mapping waste production within 30 km radius of the plant environs. The results show that the total amount of expected MSW corresponds to 51.34 tonnes daily based on per capita waste generation of 0.47 kg/day, however, only 16.62% of the expected waste generated daily is collected with the remaining waste either unofficially collected or indiscriminately disposed. With regards to quality of the waste, it was revealed that an organic waste, approximately 39% represented the major waste component and 18% plastic of the MSW generated in the municipality. Further analysis indicated that 20 MMDAs were identified within the 30 km radius of the plant location with all most populated MMDAs within the buffer zone. The immediate MMDAs that border the project site constituted 19.8% of the population translating

M. Mohammed (✉) · P. D. Dzamboe · F. B. Agyenim · E. B. Boakye-Danquah · R. Abdul-Samii · E. A. Armoo · L. Mohammed · R. Adomah · J. Yeboa
CSIR-Institute of Industrial Research, Box LG 576, Accra, Ghana
e-mail: mutbaby@gmail.com

M. A. D. Rockson · S. Ayittah
Department of Chemical Engineering, Kwame Nkrumah University of Science and Technology (KNUST), Kumasi, Ghana

S. Narra · E. A. Armoo · E. Antwi
Professorship of Material and Energy Valorisation of Biogenous Residues, Professorship of Waste and Resource Management, University of Rostock, Justus-von-Liebig-Weg 6, 18057 Rostock, Germany

K. Joshua
Tenth Technology Generation Systems, 1 Kunkun Close, East-Legon, Accra, Ghana

M. Narra and S. Narra (eds.), *Innovations in Circular Economy and Renewable Energy in Africa*, World Sustainability Series, https://doi.org/10.1007/978-3-031-68330-5_3

to approximately 189 and 72 tonnes of organic and plastic waste generation daily, respectively. Notwithstanding, about 12% each of these organic waste fraction and plastic waste is available daily for collection based on the official waste data. The analysis further showed that the feedstock needed for the biogas and pyrolysis plants could be accessed within a maximum transportation distance radius of 5 – 10 km with an additional 22 and 9 tonnes of organic and plastic waste, respectively. The analysis presented in this paper provides a true reflection of the waste generation vis-a-vis collection models and their potential for energy generation in MMDAs in Ghana, which could be used for similar communities in developing countries.

1 Introduction

Globally, energy demand is growing fast due to population increase and technological advancements. However, the negative environmental impacts of fossil fuels coupled with the rapid depletion of non-renewable resources have led to the pursue for more efficient and sustainable energy resources. The United Nations Sustainable Development Goal (SDG) number 7 is targeted at provision of affordable, reliable and modern energy services for everyone, it also aims at substantially increasing the share of energy mixed at regional and global level with renewable energy source. From circular economy perspective, waste is considered an important resource provided prevention, reuse, and recycling are prioritized in the waste management hierarchy. In developing countries, waste is still considered as "nuisance" rather than a resource and this must change in order to move towards to circular economy particularly with respect to energy generation from waste. According to the European Commission, it is only by respecting the waste hierarchy that energy from waste can maximize the circular economy's contribution to decarbonization, in line with the Energy Union Strategy and the Paris agreement (European Commission 2017; Malinauskaite et al. 2017).

Energy from waste has been highly considered as one of the options to achieve SDG 7 target and could be represented as either recovery or disposal option in the context of the waste hierarchy (Mutz et al. 2017). Utilizing Municipal Solid Waste (MSW) as an energy resource can contribute to the increase of renewable energy share in the energy mix (Shahzad et al. 2017; Nizami et al. 2017; Ouda et al. 2016; Scarlat et al. 2015; Zhang et al. 2015). Decrease in the quantity of MSW dumped in landfills with a cascading impact on carbon emissions as the process avoids methane emissions from landfills and carbon dioxide from fossil fuels as well as the provision of local source of renewable energy are some of the advantages of recovering energy from MSW (Ayodele et al. 2017; Khan and Kabir 2020). Waste-to-energy (W2E) as a source of renewable energy plays a crucial role in sustainable management of MSW at global level. It is estimated that a reduction of about 10–15% in the global Greenhouse Gas (GHG) emissions could be achieved through improved solid waste management i.e., recycling, waste diversion from landfill and energy recovery from waste.

Despite the development and adoption of the use of waste to generate energy being implemented in developed countries, Ghana is yet to implement and fully enjoy the benefits that arise from this process especially with regards to energy generation from MSW. The nation relies mainly on landfills for its MSW management which has implications for the environment and also health of citizens. However, there has been a growing need for industries to generate and use energy from the waste they produce in the face of the constant grid instabilities in the supply of electricity for production. As an economy that has been dependent on petroleum as a major share in the final energy consumption, Ghana has not yet been able to vary its energy mix to include sustainable renewable energy. The Renewable Energy provides the avenue for institutions and organizations to take advantage of the opportunities in utilizing additional resources to supplement their energy needs. The recovery of energy from MSW has a potential to become an important player in the renewable energy sector in Ghana. Currently, this potential is still far from being fully exploited as waste are dumped in landfills with no energy recovery.

Waste together with energy and transport are the largest emitters of GHG in Ghana contributing about 57% (EPA 2019). To avert this situation, Ghana has developed 20 mitigation programs of actions in seven sectors for a 10-year implementation period (2020–2030) in its Nationally Determined Contributions (NDCs) with emissions from waste been prioritized in the NDCs and Ghana's Climate Change Adaptation Strategy as well including sustainable waste management and renewable energy (MESTI 2021). With the current business-as-usual (BAU), emissions are expected to increase from 43.02 $MtCO_2e$ in 2016 to 48 $MtCO_2e$ in 2020, 59.1 $MtCO_2e$ in 2025 and 73.3 $MtCO_2e$ in 2030. Ghana aims to unconditionally lower its GHG emissions by 15% relative to a BAU scenario emission of 73.95 $MtCO_2e$ by 2030 (EPA 2019). The aforementioned documents advocate for the introduction of new and affordable technologies to treat waste. Additionally, W2E technology has been identified as one of the five key energy-related priorities stipulated in the Nationally Appropriate Mitigation Actions (NAMAs).

However, the choice of W2E technologies depends on a few factors including the quality of the waste, resource availability and local conditions of plant location as well as the purpose of energy generation. The purpose for the choice of technology could vary from electricity generation, cooling, heating to biogas for transport and biofertilizers depending on the local conditions i.e., feedstock availability, economic conditions and proximity of feedstock source to plants. Additionally, the type of substrate available for W2E plants depend on population and the economic structure of a region. Urbanization in Ghana has tripled over the last three decades from 4 million people to about 47% of the country current population of over 30 million people (World Bank 2015). Municipal solid waste (MSW) as a by-product of urbanization, population growth, and increase standard of living is increasing at a faster rate. Per capita waste generation in Ghana is estimated at about 0.47kg of solid waste daily (Miezah et al. 2015) and this translates to more than 12,000 tons daily. Considering the fact that W2E plant is dependent on renewable energy resources which is abundant in Ghana, however, inadequate data on these resources poses the biggest challenge to planning and implementation of alternative energy projects in the country.

Proper assessment of waste resources to energy potential is the first step in assessing the feasibility of establishing W2E plants; this usually depends on the availability of the primary feedstocks. Nonetheless, these resources could either be sparsely distributed or concentrated in a region. The sparsely distribution of these resources emphasis the importance of finding a suitable location for W2E plants such as biogas as a way to minimize transportation distances and associated CO_2 emission. In addition, due to the spatially varied locations of different feedstock sources, the assessment of feedstock potential for particularly biogas production and siting W2E plants in optimal locations includes the use and handling of a wide range of geographical data. Geographical Information Systems (GIS) have been considered as an appropriate platform for spatial related issues and have been studied for assessing the potential feedstock for biogas production and for site location analysis (Zubaryeva et al. 2012; Sultana and Kumar 2012; Sliz-Szkliniarz and Vogt 2012; Arifa and Amit 2012).

This study aims to assess the potential of MSW generated in Atwima Nwabiagya Municipality and 19 surrounding MMDAs (Metropolitan, Municipal and District Assemblies) within a 30 km radius from the plant location (*Gyankobaa*) in the Ashanti Region of Ghana for the installation of a hybrid W2E plant comprising of biogas and pyrolysis in addition to a solar PV system, the first of its kind in the West African sub-region to augment the energy needs of the municipality. For this purpose, a geographic information system (GIS) was used to analyze the spatial distribution of MSW resources at Atwima Nwabiagya Municipality and its surrounding districts to identify the optimal location of potential waste generation points. This study represents the first systematic effort for effective and efficient resource mapping and analysis of MSW generation in the context of circular economy in Ghana and can be extrapolated to various regions in Ghana to determine the various waste generation hotspots as well as the size of the plant to be installed.

2 Materials and Methods

2.1 Case Study–Ashanti Region

Ashanti region lies in the southern part of Ghana occupying 24,389 km^2 (9,417 mi^2) with Kumasi as its capital. It lies between longitudes 0.15W and 2.25W, and latitudes 5.50N and 7.46N. The region is the third largest administrative region in Ghana forming 10.2% of the total land area of Ghana. With regards to population, it is the second most populated region with a population of 5,432,485 accounting for 17.6% of Ghana's total population with a population density of 1,235.8 persons per sq. km (GSS 2021). According to the 2021 census, 61.6% of the populace resides in urban areas whereas 38.4% in rural areas (GSS 2021). The region shares boundaries

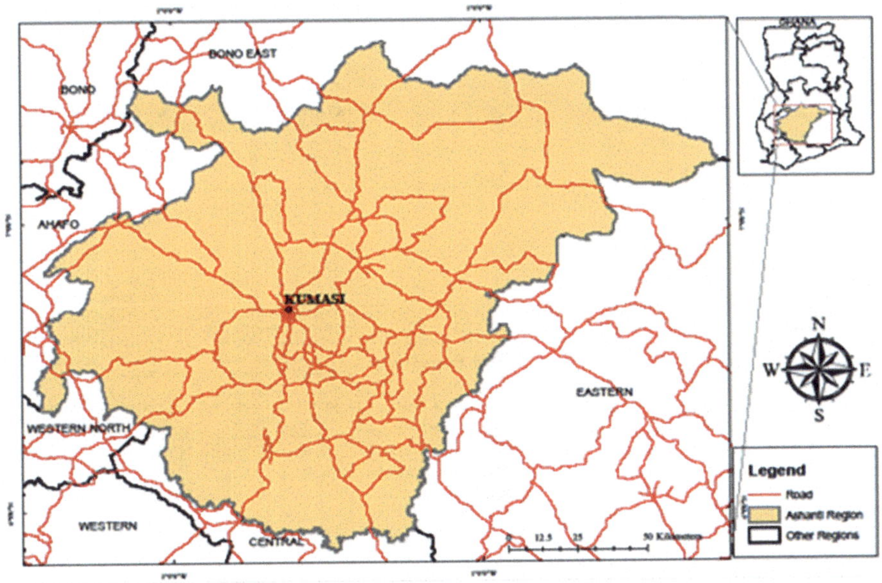

Fig. 1 Map of Ashanti region

with six of the sixteen political regions, Bono, Bono East and Ahafo Regions in the north, Eastern region in the east, Central region in the south and Western region in the southwest (Fig. 1).

Administratively the region is divided into forty-three (43) MMDAs. These are one (1) Metropolis, Eighteen (18) Municipalities and Twenty-Four (24) Districts. Figure 2 shows the population distribution in the 43 MMDAs in Ashanti Region with majority of the people residing in the Greater Kumasi (GK) area i.e., Kumasi Metropolitan Assembly (KMA) and its surrounding MMDAs (Oforikrom, Suame, Asokore, Asokore Mampong, Old Tafo, Kwadaso).

2.2 *Atwima Nwabiagya Municiaplity*

The hybrid W2E demonstration plant is sited in *Gyankobaa* in the Atwima Nwabigya Municipal Assembly (ANWA) of the Ashanti Region of Ghana. The municipality is one of the forty-three (43) administrative municipalities in Ashanti Region. It was formally Atwima Nwabiagya district until 2018 when it was upgraded to municipality status by Legislative Instrument (L.I. 2298). The municipality has about 64 settlements with Nkawie as its administrative capital. It is mainly rural with 35.3% urban and 64.7% rural. The municipality lies approximately between latitude 6° 32'N and 6° 75'N, and between longitude 1° 36' and 2° 00' West. It is situated in the western part of the Ashanti Region and shares common boundaries with Ahafo Ano

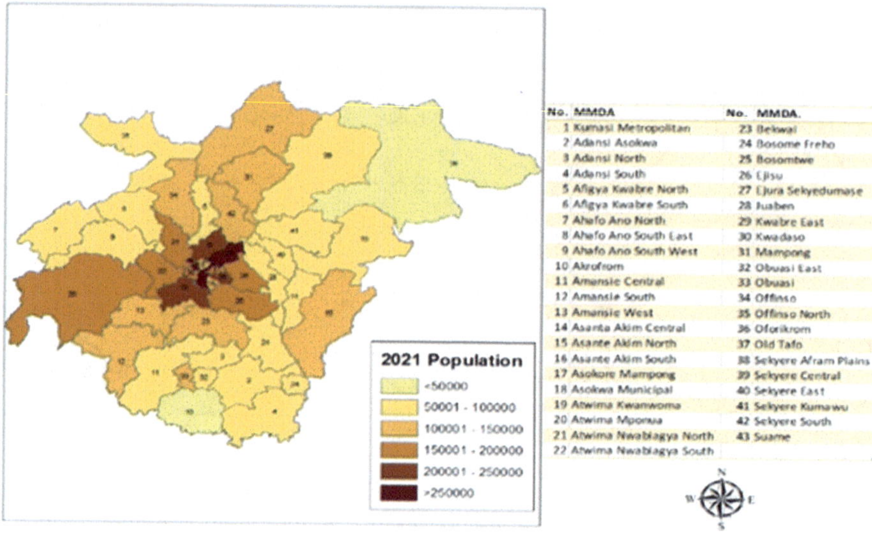

Fig. 2 MMDAs population distribution in Ashanti region

Southwest and East, and Atwima Nwabiagya North District to the north, Atwima Mponua District to the west, Amansie West and Atwima Kwanwoma District to the South and Kwadaso Municipality to the east. It covers an estimated area of 71 km^2. The project site (Gyankobaa) is situated in the southern part of the municipality, close to Amansie West and Atwima Kwanwoma (Fig. 3). The choice of location was primarily due to the availability of household waste, agricultural residues due to the agrarian nature of the local economy, and availability of an off-taker for the power to be generated. Additionally, the availability and willingness of the community to contribute 18 acres of land in kind.

The current population of the municipality is estimated at 161,893 with an annual growth rate of 2.6% (GSS 2021). The population of the municipality is densely distributed. However, the larger towns have relatively large populations as compared to the smaller towns/communities. Abuakwa and Nkawie are the two densely populated towns in the municipality with projected populations of 25,731 and 10,691 respectively. Table 1 provides a short description of the study area. Majority of the populace are engaged in farming i.e., food crop farming (cassava, plantain, cocoyam and local vegetables) and cash and food crops (cocoa, citrus, oil palm, maize, exotic vegetables, ginger, rice,) as well as livestock rearing. The municipality is also noted for the cultivation of ginger with annual production of 10,106 metric tons with the ginger produced still being sold in their raw form.

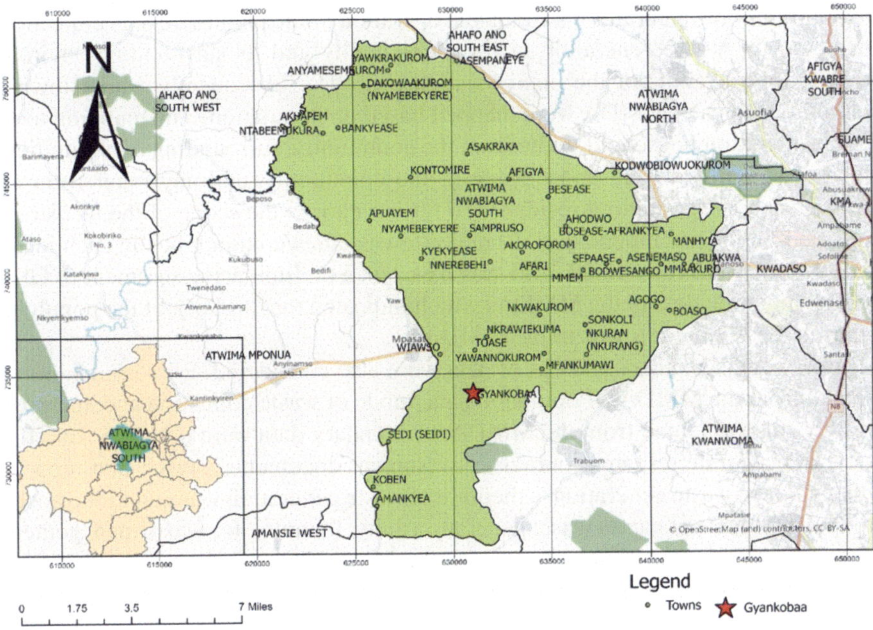

Fig. 3 Map of Atwima Nwabiagya (South) municipal assembly

Table 1 Description of the study area

Variable	Ashanti region	30 km radius
Number of MMDAs	43	20
Population (2010)	4,780,380	2,778,198
Population (2021)	5,432,485	3,880,504
Share of total population (%)	17.6[a]	65.5[b]
Area (km^2)	24,389	8,086
Share of field area of total area (%)	10.2[c]	33.2[d]

[a]Share of total population of Ghana
[b]Share of total population of Ashanti Region
[c]Share of total area of Ghana
[d]Share of total area of Ashanti Region

2.3 *Data Collection*

Data for this study was collected with a specific focus to develop a resource mapping of the potential waste available for the hybrid W2E plants i.e., biogas and pyrolysis with specific assessment on spatially optimized MSW collection areas and transportation distances. Since the location was already earmarked, proximity of waste collection sites, transportation and road network to the plant were assessed to ensure availability of raw material for sustainable operation of the plant. The data collection considered the waste generation in the municipality and its surrounding environ, within a 30 km radius of the plant location (Fig. 4). Subsequently, major towns/communities in each MMDA were mapped based on population. The mapping took into consideration household waste from the community, surrounding communities, industrial activities within and outside of the community, and any other form of biowaste such as agricultural residues. To further narrow the scope of the feedstock assessment in the catchment area, 10 major towns/cities/communities in each identified MMDA were considered. This was based on a preliminary assessment of GSS (2021) population data in the MMDAs which indicated that 80-90% of the populous resides in the 10 major towns in the MMDAs.

Primary and secondary sources of data were also considered. Primary data were obtained from the MMDAs. Population data, mode of waste collection and dumping sites were also obtained from the MMDAs. Secondary data were obtained from the review of academic research journals, government documents, books, and reports. With regards to waste generation, a theoretical waste generation was quantified based on population and per capita waste generation daily. Planning for waste management is always a daunting task for authorities at the MMDAs due to the unreliability and scarcity of data on waste generation in Ghana. There is no official government data on waste generation rate in Ghana with most available data based on academic studies. However, for the purpose of this study, 0.47kg capita^{-1} day^{-1} was used to estimate the theoretical waste generation in the catchment area. To further ascertain the amount of

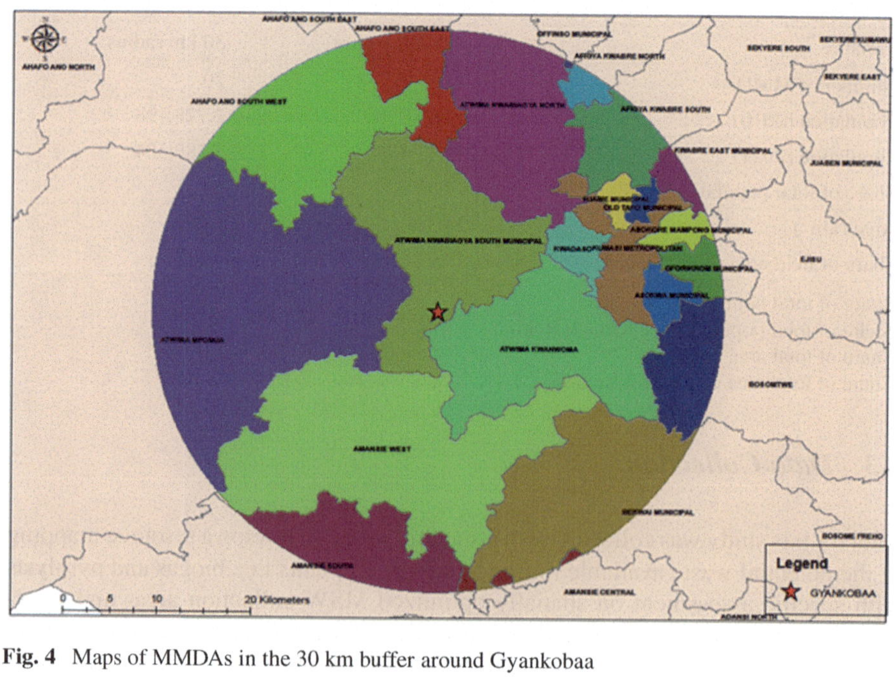

Fig. 4 Maps of MMDAs in the 30 km buffer around Gyankobaa

waste collected and disposed at dump sites, the number of communal bins provided by the MMDAs placed at designated sites including market centers, industries etc. in the communities and the frequency at which these bins are emptied were obtained from the MMDAs. A Geographical Information System (GIS) was used to identify and map central collection points and the distance from the central collection point to the processing site. Thereafter, thematic maps were generated for the various categories of items identified in the data collection process. Waste characterization was also carried out to ascertain the composition of waste generated as well as the variation in seasonal waste composition in accordance with ASTM D5231-92 (ASTM 2008).

3 Results and Discussions

3.1 Atwima Nwabiagya Municipality

3.1.1 Waste Generation and Composition

On the average, the municipality generates 51.34 tons by weight of solid waste daily with annual generation rate of 18,738 tons, based on per capita waste generation of 0.47 kg per day^{-1}. Waste generation in the municipality varies by different locations with the two densely populated communities, Abuakwa and Nkawie (municipality capital) constituting 33% of the waste generated (Fig. 5). It must be emphasized that the waste generation in the municipality was based on the official waste collection systems i.e., the use of communal containers designated at vantage points in the communities. Due to lack of data on the number of tricycles (*aboboyaa*) used for waste collection, the research could not account for waste collected by these *aboboyaa* and others. Based on field experiment, the average bulk density of a communal container in Atwima Municipality is 415 kg m^{-3} with a communal container volume of 12 m^3 implying 0.72 tons of waste collected daily per communal container. From the foregoing, only 16.62% of the expected waste generated daily is collected with the remaining waste either unofficially collected or indiscriminately disposed.

Waste characterization is very important for appropriate MSW collection, selection of transportation equipment, energy transformation and its recovery, recovery of reusable matter, as well as the proper design and implementation of optimal disposal routes and methods. Figure 6 provides the percentage of waste composition by weight of the MSW generated in the Atwima Nwabiagya Municipality. On the average, organic waste (approximately 39%) represented the major waste component of the MSW generated in the municipality. The agrarian nature of the economy in the municipality could be a contributing factor to the high fraction of organic waste generated. This is in line with other studies establishing organic waste as the largest fraction of MSW generated in Ghana (Asase 2011; Owusu-Ansah 2008; Fobil et al. 2005; Miezah et al. 2015). However, the findings differed in percentage composition

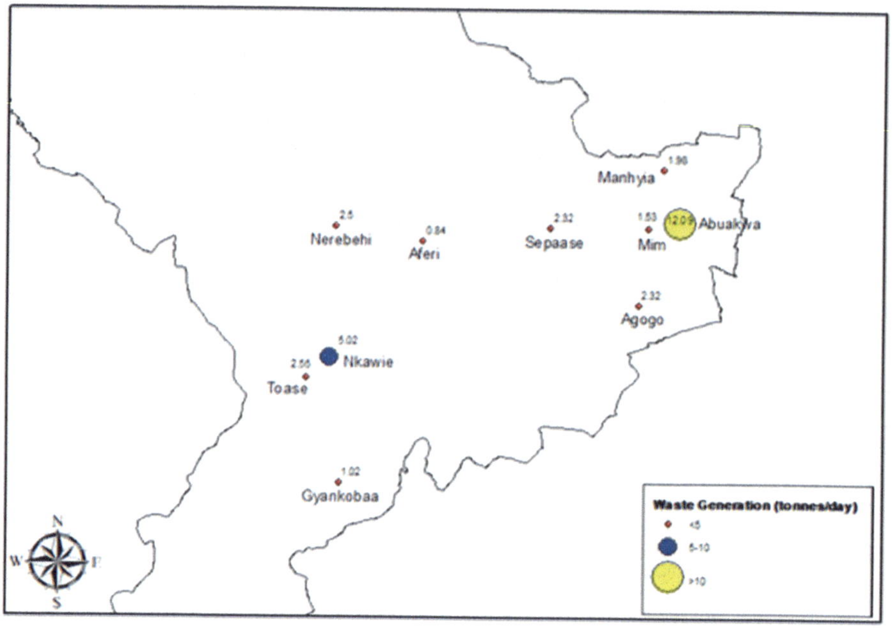

Fig. 5 Waste generation in the 10 major towns/cities in Atwima Nwabiagya Municipality

with the authors reporting 50–65% of organic fraction of waste generated in Ghana. In this study, waste sampled was from communal containers destined for final disposal site which might have undergone some form of degradation depending on the number of days the waste has been kept in the container as compared to waste sampled from households. Plastic waste was the second largest waste fraction forming 18% of the MSW generated by weight. This could be attributed to the increasing use of plastic products particularly for packaging. Additionally, there have been an increase in use of HDPEs in sachet water packaging with several water sachet companies established in the municipality. On the one hand, fine ores comprising of stones, decomposed food debris, sand, finest particles etc. contributed about 17% of the MSW generated. The recyclables (excluding plastic) including textiles, metals, glass, papers and cardboard on the other hand formed about 12% of the waste stream. Metal and papers were the least waste fractions amongst the categories of waste confirming the fact that most of these materials are recovered by itinerants who go around houses and waste collection points to collect these recyclable materials and sell them to recycling companies (Mohammed et al. 2021). It is a very lucrative business and act as source of livelihood for these scavengers.

However, seasonal changes affected the composition of waste generated in the municipality with minor variations observed during the dry and wet seasons. Higher organic waste fractions (45%) were observed during the dry season with 12% difference between dry and wet seasons (Fig. 7a and b): In addition, samples from market

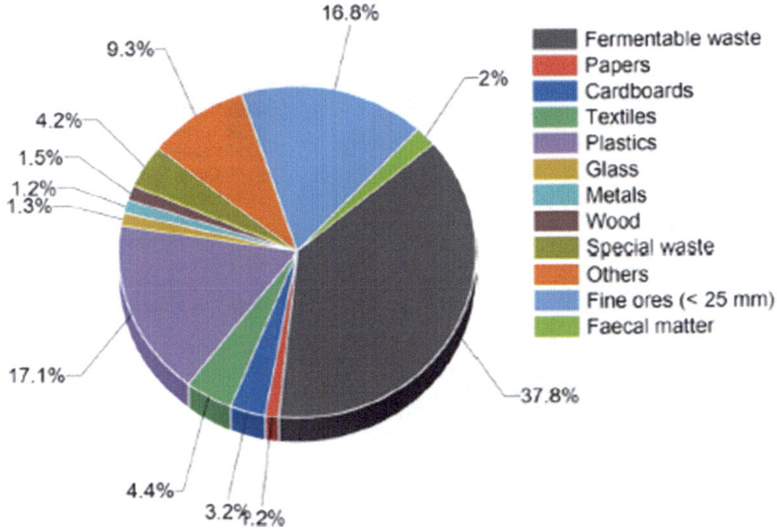

Fig. 6 Average waste composition in the municipality

centers also observed high organic fraction of waste as compared to that from township. This could be because of the nature of activities in the market centers where most fresh food products are sold. Communal containers in the market centers are usually emptied twice a week which suggest less decomposition of the materials at the time of sampling as compared to samples from township which are emptied once every two weeks in most instances. Plastic waste generation saw no variations with an average of 17%. Glass, metal and papers were the least contributing fractions to the waste generated in the municipality both in the dry and wet seasons. Several research carried out on household waste compositions in Ghana has concluded that seasonal variations have no effect on change in waste composition and generation (Adjei 2013; Asase 2011; Ketibuah et al. 2004; Anomanyo 2004; Dagadu 2005; Fobil et al. 2005). Exposure of the communal containers could be a contributing factor to the seasonal variations observed in this study which has been corroborated by Hancs et al. (2011); Kamran et al. (2015), Aslani and Taghipour (2018) that seasonal variation normally affects generation of outdoor waste.

3.1.2 Waste Collection System

The choice of W2E technologies depends on the waste characteristics, purpose of energy generation, resource availability and local conditions of plant location. Local conditions including economic conditions, availability of substrate, and proximity of substrate source to plants are key in ensuring sustainability of W2E plants. Resource availability and a clearly defined chain of production and supply are required for continual increase in the use of W2E plants. Sources of solid waste generation in

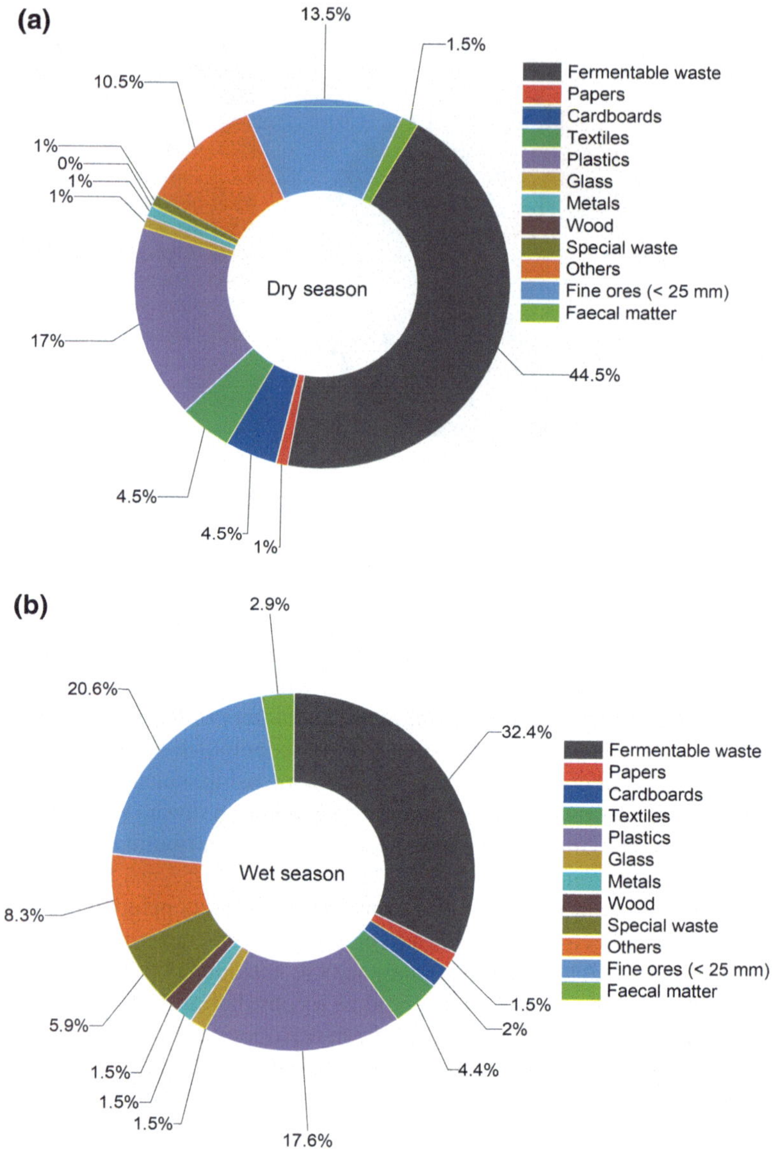

Fig. 7 **a** Waste composition–Dry season **b** Waste composition–Wet season

the municipality are mainly from domestic (household), commercial (market and trading) and institutional activities. Waste collection system in the municipality is not different from what pertains in other cities in Ghana (Mohammed et al. 2021). Waste management is usually the one service that falls completely within the local government's purview as articulated in the Local Government Law 1993

(Act 462), Section 10, and the Legislative Instrument of Establishment (L.I.) in all MMDAs. However, the collection system is handled by a private company. The municipality has a contractual agreement with Zoomlion Ghana limited, the largest private waste management company in the country, for the collection and disposal of both households (door-to-door) and communal bins designated at vantage points i.e. market centers, schools, public toilets etc. in the communities. Currently, there are 15 communal containers (10–15 m^3) placed in strategic locations in the entire municipality for waste collection as shown in Fig. 8, with 10 of them owned by the municipality and 5 by Zoomlion Ghana Limited. The collection is done with little or minimum regulatory framework or guidance from the municipality. For instance, the trucks used are not covered and usually end up littering the streets thereby complicating the waste collection arrangement.

Due to the rural nature of the municipality, the waste collection model employed by authorities is primarily through the communal container system. At the household level, waste is first stored in primary storage bins i.e., containers or used plastic sacks until it accumulates to certain quantity or full and transported to communal container collection sites. The distance from the household premises to the communal container dump site usually does not exceed 300 m. A polluter pay system applies at the communal container dump site, where waste depositors pay for the waste being disposed with varied amount depending on the quantity of waste. At the same site, there are itinerants who scavenge for recyclable materials such as metals, glasses,

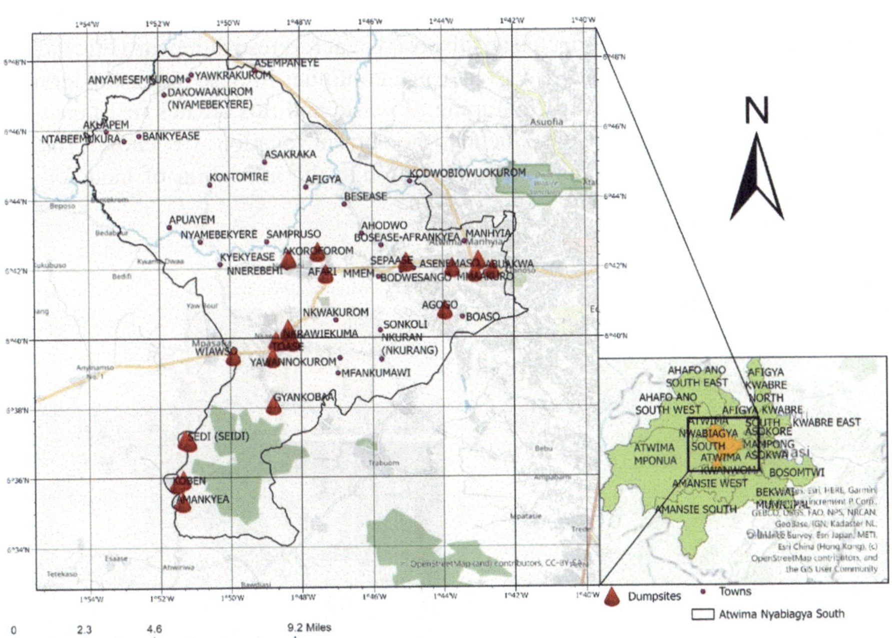

Fig. 8 Communal container dumpsites in ANMA

cardboard, plastics etc. and are sold to recycling companies. The number of days for the removal of the containers to the final disposal site are increased as the recovery of these valuable materials reduces the volume of waste. Similarly, at the institutional level such as s hospitals, educational centers, clinics and industries, communal containers are employed as mode of waste collection before being transported to final disposal site.

The municipality has two major market centers i.e., Abuakwa and Nkawie markets and the waste collection is similar to what pertains in other areas. However, the frequency of waste collection and disposal is higher than the other areas due to waste traffic at the market centers. Additionally, the communal containers at the market centers are disposed twice every week.

Depending on the population and waste traffic, some communities are allocated more containers and are usually disposed twice per week. For instance, Abuakwa and Nkawie which are the most populated communities have six and two communal containers at strategic locations respectively. On the contrary, the wastes generated in less populated communities such as Mim, Afari etc. are collected for disposal on weekly basis and in some cases once in two weeks.

3.1.3 Waste Disposal

The mode of waste disposal in the municipality includes open dumping, public dumping (container), burning, collected and buried by households, indiscriminate dumping and others. Waste collected are disposed of at Kyereso dumpsite (Fig. 9a and b). Not-In-My-Back-Yard (NIMBY) syndrome continues to pose major hinderance to authorities in locating sites for dumping of waste and this creates resistance for setting up of a W2E plant. Consequently, the current location was chosen due to the commitment shown by the community through the allocation of land for the establishing of the W2E facility.

(a)

(b)

Fig. 9 **a** Aerial view of current dumping site, **b** Skip truck offloading waste at dump site

3.2 *Environs–30 km radius*

The viability of any W2E project, critically depends upon, interalia, an assured availability of the requisite quantities and quality of the waste, implying that there will be the need to ensure proper linkages in waste management right from its generation at source to final disposal. With the planned plant at Gyankobaa, which is expected to receive 18 tons at the initial stage of the plant operation, a resource mapping was carried out to ensure constant supply of feedstock to the plant particularly in the event of the upgrade of the plant capacity in the future, mapping of household and other biowaste within an economic zone of 30 km radius from the plant location.

3.2.1 Waste Generation and Composition

As indicated earlier, it is essential to assess the quality, quantity and availability of feedstock within a certain radius of the location of the plant in identifying specific location for W2E plants. These are critical issues with regards to location and economic performances of the plant. In assessing the spatial distribution of bio-waste and plastics for the hybrid solar-biogas-pyrolysis plant, a 30 km buffer zone was considered. With the plant to be situated at *Gyankobaa*, a suburb of Atwima Nwabiagya Municipal Assembly of the Ashanti Region of Ghana, a total of 20 MMDAs were mapped out in the 30 km radius of the plant location (Fig. 10). Intriguingly, all the MMDAs in the GK area, the main economic areas in the region, fell completely within the radius, which comprises majority of the population in the region with Kumasi Metropolitan Assembly (KMA) as the most populous MMDA, constituting 26% of the population and Oforikrom, Suame and Kwadaso municipalities constituting 9.14%, 7.42 and 6.44% of the population in the catchment area, respectively. It is estimated that the GK area alone generate an approximately 1,200 tons of waste daily with majority of the waste emanating from KMA due to its population and economic activities. The immediate MMDAs that border the project site included Atwima Kwanwoma district, Amansie West district, Kwadaso Municipal assembly, Atwima Nwabiagya North district, Atwima Mponua district, Ahafo Ano Southwest district and Ahafo Ano Southeast district constituting 19.8% of the population which translate to approximately 189 and 72 tons per day of organic and plastic waste respectively. Compared to the official waste collected, about 12% each of the organic waste fraction and plastic will be available for delivery to the plant daily. Further analysis revealed that an estimated 1,824 tons of waste based on population is expected to be generated daily in the 20 MMDAs in the catchment area with Afigya Kwabre North district as the least contributing district. Atwima Nwabiagya Municipal assembly contributes 2.8% of the waste generation in the catchment area with 25 tonnes of bio-waste daily. With an estimated 19% of plastic waste in the composition of waste stream in Ghana, 347 tons of plastic waste generation are expected daily in the catchment area.

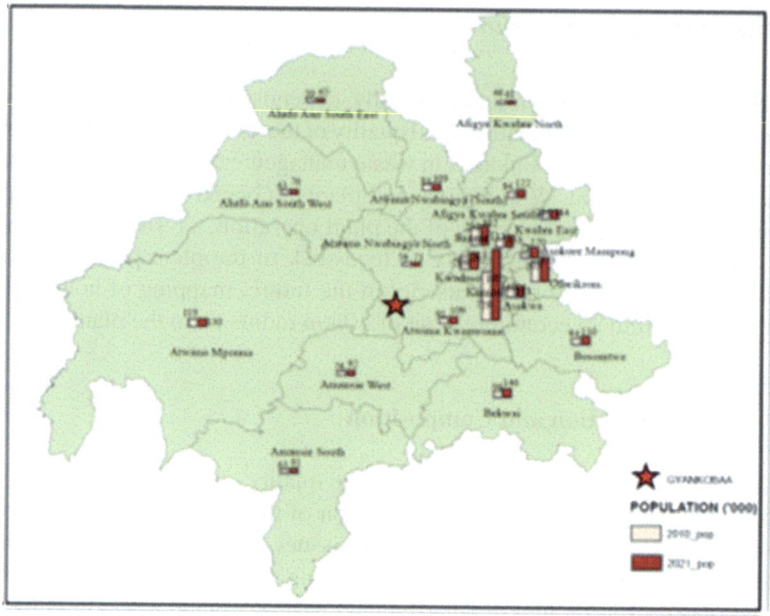

Fig. 10 Projected populations of the 20 MMDAs in the study area

However, further analysis on the waste generation based on the average bulk density of waste in each MMDAs revealed differently (Fig. 11). A total of 1,954.81 tons of waste generated weekly and 279.26 tons per day. Asokwa and KMA alone contribute 33% of the daily waste generation in the study area with Ahafo Ano Southwest as the least contributing district. The immediate 7 neighboring MMDAs comprising of Atwima Kwanwoma, Atwima Mponua, Atwima Nwabiagya North, Ahafo Ano Southeast, Ahafo Ano Southwest and Amansie West and Kwadaso municipal contributes 21% of the total waste generation in the catchment area. Atwima Nwabiagya Municipal is the highest waste generation contributor after Kwadaso municipal (9.6%). In comparison to the established waste generation of over 1,800 tons of waste generated daily based on population, only 15% of the waste generated in the catchment are collected based on the established waste collection centers in the various MMDAs. Similarly, it must be emphasized that an appreciable amount of the waste from households is also collected and disposed by tricycles (*aboboyaa*).

The composition of waste generated in the 30 km environs did not differ much from that of the Atwima Nwabiagya Municipality. Analysis of waste composition showed predominance of organic or biodegradable materials (37%) such as vegetables, left-over food etc. lower than the 61% reported by Miezah et al. (2015) and 55% reported by Ketibuah et al. (2004) on household waste characterization in Ghana and Kumasi, respectively. The difference could be a result of the sampled waste as the waste characterized in this study was mainly from communal containers which might have undergone some of decomposition prior to its final disposal. On the average,

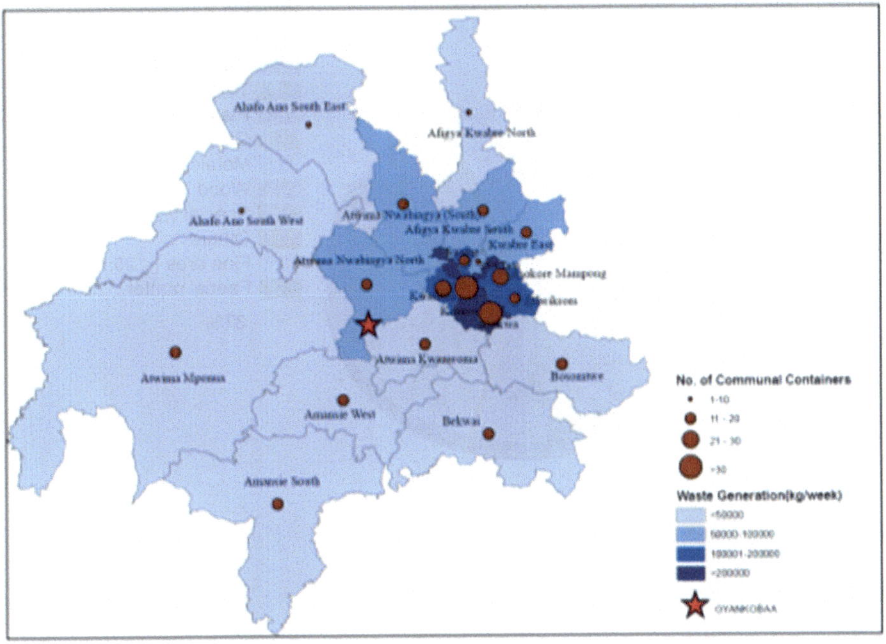

Fig. 11 Waste generation in the MMDAs on bulk density basis (kg m^{-3})

recyclables including plastic, textiles, metals, glass, paper and cardboard accounted for 30% in this study (Fig. 12). Notwithstanding the high quantity of municipal residues in the catchment area, there are few agri-industrial waste generated specifically in the GK area. Table 2 provides the agri-industrial companies in the 30 km buffer zone. Intriguingly, no waste is left unutilized in most of the industries sited in this study. Waste generated during processing are either converted into other material form for usage at the factory or sold to farmers.

3.2.2 Waste Collection and Transport

Waste collection system in the study area is inadequate. The waste collection service in the MMDAs is performed by the private sector under various agreements with the MMDAs, as well as the use of communal bins provided by private contractors. It is estimated that over 80% of waste services in all MMDAs of Ghana are provided by the private sector. The impression is that the shift towards the private sector has effectively improved waste collection services in the major cities in the country. In most of the MMDAs, particularly in the GK area, there are averagely two to three licensed private solid waste companies operating in the catchment area. The major cities and towns within the MMDAs in the Greater Kumasi area have been

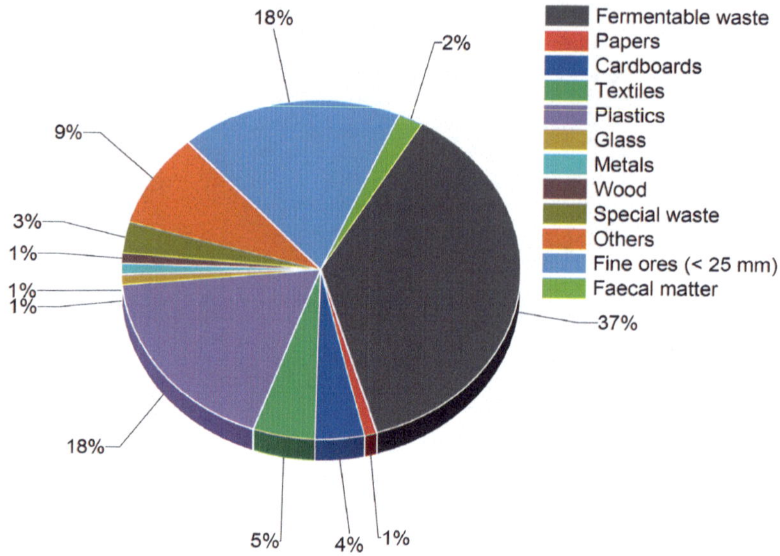

Fig. 12 Average waste composition in the 30 km radius

Table 2 Types of residual resources available in the study area

Municipal residues	Agri-industrial residues
Municipal solid waste	Oil palm processing
Municipal sewage sludge	Fruit processing
Food processing residues	Cocoa processing
Market residues	Starch production
Household waste	Breweries
Institutional/establishment waste	Livestock farming
Manufacturing residues	Poultry farming
	Slaughterhouses
	Wood processing (sawmill)

zoned into units with each company enjoying a zonal monopoly to operate house-to-house collection in a delineated zone, except Zoomlion Ghana Limited, which unofficially extends its house-to-house collection to zones of other companies. The companies are responsible for collecting and transporting the waste from the various zones to the final disposal site. During the study, Zoomlion was identified as the major waste management service provider in all the MMDAs with Asadu Waste Company engaged in some MMDAs within the GK area i.e. Kwadaso municipal Assembly, Oforikrom Municipal Assembly etc. Other waste management service providers including Venmark and Kumasi Waste Management Limited (KWML) are contracted for waste collection in the catchment area. However, there is no data in terms of waste generation, frequency of collection and disposal sites in the MMDAs.

Similarly, the study revealed three modes of waste collection across the MMDAs i.e. door-to-door in urban centers particularly high-income communities, communal container system for waste collection in low income communities and indiscriminate/crude dumping in rural or low income communities. The private companies are responsible for the door-to-door collection within the high- and middle-income communities utilizing compactor trucks. The waste is stored in closed plastic containers and picked up by the trucks once a week. With the communal waste container system, metal containers or skips are placed at designated sites commonly known as transfer stations, which are shared by a number of households within the community. These are then transported and emptied at final disposal site by skip loading trucks once every week or when the containers are full. Due to the exposure of these containers to rain, they produce leachate with unpleasant smell as well as causing pollution to nearby water bodies. Fig. 13 shows the towns in the 30 km radius from the project site (Gyankobaa).

In some MMDAs, particularly in KMA, 90% of the communal containers had shed under which the communal containers are placed and operated by a caretaker, responsible for manning the place. Over the last decade, MMDAs have adopted the polluter-pay system where waste generators pay before dumping waste. Despite the positive aspect of the initiative, majority of the citizens prefer to dispose their waste indiscriminately as a way to avoid paying for dumping which has negative consequences on human health and the environment. Additionally, bad road conditions in densely populated poor neighborhoods make waste collection a challenge

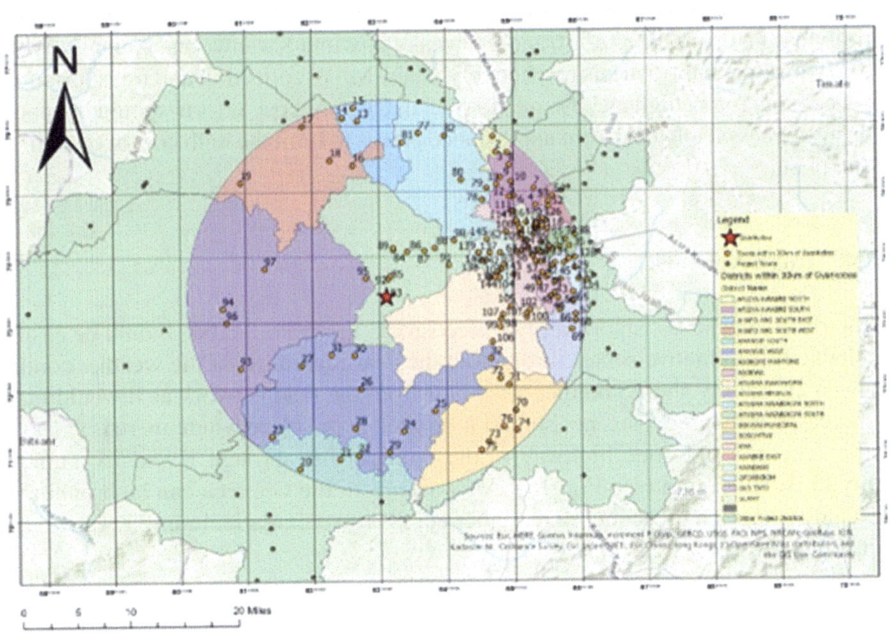

Fig. 13 30 km buffer from Gyankobaa and the towns that fall within

resulting in the utilization of tricycle *(aboboyaa)*. These tricycles operate from door-to-door for waste collection and subsequent disposal to nearby dumpsite. Though the tricycles operators are seen to be complementing the work of the MMDAs in waste management, some of the operators seem to be creating problems for the assembly as they turn to dispose their waste on roadsides instead of the dump site due to distance. Communal collection system represents majority (>50%) of waste collection method whilst the door-to-door or curbside system makes up 33%. The remaining populace indiscriminately dumps their waste in open places, burned at the backyards, dumped on the streets or abandoned in water bodies causing flooding of rivers and other water bodies in the city.

3.2.3 Distance to Project Site

Siting of W2E plants in strategic locations is a major means of combating some of the environmental challenges of bio-waste generation, both convenient and economically advantageous. Supply chain analysis is therefore undisputable in the realization of a sustainable waste management system. Feedstock supply is paramount to the sustainability of W2E plants. It is recommended that W2E plant should be close to the source of feedstock to overcome the difficulties of fuel transportation and rural waste disposal. Figure 14 pictorially describes the distance of the major townships in the MMDAs to the W2E plant. Neighbouring MMDAs that might contribute to feedstock supply within 10 km radius included Amansie West (Ahwerewa) and Atwima Mponua (Mpasatsia). Out of 25 towns within 10-20 km, Kwadaso Municipality falls entirely within in addition to 7, 6 and 2 towns of Atwima Kwanwoma, Amansie West and Atwima Mponua districts respectively. Since MSW collection and transportation cost accounts for as high as 70% of MSW management cost, efforts should be made to ensure access feedstock from not distance places to limit the high cost of operation of the plant.

3.2.4 Road Network

Similarly, to ensure constant supply of feedstock, the W2E facility should be easily reachable by alternative routes throughout the year regardless of the weather conditions. W2E facility should neither be too far from the road network as it would lead to increased transport costs, nor should it be sited at the place where there is a heavy traffic. There is only one highway to the municipality through Kwadaso municipality to Bono Region (see Fig. 15). All waste from the GK area can be brought to the facility through the same route. Other access route exists including Kwanwoma and Barekese.

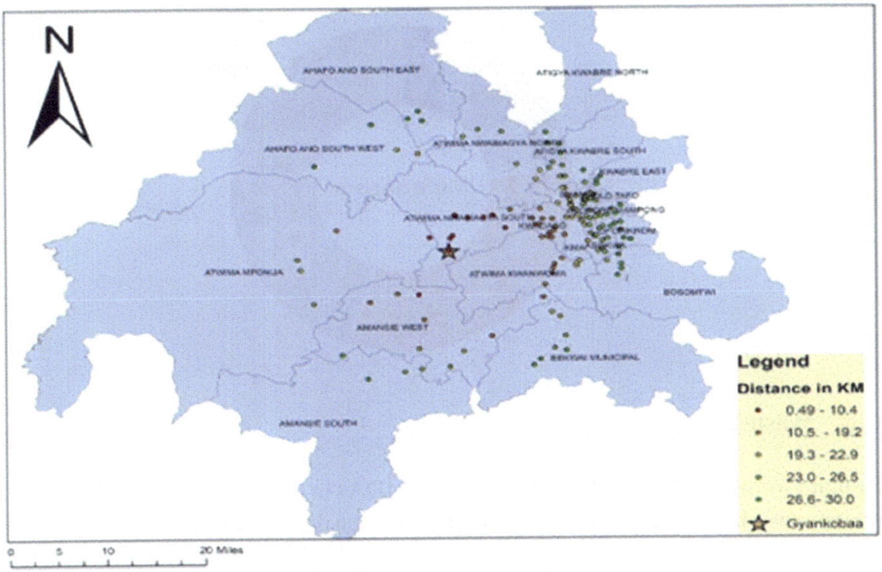

Fig. 14 Distance of towns to Gyankobaa (plant site)

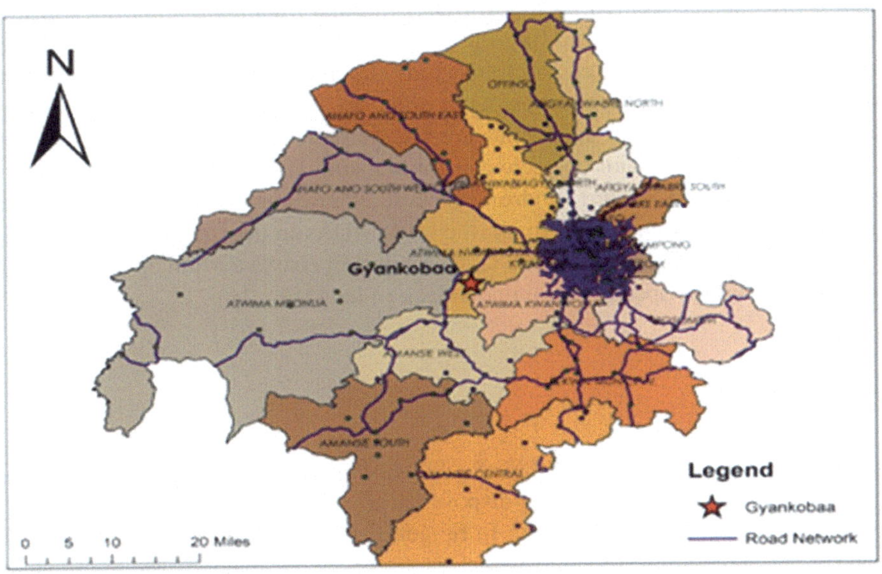

Fig. 15 Road network map

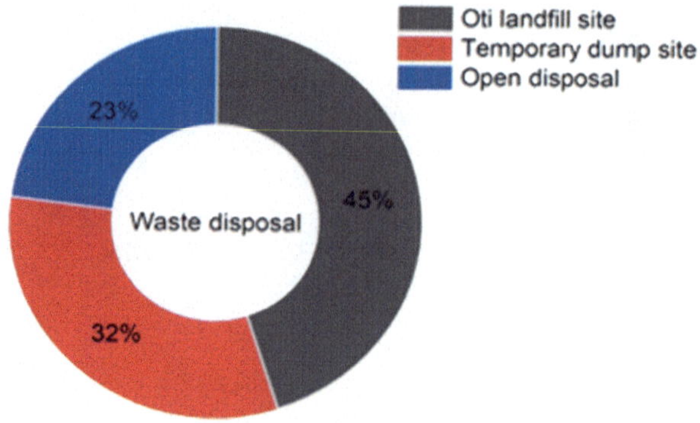

Fig. 16 Mode of waste disposal in the MMDAs

3.2.5 Waste Disposal

The main form of waste disposal by various MMDAs in Ghana is by landfill, whereas also in many places waste is disposed of through open dumps. During the study, it was revealed that majority of the institutions including health and educational facilities as well as industries employ the use of incinerators for waste disposal. All waste collected within the MMDAs are disposed of at a non-engineered landfill site. The landfills are usually secured by the MMDAs through traditional authorities in the communities at a fee to be paid by the private companies before dumping at the site. However, it was gathered from this study that these private companies do not adhere to this agreement which results in difficulty in securing location for waste dumpsite as the traditional authorities are always reluctant to make land available as dumpsite. Another major obstacle confronting authorities at the local level is the issue of not-in-my-back-yard syndrome as communities agitate against location of dumpsite in their communities. In all, the 20 MMDAs in the catchment area, 43% of the MMDAs disposed their waste at Oti landfill in Asokwa Municipal Assembly with 23% opting for open disposal due to lack of land (Fig. 16).

4 Conclusion

Averagely, a total of 8.5 tons per day of solid waste generated from Atwima Nwabiagya Municipality is collected and disposed at a temporary dumpsite representing 16.62% of the expected total waste to be generated based on population and per capita waste generation. Organic waste fraction was the major waste component generated in Atwima Nwabiagya Municipality followed by plastic waste. Municipal solid waste management in the municipality is characterized by ineffective collection methods, inadequate coverage of the collection system, and improper disposal of municipal solid wastes with communal containers placed at vantage points for waste collection. Waste collection in the municipality is dominated by the private sector that are contracted by the municipality. The additional feedstock needed for the hybrid renewable energy plant could be accessed within 10 km radius from the plant location. Similarly, the waste composition in the neighboring environment is similar to the Atwima Nwabiagya Municipality with organic waste constituting 37% of the waste generated and 18% plastic waste. About 45% of the waste in 30 km radius from the plant location is collected and transported to Oti landfill, the only sanitary landfill site in the Ashanti Region of Ghana.

References

Adjei, S. M. (2013). Municipal solid waste minimization through household waste segregation in Bantama, Kumasi. A thesis submitted to the Department of Environmental Science, Kwame Nkrumah University of Science and Technology, Kumasi, in partial fulfillment of the requirement for the award of Master degree in Environmental Science. http://ir.knust.edu.gh/bitstream/123456789/5587/1/MARY%20SERWAAH%20ADJEI.pdf Accessed 30 November 2021

Anomanyo, E.D. (2004). Integration of municipal solid waste management in Accra (Ghana) Bioreactor treatment technology as an integral part of the management process. A thesis submitted to the Lund University in partial fulfillment of the requirements for the award of the Master of Science (MSc) degree in International Environmental Science. https://www.lumes.lu.se/sites/lumes.lu.se/files/anomanyo_edward.pdf Accessed 15 June 2021

Arifa, S., Amit, K. (2012). Optimal siting and size of bioenergy facilities using geographic information system. Appl. Energy. 94, 192–201. https://doi.org/10.1016/j.apenergy.2012.01.052

Asase, M.A.D (2011). Solid waste separation at source. A case study of Kumasi Metropolitan Area. PhD thesis submitted to Faculty of Chemical and Materials Engineering, College of Engineering, Kwame Nkrumah University of Science and Technology, Kumasi, Ghana. http://ir.knust.edu.gh/xmlui/bitstream/handle/123456789/6304/MIZPAH%20AMA%20DZIEDZORM%20ASASE.pdf?sequence=1 Accessed 4 December 2021

Aslani, H., Taghipour, H. (2018). Seasonal characterization and quantification of municipal solid waste: energy content and statistical analysis. J. Adv. Environ. Health Res. 6, 34-43. https://doi.org/10.22102/JAEHR.2018.105728.1053

ASTM D5231-92 (2008). Standard test method for determination of the composition of unprocessed municipal solid waste.

Ayodele, T.R., Ogunjuyigbe, A.S.O., Alao, M.A. (2017). Life cycle assessment of waste-to-energy (WtE) technologies for electricity generation using municipal solid waste in Nigeria. Appl. Energy 201, 200–218. https://doi.org/10.1016/j.apenergy.2017.05.097

European Commission (2017). Communication from the commission to the European parliament, the council, the European economic and social committee and the committee of the regions - the role of waste-to-energy in the circular economy. COM/2017/0034 Final. http://ec.europa.eu/environment/waste/waste-to-energy.pdf.

Dagadu, P.K. (2005). Municipal solid waste source separation at the household level: A case study of the Accra Metropolitan Area. MPhil. Thesis, Environmental Science, University of Ghana. https://iess.ug.edu.gh/sites/iess.ug.edu.gh/files/Student_Projects/2005%20IESS%20-%20Municipal%20Solid%20Waste%20Source%20Seperation%20at%20the%20Household%20Level%20%20A%20Case%20Study%20of%20Accra%20Metropolitan%20Area_0.pdf Accessed 20 April 2022.

EPA (2019). Ghana's Fourth National Greenhouse Gas Inventory Report, Environmental Protection Agency, Ministry of Environment, Science, Technology and Innovation, Accra. https://unfccc.int/sites/default/files/resource/gh_nir4-1.pdf Accessed 20 November 2021

Fobil, J.N., Carboo, D., Armah, N.A. (2005). Evaluation of municipal solid wastes (MSW) for utilisation in energy production in developing countries. Int. J. Environ. Technol. Manag. **5**(1),76–86. https://doi.org/10.1504/IJETM.2005.006508

Ghana Statistical Service (GSS) (2021). Population and Housing Census. 2021 Population and Housing Census Report.

Kamran, A., Chaudhry, M.N., Batool, S.A. (2015). Effects of socio-economic status and seasonal variation on municipal solid waste composition: a baseline study for future planning and development. Environ. Sci. Eur. 27, 16. https://doi.org/10.1186/s12302-015-0050-9

Ketibuah, E., Asase, M., Yussif, S., Mensah, M.Y., Fischer, K. (2004). Comparative analysis of household waste in the cities of Stuttgart and Kumasi – Options of waste recycling and treatment in Kumasi. In: Proceedings of the 19th International CODATA Conference Berlin, Germany, 7–10 November.

Khan, I., Kabir, Z. (2020). Waste-to-energy generation technologies and the developing economies: a multi-criteria analysis for sustainability assessment. Renewable Energy 150, 320–333. https://doi.org/10.1016/j.renene.2019.12.132

Malinauskaite, J., Jouhara, H., Czajczyńska, D., Stanchev, P., Katsou, E., Rostkowski, P., Thorne, R.J., Colon, J., Ponsa, S., Al-Mansour, F., Anguilano, L., Krzyzynska, R., Lopez, I.C., Vlasopoulos, A., Spencer, N. (2017). Municipal Solid Waste Management and Waste-to-Energy in the Context of a Circular Economy and Energy Recycling in Europe. Energy 141, 2013–2044. https://doi.org/10.1016/j.energy.2017.11.128

Miezah, K., Obiri-Danso, K., Kadar, Z., Fei-Baffoe, B., Mensah, M.Y. (2015). Municipal solid waste characterization and quantification as a measure towards effective waste management in Ghana. J. Waste Manag. 46, 15-27. https://doi.org/10.1016/j.wasman.2015.09.009

Mohammed, M., Ozbay, I., Donkor, A.K., Olalekan, I.E., Nyankson, E.A., Bayitse, R., Mohammed, L., Fahanwi, A.N. (2021). Inter-continental differences in waste management in developing countries: Turkey and Ghana. Proc. Inst. Civ. Eng.: Waste Resour. 175, 14-28. https://doi.org/10.1680/jwarm.20.00029

Mutz, D., Hengevoss, D., Hugi, C., Gross, T. (2017). Waste-to-energy options in municipal solid waste management: A guide for decision makers in developing and emerging countries, Deutsche Gesellschaft für Internationale Zusammenarbeit (GIZ) GmbH. Eschborn, Germany. https://www.giz.de/en/downloads/GIZ_WasteToEnergy_Guidelines_2017.pdf. Accessed 15 March 2021

Nizami, A.S., Rehan, M., Waqas, M., Naqvi, M., Ouda, O.K.M., Shahzad, K., Miandad, R., Khan, M.Z., Syamsiro, M., Ismail, I.M.I., Pant, D. (2017). Waste Biorefineries: Enabling Circular Economies in Developing Countries. Bioresour. Technol. 241, 1101–1117. https://doi.org/10.1016/j.biortech.2017.05.097

Ouda, O.K.M., Raza, S.A., Nizami, A.S., Rehan, M., Al-Waked, R., Korres, N.E. (2016). Waste to energy potential: A case study of Saudi Arabia. Renew. Sust. Energ. Rev. 61, 328-340. https://doi.org/10.1016/j.rser.2016.04.005

Owusu-Ansah, K. (2008). Evaluation of household solid waste processing options in Accra, Ghana. MSc. Thesis, UNESCO-IHE Institute for Water Education, Delft, the Netherlands.

Scarlat, N., Motola, V., Dallemand, J.F., Monforti-Ferrario, F., Mofor, L. (2015). Evaluation of energy potential of Municipal Solid Waste from African urban areas. Renew. Sust. Energ. Rev. 50, 1269–1286. https://doi.org/10.1016/j.rser.2015.05.067

Shahzad, S., Butt, A., Anwar, S., Ahmad, S.R., Sarwar, T., Asghar, N. (2017). Municipal solid waste as a renewable source of energy: An Overview from the Lahore District in Punjab, Pakistan. Pol. J. Environ. Stud. **26**(6), 2721–2729. https://doi.org/10.15244/pjoes/69143

Sliz-Szkliniarz, B., Vogt, J. (2012). A GIS-based approach for evaluating the potential of biogas production from livestock manure and crops at a regional scale: a case study for the Kujawsko-Pomorskie Voivodeship. Renew. Sust. Energy Rev. 16, 752–63.

Sultana, A., Kumar, A. (2012). Optimal siting and size of bioenergy facilities using geographic information system. Appl. Energy. 94, 193–201. https://doi.org/10.1016/j.apenergy.2012.01.052

World Bank (2015). Rising through Cities in Ghana, Ghana Urbanization Review, Overview Report, The International Bank for Reconstruction and Development, Washington DC, USA. https://dokumen.tips/documents/rising-through-cities-in-ghana-the-world-rising-through-cities-in-ghana-ghana.html?page=2 Accessed 2 January 2022

Zhang, D., Huang, G., Xu, Y., Gong, Q. (2015). Waste-to-Energy in China: Key Challenges and Opportunities. Energies 8(12), 14182–14196. https://doi.org/10.3390/en81212422

Zubaryeva, A., Zaccarelli, N., Giudice C.D., Zurlini, G. (2012). Spatially explicit assessment of local biomass availability for distributed biogas production via anaerobic co-digestion – mediterranean case study. Renew. 39, 261–70. https://doi.org/10.1016/j.renene.2011.08.021

Assessment of Knowledge, Attitudes and Practices Towards Waste Management in Ghana: Implications for Energy Production

Roland Asare, Abdalla Mahama, Portia Adade Williams, Julius Ahiekpor, Alberta Aryee, and Rankine Asabo

Abstract Waste management is a major environmental concern in Ghana. Converting municipal solid waste into energy thus presents a workable pathway to solving Ghana's environmental challenges and bridging the gap of the country's excess energy demand gap. With the introduction of Hybrid Waste-to-Energy plant in the research area, this study aimed at providing empirical evidence into understanding the Knowledge, Attitudes and Practises (KAP) of waste management and the factors which influence willingness to pay for waste disposal for potential operationalisation of a Waste to Energy plant in Atwima Nwabiagya Metropolitan Assembly, Ghana. A cross sectional study was conducted among 614 randomly selected respondents. The data applied mixed methods. The KAP was analysed using descriptive statistics whilst respondents' willingness to pay was estimated using a logistic model. The results show that respondents have adequate general KAP in waste management. Although very few knew of alternative waste management practises different from what they currently use. The study further showed that organic waste and plastic were the most generated types of waste which households mostly stored in rubber bins/bowls. Gender, knowledge about waste conversion, income, current payment for waste disposal status and the zonal council relative to the plant site are factors that significantly influence the willingness to pay ($p < 0.05$). Respondents with no knowledge on waste conversion were significantly less willing to pay compared to respondents with knowledge of waste to energy. Whilst Respondents living in the communities of the Waste to Energy plant site are generally more willing to pay. Albeit the low knowledge, respondents perceived that a waste to energy plant will promote sanitation and health as well as employment opportunities in their municipality. Thus, for effective waste to energy plant operationalization and acceptance,

R. Asare (✉) · A. Mahama · P. A. Williams · R. Asabo
Science and Technology Policy Research Institute, CSIR, Accra, Ghana
e-mail: nryasare14@gmail.com

J. Ahiekpor
Chemical Engineering Department, Kumasi Technical University, Kumasi, Ghana

A. Aryee
West Africa Science Service Centre on Climate Change, Accra, Ghana

M. Narra and S. Narra (eds.), *Innovations in Circular Economy and Renewable Energy in Africa*, World Sustainability Series, https://doi.org/10.1007/978-3-031-68330-5_4

education should be increased on waste segregation while the net social and economic benefits should be widely publicised to the target communities.

1 Introduction

Waste management includes services such as proper sanitation provision of potable water and energy and is important for ensuring a sustainable environment (Wilson et al. 2015). Efficient waste management directly influences achieving a circular economy while playing a critical role in the attainment of 2030 Agenda (United Nations, 2015). Weitz et al. (2019) notes that the Sustainable Development Goal (SDG) target 12.4, which focuses on responsible management of waste and chemicals has highest "synergistic potential" among SDG targets to make progress on SDGs implementation.

Solid waste generation in sub-Saharan Africa is estimated at approximately 62 million tons per year with a household estimation of average of 0.65 kg per capital per day (Hoornweg et al. 1999). In Ghana, the Metropolitan Municipal and District Assemblies MMDAs is responsible for managing solid waste. The country generates an estimated 3.0 million tons of solid waste annually with a daily output of 13,000 tons. National average per capita daily waste generation rates are also 0.5kg (Weitz et al. 2019, Hoornweg et al. 1999, Puopiel (2010). According to Hoornweg and Bhada-Tata (2012) in a World Bank report on waste management systems in urban areas, global municipal waste generation of about 1.3 billion is projected to increase to 2.2 billion tons by 2025.

Trajectory for efficiently managing refuse generation has necessitated the demand for effective waste management services. Adoption of environmentally friendly waste-to-energy technologies that allows treatment and potential processing of wastes into energy could mitigate challenges of waste management. Waste-to-energy is a win-win endeavor. It is a sustainable waste management system approach that produces energy for economic gains and green jobs. Renewable energy has become mainstream energy source and a viable option in countries at all levels of income globally (Agbelie et al. 2015). Meanwhile, how to meet energy needs without impairing both the economy and the environment remains an urgent problem for every country (Tsiboe and Marbell 2004),

The persistent increase in human population and rapid industrialization has caused the continuing global problems of improper wastes disposal (ATIeNZA 2008). Ineffective waste collection strategies and the lack of disposal sites can be cited as the major challenges concerning wastes management, especially in the developing countries (Reyes and Furto 2013). Licy et al. (2013) suggests, that the practices of basic solid waste management (SWM) at the individual level are often neglected. That is, while most people are often aware of the negative effects of mismanaging wastes on the environment, a combination of their negative attitude and insufficient environmental knowledge usually corresponds to poor practices towards maintaining good environmental conditions. Enactment of certain policies relevant to improving environmental sanitation and community-perception on waste management is a matter of national urgency to minimize imminent outbreaks of diseases and adverse impacts on the economy due to loss of workdays, treatment cost, and clean-up activities (Joseph 2006).

In most developing economies including Ghana, achieving such efficient management of waste has been a difficult objective for local authorities due to factors such as rapid population growth leading to increased waste (Oteng-Ababio et al. 2013). In Accra, Ghana, nearly 900,000 metric tons of solid waste per year is generated of which approximately 67% is organic (CCAC 2014).

Notwithstanding the issues with waste management, Ghana's electricity sector is saddled with many supply challenges such as an over reliance on non-renewable energy sources for power generation. Existing power plants are also not able to attain full generation capacity due to constraints such as fuel supply, the uncertainty of rainfall and water inflows into hydroelectric power facilities (Kuamoah 2020). The waste-to-energy has proven itself to be an environment friendly solution for the disposal of municipal solid waste. This technology for conversion of MSW into power generation would not only be beneficial to meet the power demand but also reduce the environmental pollution to a certain extent. Given the positive correlation between energy access and human development, Ghana needs to improve energy security to provide insulation against external shocks such as price spikes of fossil fuels by integrating a broad range of non-conventional energy technologies into its generation portfolio. Sustainable energy systems provide such an opportunity. Transition to renewable energy technologies however requires long-term planning and commitment on the part of government (Puopiel 2010).

This study therefore seeks to assess the current Knowledge, Attitudes and Practices (KAP) of individuals regarding waste management; to identify environmental practices concerning waste management that individuals engage in; to examine whether beneficiaries will be willing to pay for waste management and to examine perceptions of individuals on benefits and challenges with waste to energy technology.

2 Methodology

2.1 Study Area

The study was conducted in the Atwima Nwabiagya Municipality in the Ashanti Region of Ghana, West Africa. Nkawie/Toase, Afari and Abuakwa are the three (3) zonal councils within the municipality with Nkawie as the capital. Nkawie is located 27.3 km west of the regional capital and 273 km from the national capital. The population of the municipality according to the 2017 projected population size according to the Atwima Nwabiagya Municipal Assembly Physical Planning Department is 106,569. Agriculture is the mainstay of residents in the municipality.

2.2 Study Design, Sampling Method and Data Collection

The study employed a cross sectional design to obtain both qualitative and quantitative data. *A* multistage sampling method was adopted with cluster and simple random sampling methods. Fifteen communities from the three zonal councils (constituting the study area) were randomly selected. Study respondents, who were basically adults, were also randomly selected at the household level. Ghana Living Standard Survey's definition of a household was adopted and applied for the study. Here, a household was defined as a person or group of persons living together in one housing unit, who share the same housekeeping and cooking arrangement and acknowledge a male or female adult as a household head (Ghana Statistical Service 2018). A total of 614 respondents were interviewed from the population size of 160,569. In the Nkawie/Toase Zonal Council, 37 respondents were interviewed in Gyankoba, 37 respondents in Toase, 50 respondents in Kobeng, 39 respondents in Nkawie, 40 respondents in Amanchia and 36 respondents in Wioso, summing up to 239 respondents in the zonal council. A total of 160 respondents were interviewed in Afari zonal council area. Each community in the zonal council; Afari, Mim, Kyereyaase and Nkontomire were represented by 40 respondents. Also, 215 respondents were interviewed in the Abuakwa Zonal Council with 46 respondents from the Abuakwa community, 43 from Manhyia, 42 from Nerebehi, 47 from Sepaase and 37 respondents from Adankwame.

Data collection was done using a pretested structured questionnaire and was conducted between September 2020 and February 2021. Validity of the questionnaire was achieved through the review of existing published research literature and project team member's input. Recommended changes suggested through the questionnaire validation process and initial pretesting was incorporated into the final questionnaire administered. The data was organized under the following themes: demographic characteristics of respondents; current waste management practices; knowledge about waste management and knowledge and perceptions about converting waste to energy.

The questionnaires were administered by trained enumerators through face-to-face interviews in both English Language and paraphrased into local languages for respondents with low literacy levels. The respondents were informed about the purpose and objectives of the study and assured of the confidentiality of their participation and any information given in the interviews. The respondents thus consented to their willingness to participate in the survey.

2.3 Data Analysis

Data from the survey was analyzed using STATA 15 software. The data analysis involved descriptive statistics comprising demographic characteristics of the respondents interviewed, their knowledge, attitudes and practices on waste management and perceptions of the benefits, challenges and expectations from conversion of waste to energy. The descriptive analysis was done by using descriptive statistical analysis and the chi-squared tests of association. The descriptive analysis was further used to understand perceptions of the benefits and challenges of waste to energy within the study area. The Garret ranking technique (Dhanavandan, 2016) was used to rank in percentages, the various types of waste generated in the households of respondents. Additionally, logistic regression was applied to estimate the probability relationships between willingness to pay for waste disposal or collections and social demographic factors as employed by Adebo and Ajewole (2012).

2.4 Estimation Procedure

To estimate factors, which influence respondents' willingness to pay for waste management in a waste to energy conversion process, the study adopted the logistic regression model because the dependent variable, thus willingness to pay, is binary (either Yes or No). The fitted dependent variable is expressed as a function of the independent variables chosen based on theory, information from literature and knowledge of the communities from which data was gathered.

In the logistic model, WTP represents the dichotomous variable of respondents' willingness to pay which is a latent variable in the model specification (Wilson et al. 2015). With X representing a matrix of explanatory variables β, vectors of the parameters and ε is the normally distributed error term.

$$WTP_i^* = \beta X_i + \varepsilon_i \quad (1)$$

The probability that a respondent is willing to pay dependent on the explanatory variables (X_i) since WTP^* is a latent variable is expressed as:

$$\Pr(\mathrm{WTPi} = 1|\mathrm{X}) = \varphi(\beta \mathrm{X}i) \quad (2)$$

$$WTP = \begin{cases} 1\ if\ WTP^* > 0 \\ 0\ if\ WTP^* = 0 \end{cases} \quad (3)$$

$$WTP^* = \ln\left(\frac{p}{1-p}\right) = \beta_0 + \beta_1 GEN + \beta_2 AGE + \beta_3 EDU + \beta_4 ZNC + \beta_5 EMP + \beta_6 INC + \beta_7 FRQ + \beta_8 DIST + \beta_9 PAY + \beta_{10} DISP + \beta_{11} WCON + \beta_{12} CHA + \beta_{12} LOC + \varepsilon_t \quad (4)$$

The study further estimates the average marginal effects of the explanatory variables on the explained variable to examine the factors associated with the probability of WTP of respondents. The marginal effect is the effect of a unit change in an explanatory variable on the probability of willingness to pay of respondents. Since the logistic model is binary the effect of a change in the explanatory variable, all things being equal, will differ with each individual. The average marginal effect thus estimates the marginal effect on average for individuals in the sample (Cameron and

Trivedi 2010).

$$\frac{\partial \Pr(WTP_i = 1|X_i)}{\partial X_i} = \varphi(X_i^*) \tag{5}$$

Hosmer-Lemeshow tests of goodness of fit and a Link test check for specification errors were estimated as diagnostic measures for the fitted model. For respondents who were willing to pay, the study compared the means of the maximum amount they were willing to pay and the how much they currently pay by estimating the significance of the t-value. Table 1 shows the measurement of each variable used and the a-priori expectations of the model.

Table 1 A-priori expectation of variables

Variable	Symbol	Measurement	A-priori expectation
Gender	GEN	Gender. Dummy (1 = Male 0 = Female)	+/−
Age	AGE	Years	+/−
Educational level	EDU	Number of years in schools (categorical)	+
Zonal council	ZNC	Zonal Councils in the Atwima Nwabiagya Municipality	+
Employment status	EMP	Current employment status	+
Total Household income	INC	Total monthly household income (categorical)	+
Frequency of disposal	FRQ	How many times do respondents dispose refuse	−
Distance to the community dump	DIST	Measures how far, in km, the distance of the household to the community dumpsite	+
Current payment for waste disposal	PAY	Measures if respondents currently pay for waste disposal. Dummy (1 = Yes, 0 = No)	+/−
Current disposal method	DISP	Current disposal method. Categorical	+/−
Waste conversion	WCON	Measures respondents' knowledge of waste conversion. Dummy (1 = Yes, 0 = No)	+
Willingness to change current practises	CHA	Measures respondents' willingness to change current waste management practises. Dummy (1 = Yes, 0 = No)	+
Local institution in converting waste to other products	LOC	Measures respondents' knowledge of waste conversion institutions. Dummy (1 = Yes, 0 = No)	+
Willingness to pay	WTP	Dummy (1 = Willing to pay; 0 = Not willing to pay)	

3 Results and Discussion

3.1 Socio-Demographic Characteristics

In total, more female (79%) respondents compared to male respondents (21%) were interviewed. The mean age group was 40–49 years. The results showed that a fifth (17%) of the respondents have no formal education. More females than males proportionately had received Primary/Junior secondary education as well as Technical/ Vocational education. The males also had proportionately more Senior Secondary and Tertiary education than females. This educational pattern is in line with GLSS 7 where lesser percentages of females than males had University education. Higher percentage of females (23.9%) than males (12.1%) also had no education. About ninety-three (93%) of respondents who had no form of education were above 30 years old (Table 2).

Table 2 Socio-demographic characteristics of the study respondents

Variable	Total	Male	Female
Age			
Younger than 20 years	19 (3.1)	3 (2.29)	16 (3.33)
20–29	88 (14.38)	19 (14.5)	69 (14.35)
30–39	169 (27.61)	27 (20.61)	142 (29.52)
40–49	147 (24.02)	29 (22.14)	118 (24.53)
50–59	96 (15.69)	22 (16.79)	74 (15.38)
Above 60	93 (15.20)	31 (23.66)	62 (12.89)
Highest education level			
No education	104 (16.97)	8 (6.11)	96 (19.92)
Primary/JHS	337 (54.98)	59 (45.04)	278 (57.68)
SHS/Six form	124 (20.23)	48 (36.64)	76 (15.77)
Technical/Vocational	10 (1.63)	1 (0.76)	9 (1.87)
University/Polytechnics	38 (6.20)	15 (11.45)	23 (6.20)
Marital status			
Single	140 (22.88)	35 (26.92)	105 (21.78)
Married	404 (66.01)	89 (68.46)	315 (65.35)
Divorced	24 (3.92)	3 (2.31)	21 (4.36)
Widowed	44 (7.19)	3 (2.31)	41 (8.51)
Employment status			
Private sector employee	9 (6.87)	12 (2.48)	21 (3.42)
Public sector employee	10 (7.63)	17 (3.52)	27 (4.40)
Self employed	90 (68.70)	356(72.71)	446 (72.64)
Retired	12 (9.16)	15 (3.11)	27 (4.40)
Unemployed	10 (7.63)	83 (17.18)	93 (4.40)

Frequency (Row percentages), *Source* Authors calculations from data

Two-thirds of the respondents were married at the time. The results showed that only 5% of respondents younger than 20 years were married, as such respondents were either in school or graduates. In conformation with national statistics, the majority of the widowed respondents were above the age of 60 years and proportionately dominated by females.

Household size ranged between 1 and 30 persons. The majority (80.5%) of the respondents had an economic activity they were engaged in but mainly as small-scale farmers and traders. Only respondents less than 20 years (mostly students) were unemployed. More than half of the respondents were mostly self-employed (e.g. farmers). Unemployment rate was higher for female respondents (17.2%), compared to their male counterparts (7.6%). The majority of the respondents interviewed were largely (95%) responsible for decisions on waste management in the household. Summary of the socioeconomic characteristics is shown in below (Table 2).

3.2 *Respondents Knowledge, Attitudes and Practices (KAP) Towards Waste Management*

Knowledge towards waste disposal and waste conversion

The majority (81%) of the respondents confirmed that they were aware of how waste collected was finally disposed-off. About 15% had no knowledge while 4% of the respondents were indifferent. Less than 3% practice open spaces waste disposal methods and other specified methods such as using self-dug pits as final disposal methods. Further, 88% attested to using and/or knowing community disposal sites (disposal site/bins are at least 0.01 km and at most 3.5 km from households). In estimating the knowledge about alternative methods for waste disposal, over a third of the respondents answered yes to having knowledge of alternative waste disposal methods while the remaining respondents were unaware of alternative methods for waste disposal. Adding a chi-squared test revealed a strong positive statistical relationship between respondents' knowledge of alternative methods for waste disposal and their level of education ($\chi 2$ (4) = 14.5, p-value = 0.01).

About knowledge on waste conversion, two thirds of the respondents confirmed that they know waste could be converted to other uses including energy generation. Primary source of information about waste conversion to other uses include television (33.5%), radio (26.3%), social media (Facebook, twitter) (1.02%) and other specified sources of information by 15.1% of the respondents include friends, agricultural extension officers, NGOs and private recyclers. A cross tabulation with gender and a significant chi squared test, ($\chi 2$ (1) = 22.5, p-value = 0.00), showed that male respondents were more likely to know that waste can be converted into other resourceful products than female respondents.

Respondent's attitude towards payment for waste disposal.

With regards to attitudes of respondents concerning with payment for waste disposal, some of the respondents found payment for waste disposal as extremely expensive (9%), moderately expensive (9%), expensive (18%) and a little expensive (17%). The rest (47.3%) reported that the cost of waste disposal was not at all expensive and felt extremely satisfied with charges for waste disposal. About 15% expressed that they were very satisfied, a third of the respondents were moderately satisfied whilst a tenth of the respondents who pay for disposing waste were not at all satisfied with the charges for waste disposal.

3.3 *Waste Generation, Segregation, Storage and Disposal Practices*

Applying the Garret ranking technique (Dhanavandan 2016), organic waste (37%) (kitchen waste, decomposable waste, etc.), plastic waste (32%), glass and ceramic waste (10%), paper waste (15%) and other specified waste products (6%) like metallic tin waste were ranked in proportions as the various waste types usually generated in the household. The waste generated is mostly (52.5%) stored in plastic bowls and bins. Plastic bags were also used by 17.3% of the respondents as storage for waste generated while 15.3% used metallic bowls/basin. Other specified waste storage systems (13.2%) such as dugouts and pits were used in storing household generated waste. Paper boxes constituted the least (0.8%) waste storage method used. A chi square test used in testing the dependence of relationships showed that the community of the respondents significantly, ($\chi 2\ (56) = 269.6$, p-value $= 0.00$) influence the waste storage type statistically.

In assessing the waste disposal methods particular to the respondents and their respective communities, two-thirds of the respondents cited community dumps as the main avenue for waste disposal. While the same proportion, 14%, disposed their waste through private waste collectors and through burning their waste generated in the households respectively. About 45.8% of the respondents disclosed that they pay for each waste disposal while the remaining acknowledged waste disposal is free. The chi-square showed significant relationship between how much respondents paid, their level of education and their total household income ($\chi 2\ (100) = 146.3$, p-value $= 0.00$ and $\chi 2\ (72) = 101.5$, p-value $= 0.01$).

Further on the respondents' current practices of waste segregation, only a quarter of respondents answered yes, to separating the different types of waste generated at the household level. The proportion of male to female respondents showed that female respondents who separate their waste were almost twice (28.4%) as much as male respondents (19.8%). Of the respondents who separate waste, 71% separate the different types of waste into different bins/sites at the household level. The remaining 19% of such respondents explained that they burn the separated waste immediately. All respondents confirmed that there were no different waste bins/sites for segregated waste generated at the community level. A chi-test showed that respondents' level of household incomes and education respectively, had significant influence on the percentage difference between households that separate their waste or households which do not separate their waste. The results showed that as household income increases or the level of education increases, the percentage difference between households separating and not separating their waste decreases $\chi 2$ (3) = 9.1, p-value = 0.03).

3.4 Willingness to Pay for Waste Management

With regards to willingness to pay for waste generated, the results showed that female respondents were more likely to pay for waste generated than male respondents, all things being equal. Using the marginal effect, the model predicted females were 21.5% more likely to pay for the waste generated than males. A significant effect of the employment status on willingness of respondents to pay for waste management was identified. Unemployed respondents seem less willing to pay compared to respondents who were employed. The location of the respondents in terms of their zonal council categorization also showed statistically significant influence on the willingness to pay for waste generated and subsequently waste management services like conversion of waste to energy. The model revealed that using Nkawie/Toase Zonal Council, there was statistically significant difference with respondents in Afari Zonal Council while there existed no statistically significant difference between the base zonal council and Abuakwa Zonal council all things being equal. Thus at 1% significance level, respondents in Afari were 32.7% less likely to want to pay for the waste generated compared to respondents in Nkawie/Toase Zonal Council. This is confirmed by the low level of knowledge about the waste to energy projects in zonal councils or communities farther from the Nkawie/Toase Zonal Council or Gyankoba.

Other demographic characteristics such as the age, level of education, marital status and household income had no significant influence on the willingness of respondents to pay for waste management for the waste to energy plant. Respondents who do not currently pay the waste they generate were also less likely (23%) willing to pay for the technology. The results further showed that respondents with prior knowledge about waste conversion to other products were 10% more likely to pay for the waste management and waste to energy technology, as depicted in Table 3.

However, the study found that neither the distance of the waste disposal sites nor how frequent respondents dispose of their waste had significant effects on their willingness to pay. Similarly, the willingness of respondents to change their current waste if waste as a resource is explained to them or respondents' knowledge of any local institutions which convert waste to resourceful products had no statistically significant influence on respondents willing to pay for waste management for the waste to energy technology.

Hosmer-Lemeshow goodness of fit test was run on 10 groups to test the fitness of the model. The Hosmer-Lemeshow chi square proved that the model was well fitted (Hosmer-Lemeshow chi2(8) = 9.99, Prob > chi2 = 0.2657). A Linktest was also

Table 3 Determinants of willingness to pay for waste management (Average marginal effects)

Variable	dx/dx	Delta-method standard errors	z	Prob > z
GEN				
Male	−0.110	0.054	−2.030	0.042**
AGE	−0.010	0.018	−0.550	0.586
EDU				
Primary/JHS	−0.018	0.053	−0.330	0.743
SHS/Six Form	0.012	0.071	0.160	0.870
Technical/Vocational	−0.114	0.170	−0.670	0.503
University/ Polytechnic, teacher training, post-secondary nursing	0.217	0.125	1.750	0.081
EMP				
Unemployed	−0.139	0.059	−2.340	0.019**
MAR	−0.032	0.032	−0.980	0.329
INC	−0.030	0.025	−1.210	0.228
ZNC				
Afari Zonal Council	−0.310	0.046	−6.740	0.000***
Abuakwa Zonal Council	−0.015	0.053	−0.290	0.773
DIST	0.025	0.029	0.860	0.391
FRQ	0.000	0.024	0.010	0.991
PAY				
No	−0.233	0.049	−4.780	0.000***
DISP	−0.023	0.024	−0.940	0.346
WCON				
No	−0.103	0.040	−2.550	0.011**
CHA	−0.085	0.084	−1.010	0.314
LOC	−0.128	0.129	−0.990	0.322
Number of obs. = 505 Chi-square = 84.67 Prob > chi2 = 0.0000		Pseudo R2 = 16.79 Model VCE: Robust AIC = 626.1849 BIC = 736.0234		

***, **, *; Significance at 1%, 5% and 10% respectively
dy/dx for factor levels is the discrete change from base level
Source Field survey

carried out to check for specification errors. The significant _hat and insignificant _hatsq implied that the model is with specification errors (Nomenclature).

Additionally, analysis was carried out on the mean price differences between how much respective respondents are willing to pay for waste disposal and how much they are currently paying using the t-test (one-tailed and two-tailed). The results (Table 4) showed that there was no statistically significant difference between

Table 4 Paired t test

Variable	Obs.	Mean (standard deviation)	t-value	Prob.
Maximum amount willing to pay	220	1.593 (1.29)	−1.56	0.1205
Current amount being paid	220	1.752 (1.62)		

Source Field survey

how much respondents were averagely willing to pay (M = 1.6) and how much respondents averagely currently paid (M = 1.8) for waste disposal charges at 5% significance level, $t = -1.55$ $p = 0.12$. This implies that respondents who are willing to pay set the average amount to be paid for the Waste to Energy project waste collection. This is not different from the average amount for current waste disposal in communities.

3.5 *Perception About Conversion of Waste to Energy*

Although almost two-thirds of respondents had no knowledge of what waste to energy entailed, respondents opined on the benefits, concerns and expectations of a plant that converts waste to energy in their municipality. According to the respondents, some benefits of transforming waste into useful products, included sanity to the community (26%), job creation through the operationalization of waste to energy technology (9%), conservation of the environment and natural resources from switching from biomass (5%), producing co-benefits (recycled plastics [21%]; fertilizer [18%]; energy cost savings from relying on convention energy supplies such as electricity and LPG [18%]) while 3% cited economic productivity as a benefit from operating waste to energy technologies.

Concerns of respondents on the construction bordered on health and sanitation, employment of community members, general plan for waste management and sustainability of the project and its goals. A quarter of the respondents were concerned about the health and environmental issues from operating a waste to energy plant within their municipality. Considering that study communities fall within high rainfall regimes, unprocessed waste could produce bad odor and spread diseases. Employment approach that would ensure residents is offered managerial positions within the plants were also raised. Culturally, waste is not meant to be kept in households for more than a day, hence respondents were concerned about how regular waste collection to the plant would be done. Availability of waste bins to ensure effective waste segregation at source (households) was also raised. Additionally, sustainability of the plant as most fold up at the end of government tenures because of neglect from newly elected government was additionally raised as a concern.

Respondents (31%) however expect that with the establishment of a Waste to Energy plant, there will be increased employment opportunities particularly among the youth. Others (27%) expressed that they expect the commencement and operationalization of a waste to energy plant to result in the provision of certain basic social amenities and infrastructure like potable water, schools, toilet facilities and roads by the government as well as corporate social responsibility of the Waste to Energy project. About 15% of the respondents also expect that converting waste to energy will lead to a positive shift in the environment's sanitation and health, while 27% of the respondents expect that the by-products from operation of the Waste to Energy plant would be accessible to the locals.

4 Discussion

4.1 *Knowledge, Attitudes and Practices Towards Waste Management in Ghana*

There is a generally good impression in the current study that participants' knowledge of the kind of waste generated and treatment methods is high. This includes kind of waste generated, waste collection, what happens to waste at the final disposal site, kind of waste treatment method used at the final disposal site among others at the community level. Few respondents lacked such knowledge implying education and awareness creation on alternative waste management should be continually pursued as lack of understanding and/or knowledge from such people contributes to improper solid waste management. In semi-urban areas of Nigeria, it was seen that knowledge gap between the level of respondents' sanitation practices and attitude of the respondents' sanitation practices resulted in illegal disposal of waste which was further considered to be cheaper (Duru et al. 2017). Increasing education has the potential of increasing respondent's knowledge and perhaps attitude towards waste management.

This study assessed the knowledge, attitudes and practices towards waste management by residents in Atwima Nwabiagya Municipality of Ghana where a pilot waste to energy conversion plant is to be built. Results of this study revealed biodegradable composition of waste making up more than half of household waste generated. Other studies have also shown similar waste composition in Ghana (Gyimah et al. 2016, Miezah et al. 2015). Ability to separate the various waste generated before final disposal is essential. Waste separation is a key process for waste management, yet the study identified that only 25% practice some level of waste separation.

Absence of extra bins posed a major challenge for waste segregation hence improving access to waste collection bins are recommended. In other studies in Ghana, such as Asare et al. (2020), it was shown that the low level of waste segregation was influenced by lack of motivation (prize addition), level of income, lack of waste bins and lack of education. Gyimah et al. (2021) suggested in a study conducted in the Cape Coast Metropolitan of Ghana that 81% of residents were willing to perform source separation.

Temporary storage of solid waste by households for collection is an essential stage during solid waste management, which depicts some improvement in waste management by households. Low percentage of waste storage in dug-out/pits, which is mostly uncovered, is a good sign for prevention of land and air pollution in the Municipality. Also, it was found that most of the waste generated in households are already properly stored in the various collection system (excluding dug-out/pit), and therefore building and increasing communal collection points for recycling is believed to be possible to collect most waste generated. Without reviewing the current waste management practices in some communities, especially those with low storage system depicting poor waste management, collection of solid waste generated from households for processing could be a major challenge. Although Seng et al. (2018) observed that households that do not use proper waste collection/storage systems continue to use the usual illegal dumping method because it is cheaper this often results in land pollution. According to the Ministry of Sanitation and Water Resources (2020), most community wastes are emptied in uncontrolled dumpsites that are not fenced leading to pollution and poor waste management. Practicing waste reduction and recycling (for example waste to energy conversion) could reduce costs. Respondents, however, do not mostly pay for waste disposal with mixed feelings on level of waste disposal satisfaction. More than half of the study's responses disclosed the need for waste disposal to be free.

4.2 Conversion of Waste to Energy

Results from this study showed study participants' have limited knowledge and awareness about waste to energy technology. Meanwhile, the respondents perceived some expectations (benefits) that could be derived from such technology if established in their locality. These included improved sanitation, energy saving, job creation, conservation of the environment and natural resources, as well as economic productivity are the main benefits mentioned. The majority (83%) of the responses were related to improved sanitation implying the current sanitation and recycling system needs to be improved as well as showing the acceptance of such technology. In Ghana, a study by Bensah et al. (2011) highlighted similar benefits from a waste to energy technology, thus biogas production.

Concerns were, however, raised with health and sanitation, employment of community members, a general plan for waste management, and sustainability of the project and its goals. This suggests that while the respondents believed the existence of such technology would transform waste management challenges in the Municipality, the health and sanitation of harboring waste in the municipality for processing needs to be critically considered. Strategies to ensure structures are put in place to improve sanitation in the area should also be critically looked at. A high number of the respondents (95%) showed willingness to use proper waste management alternates and also willing to give out the waste for recycling, which is a satisfactory attitude to improved waste management in the Municipality.

4.3 Ensuring Sustainability of Waste to Energy Production in Ghana

In terms of willingness to pay for waste management, females form the majority group who show greater willingness to pay compared to males. This could be because females are particularly responsible for waste management, collection and monthly communal waste disposal site management, hence more willing to pay for even external parties to efficiently manage communal waste as shown in other studies (Awunyo-Vitor et al. 2013, Addai and Danso-Abbeam 2014). The study further showed that unemployed respondents were 3% less willing to pay compared with employed respondents. This could be explained by the fact that employed individuals all things being equal, have comparatively more disposal income to pay for improved waste management services as corroborated by a study by Owusu, Boaheng and Sundberg (2011). A similar unemployment scenario was observed in another study in Ghana where urban households' willingness to pay for improved solid waste disposal services in another Metropolis was 57% (Awunyo-Vitor et al. 2013). On the contrary, a study by Seth et al. (2014), in Ghana showed that 62% of the respondents were unwilling to pay for waste management. The findings indicate that, willingness to pay for waste should be critically examined contextually before investment in a waste to energy plant to ensure sustainability of such facility.

It was also recorded that age, level of education, marital status, and household income had no significant influence on the willingness to pay for waste management. This is encouraging and shows the support participants have for the waste to energy plants to ensure its continuous existence. It was additionally identified that currently non-paying waste generators and those with prior knowledge of the waste conversion technology were 23% less likely willing to pay and 10% more likely to be willing to pay respectively. This is because non-paying respondents use alternatives such as burnings of waste and therefore find it expensive to pay for the waste generated. Furthermore, it also indicates that switching from free to paying is uneasy to conform especially where there are no illegal waste disposal sanctions. More education is needed to ensure there is increased willingness for sustainability of waste to energy facilities while ensuring policy guidelines are developed to guide implementation of such facilities in future.

5 Conclusion

Converting waste to energy is an effective way of conserving the environment and closing the domestic energy supply gap which is dominant in most Sub-Saharan African countries including Ghana. The study aimed to examine the knowledge, attitudes and practices of waste management, respondents' perception on waste to energy conversion as well as analyze the factors which influence the willingness of members of the Atwima Nwabiagya Municipal Assembly to pay for disposing waste for the operationalization of a Waste to Energy plant in their municipality. The study showed that, majority of the respondents know the final waste disposal sites as well as the kind of waste disposal methods at the site. Very few knew of alternative waste management practices different from what they currently use. Respondents, nonetheless, largely found the cost of waste disposal moderately expensive and were extremely satisfied with the costs at their respective waste disposal sites. The study further showed that organic waste and plastic were the most generated types of waste which households mostly stored in rubber bins/bowls. Although a test of significance showed that the community of the respondents significantly influences the type of waste storage. However, two-thirds of the respondents cited the community dumps as the main avenue for waste disposal. Moreover, only a quarter of the respondents separate the waste they generate in their households whilst there exists no form of waste segregation at the community level.

Using the logistic regression model, the study found that gender significantly influences respondents' willingness to pay for waste disposal for the Waste to Energy plant, since more females according to the results are responsible for waste management at the household level. The employment status of the respondents significantly effects on their willingness to pay as unemployed respondents were comparatively less willing to pay for waste disposal for the Waste to Energy plant to employed respondents. The low levels of respondents' knowledge on waste conversion revealed to also have a significant negative effect on their willingness to pay. Respondents with no knowledge on waste conversion were significantly less willing to pay compared to respondents with knowledge of waste to energy. Also, the location of respondents in relation to the Waste to Energy plant also affects the willingness to pay. Respondents living in the communities of the Waste to Energy plant site are generally more willing to pay. Albeit the low knowledge about the conversion of waste to energy, respondents perceived that a waste to energy plant will pro-mote sanitation and health as well as employment opportunities in their municipality. Although, they were generally concerned about the smell and health implications of operating a Waste to Energy plant in the communities.

In the setting and operation of the waste to energy plant, since the concept of waste segregation is critical to the successful running of the plant, education on waste segregation should be widely publicized. Adequate waste bins can be provided which should be collected regularly to prevent a cultural friction pertaining to waste storage at the house-hold level indicated by respondents' high numbers of waste disposal in a week. Lastly, to achieve the acceptance of a price mark-up for waste disposal, decision makers should widely disseminate to the communities the social and economic benefits the members of the communities will enjoy from the price trade-off.

This study is not without limitation. It was conducted in one municipality (Atwima Nwabiagya Municipality in the Ashanti Region of Ghana) making it not representative of Ghana at the national level. Waste for the operations of the Waste to Energy plants would however also be sourced from other prominent metropolitan and municipal assemblies therefore the gap identified can serve as a baseline to design effective strategies and interventions for collection and operationalization of Waste to Energy projects in the selected areas. As well, the study provides insight on areas for improving the knowledge, attitudes and practises of waste management such as education and awareness of waste segregation to targeted geographic regions.

Nomenclature

Linktext for specification errors

Number of obs = 505

LR chi2(2) = 113.63

Prob > chi2 = 0.0000

Log likelihood = −288.21844 Pseudo R2 = 0.1647

q42	Coef.	Std.Err.	z	P>z	[95% Conf.	Interval]
_hat	1.072	0.124	8.670	0.000	0.830	1.315
_hatsq	−0.125	0.080	−1.580	0.115	−0.281	0.031
_cons	0.096	0.121	0.790	0.430	−0.142	0.333

Logistic model for q42, goodness-of-fit test

Number of observations = 505

Number of groups = 10

Hosmer-Lemeshow chi2(8) = 9.99

Prob > chi2 = 0.2657

References

Addai, K. N., & Danso-Abbeam, G. (2014). Determinants of willingness to pay for improved solid waste management in Dunkwa-on-Offin, Ghana. Journal of Agriculture and Environmental sciences, 3(1), 01–09.

Adebo, G. M., & Ajewole, O. C. (2012). Gender and the urban environment: Analysis of willingness to pay for waste management disposal in Ekiti-State, Nigeria. American International Journal of Contemporary Research, 2(5), 228–236.

Agbelie, I., Bawakyillenuo, S., & Lemaire, X. (2015). Waste-to-energy: African cities can transform their energy landscapes. Retrieved January, 30, 2017.

Asare, W., Oduro Kwarteng, S., Donkor, E. A., & Rockson, M. A. (2020). Recovery of municipal solid waste recyclables under different incentive schemes in Tamale, Ghana. Sustainability, 12(23), 9869.

ATIeNZA, V. A. (2008). A breakthrough in solid waste management through participation and community mobilization: The experience of Los Baños, Laguna, Philippines. Ritsumeikan Asia Pacific University.

Awunyo-Vitor, D., Ishak, S., & Seidu Jasaw, G. (2013). Urban Households' willingness to pay for improved solid waste disposal services in Kumasi Metropolis, Ghana. Urban Studies Research, 2013.

Bensah, E. C., Mensah, M., & Antwi, E. (2011). Status and prospects for household biogas plants in Ghana–lessons, barriers, potential, and way forward. International journal of energy and environment, 2(5), 887–898.

Cameron, A.C. and Trivedi, P.K., 2010. Microeconometrics Using Stata Second Edition CHAPTER 28 DRAFT. Accessed: Aug. 13, 2022. [Online]. Available: http://cameron.econ.ucdavis.edu/sfu2022/mus2_chapter28.pdf.

CCAC (2014) Solid Waste Management in Accra Clim. Clean Air Coalit. Munic. Solid Waste Initiat., pp. 153–182, [Online]. Available: http://waste.ccac-knowledge.net/sites/default/files/files/city_fact_sheet/Accra_MSW_%5CnFactSheet_0.pdf%5Cn.

Dhanavandan, S. (2016). Application of garret ranking technique: practical approach. International Journal of Library and Information Studies, 6(3), 135–140.

Duru, C. B., Iwu, A. C., Diwe, K. C., Uwakwe, K. A., Merenu, I. A., Madubueze, U. C., ... & Nwaigbo, E. (2017). Environmental sanitation practices: A case study of solid waste management in semi-urban communities in Orlu, Imo State Nigeria. Occupational Diseases and Environmental Medicine, 5(04), 88–105.

Ghana Statistical Service (GSS) (2018): Ghana Living Standard Survey (GLSS 7) 2017.

Gyimah, P., Amuah, P., & Berko, E. S. (2016). Assessment of teaching and learning resources in geography in senior high schools. Dama International Journal of Researchers (DIJR), 1(1), 217–226.

Gyimah, P., Mariwah, S., Antwi, K. B., & Ansah-Mensah, K. (2021). Households' solid waste separation practices in the Cape Coast Metropolitan area, Ghana. GeoJournal, 86(2), 567–583.

Hoornweg, D & Bhada-Tata, P. (2012). What a Waste : A Global Review of Solid Waste Management. Urban development series;knowledge papers no. 15. World Bank, Washington, DC. © World Bank. https://openknowledge.worldbank.org/handle/10986/17388 License: CC BY 3.0 IGO."

Hoornweg, D., Thomas, L., & Otten, L. (1999). Composting and its applicability in developing countries. World Bank working paper series, 8, 1–46.

Joseph, K. (2006). Stakeholder participation for sustainable waste management. Habitat International, 30(4), 863-871. https://www.sciencedirect.com/science/article/pii/S0197397505000524.

Kuamoah, C. (2020) Renewable Energy Deployment in Ghana: The Hype, Hope and Reality' journals.sagepub.com, vol. 12, no. 1, pp. 45–64, Jan. 2020, https://doi.org/10.1177/0975087819898581.

Licy, C. D., Vivek, R., Saritha, K., Anies, T. K., & Josphina, C. T. (2013). Awareness, attitude and practice of school students towards household waste management. Journal of environment, 2(6), 147–150.

Miezah, K., Obiri-Danso, K., Kadar, Z., Fei-Baffoe, B., Mensah, M.Y. (2015). Municipal solid waste characterization and quantification as a measure towards effective waste management in Ghana. J. Waste Manag. 46, 15–27. https://doi.org/10.1016/j.wasman.2015.09.009.

Ministry of Sanitation and Water Resources (2020). National Solid Waste Management Strategy for Ghana, no. June, p. 42.

Oteng-Ababio, M., Arguello, J. E. M., & Gabbay, O. (2013). Solid waste management in African cities: Sorting the facts from the fads in Accra, Ghana. Habitat International, 39, 96–104.

Owusu, V., Boaheng, J. M., & Sundberg, C. (2011). Are households willing to pay for full-cost solid waste collection? Evidence from Kumasi Metropolis of Ghana. Journal of Environmental Studies and Sciences, 1, 144–153.

Puopiel, F. (2010). Solid waste management in Ghana: the case of Tamale Metropolitan Area (Doctoral dissertation).

Reyes, P. B., & Furto, M. V. (2013). Greening of the solid waste management in Batangas City. Journal of Energy Technologies and Policy, 3(11), 187–194.

Seng, B., Fujiwara, T., & Spoann, V. (2018). Households' knowledge, attitudes, and practices toward solid waste management in suburbs of Phnom Penh, Cambodia. Waste Management & Research, 36(10), 993–1000.

Seth, K., Cobbina, S. J., Asare, W., & Duwiejuah, A. B. (2014). Household demand and willingness to pay for solid waste management service in Tuobodom in the Techiman-North District, Ghana.

Tsiboe, I. A & Marbell, E. (2004), A Look at Urban Waste disposal problems in Accra, Ghana.

United Nations, A/RES/70/1. (2015). Transforming our world: The 2030 agenda for sustainable development transforming our world: The 2030 agenda for sustainable development preamble. United Nations Gen. Assem. Resolut.

Weitz, N., Carlsen, H., Skånberg, K., & Dzebo, A. (2019). SDGs and the environment in the EU: A systems view to improve coherence.

Wilson, D. C., Rodic, L., Modak, P., Soos, R., Carpintero, A., Velis, K., ... & Simonett, O. (2015). Global waste management outlook. UNEP.

Business Case for Hybrid Waste to Energy System for Ghana

Ekua Afrakoma Armoo, Satyanarayana Narra, Mutala Mohammed, Essossinam Beguedou, and Francis Kemausuor

Abstract This study aims to develop a business case for a hybrid renewable energy plant located in Ghana which has the capacity to treat 50 tons of municipal solid waste daily through Anaerobic digestion and pyrolysis and also integrates a solar photovoltaic (PV) system. The plant is estimated to generate 810 MWh of electricity annually. The study developed a techno-economic model for assessing the viability using RETSCREEN software. The plant shows a positive net present value (NPV) and an internal rate of return (IRR) of 20.1%, despite a levelized cost of energy (LCOE) of €0.331/kWh, which exceeds the average End User Tariff (EUT) of €0.887/kWh. Sensitivity analysis identifies key variables, including electricity exported to the grid and debt financing parameters. Scenario 2 introduces grant-based financing and reduced capital expenditure (CAPEX) by eliminating the segregation system, reducing the LCOE to €0.145/kWh, but still above current tariffs at the same capacity. A business case for this waste-to-energy plant under the current worse case economic conditions of the country will rely mainly on a large plant with a minimum capacity of 1.6 GWh/yr. With decreasing global costs of solar PV, the technical plant design should have at least 50% solar PV capacity.

E. A. Armoo (✉) · S. Narra · E. Beguedou
Professorship of Material and Energy Valorisation of Biogenous Residues, Professorship of Waste and Resource Management, University of Rostock, Rostock, Germany
e-mail: ekua.armoo2@uni-rostock.de

S. Narra
e-mail: satyanarayana.narra@uni-rostock.de

E. Beguedou
e-mail: essossinam.beguedou@uni-rostock.de

E. A. Armoo · M. Mohammed
Council for Scientific and Industrial Research – Institute of Industrial Research, Accra, Ghana
e-mail: mutbaby@gmail.com

F. Kemausuor
Department of Agricultural and Biosystems Engineering, Kwame Nkrumah University of Science and Technology (KNUST), Kumasi, Ghana
e-mail: kemausuor@gmail.com

M. Narra and S. Narra (eds.), *Innovations in Circular Economy and Renewable Energy in Africa*, World Sustainability Series, https://doi.org/10.1007/978-3-031-68330-5_5

1 Introduction

Recent health and geopolitical tensions have negatively impacted the global energy system which depends heavily on fossil fuels. Price fluctuations for key fuel commodities, on the one hand, is commonplace but has plummeted within the last year which may lead to over 75 million vulnerable people potentially losing access to electricity. On the other hand, renewable energy is fast becoming an important energy commodity in most countries (International Energy Agency 2022). The capital cost for renewable energy projects is generally regarded relatively high with low operation and maintenance costs as compared to alternatives such as crude oil and coal (Pueyo et al. 2016). However, over the last decade, technological advances have resulted in renewable energy technologies becoming more cost competitive. The global weighted levelized cost of energy (LCOE) for solar PV for instance dropped from 700% more than fossils in 2010 to 29% less as at 2021 (International Renewable Energy Agency 2023). The slow growth within the renewable energy sector is affected by actors such as interest rate fluctuations, uncompetitive tariff regime, high cost of capital, fragile economic growth patterns, and inadequate interest and government funding into research and development. The factors affecting feasibility are context specific and include the seasonal availability of feedstock and governmental subsidies (van den Berg and Duong 2018).

Financing for energy projects may be sourced from internal resources of the investor, external equity or debt from a lender. Often, a combination of any of these is used (Harvey 2020). To consider the feasibility of investments in the energy sector, the LCOE is one of the most important indicators. It is calculated as the average generation cost throughout the project's lifespan, determines the project's viability relative to other projects and is generally linked to the capacity of the plant (Weinand et al. 2023).

Hybrid energy systems which integrate two or more technologies have been studied by various authors and at different scales. Examples of various combination which have been studied include: anaerobic digestion (AD) and solar PV (Odoi-Yorke et al. 2023; Buragohain et al. 2021; Agyenim et al. 2020); AD and gasification (Mabalane et al. 2021; Hadidi and Omer 2017; Esfilar et al. 2021; Asante et al. 2023); AD, aerobic composting, gasification and incineration (Xiong et al. 2016); solar PV and biodiesel (Adaramola et al. 2017); Concentrating solar power (CSP) and AD (Petrollese and Cocco 2020) as well as solar thermal and flue gas condensation through waste incineration (Behzadi et al. 2021). In all these studies electricity is the main product with some studies also considering compost and heating. Most studies have focused on technoeconomic feasibility assessment. For instance, in Moroccan cities, a hybrid anaerobic digestion and solar thermal system was found to be the most feasible option with an Internal Rate of Return (IIR) of 10.14% (El Ibrahimi et al. 2021). In three North African countries (Tunisia, Algeria and Morocco), co-digesting agro-industrial waste with organic fraction of MSW (OFMSW) in a hybrid solar- biogas plant was found to be feasible with IRR between 10 and 20% based on seasonal availability of feedstock. Although the estimated CO_2-eq emissions during the lifetime of the project was quantified (26–411 MT), the economic returns were not incorporated into revenues and feasibility which could be due to the project being economically feasible from the sale of the primary product (electricity) (El Ibrahimi et al. 2022).

One important source of energy is MSW. Conversion of waste to energy is known as waste to energy (WtE) or Energy from Waste (EfW) (Hoang and Fogarassy 2020). The recovered energy is considered a form of renewable energy and thus has positive environmental impacts as compared to fossil fuels and creates a bioeconomy for economic gains as well (Anyaoha and Zhang 2022). WTE technologies are available depending on the type, quantity and characteristics of feedstock, required energy method, economic conditions, environmental standards and specific factors (Mutz et al. 2017). The most commonly used WtE technologies are thermal, biochemical and chemical technologies although thermal technologies dominate (incineration, gasification/pyrolysis and landfill gas recovery) (Tozlu et al. 2016).

Like conventional power plants, the investment cost for waste energy power mainly consists of engineering, construction, and equipment, as well as infrastructure costs (e.g., fuel supply system and grid connection). Some studies have revealed that WTE systems are not financially sustainable without subsidies and external support. Revenues to be obtained from electricity sales does not directly cover both CAPEX and OPEX. Additional products such as Carbon Credits may be important to achieve this (Mutz et al. 2017). A landfill gas to energy project was modelled in South Africa and it was concluded that although the project had the potential to produce 6.6GWh of electricity to 2952 households, the plant was not economically feasible. Feasibility could only be attained through the sale of Certified Emission Reduction (CER) which could bring in revenues of over $15,000,000 as at 2020 (Njoku et al. 2020).

Several large WtE plants are springing up in Africa. Ethiopia has recently commissioned a WtE plant which processes 1,400 tons of MSW per day into electricity to cover 30% of the Addis Ababa's household energy needs (UNEP 2017). Also, Safisana is a 100 kW hybrid solar-biogas plant in Ghana providing grid connected electricity and compost (Agyenim et al. 2020). Through funding support from the German Government, a 400 kW pilot WtE plant is being constructed in Ghana with capacity to treat 50 tons/day of mixed MSW through a hybrid solar PV-biogas-pyrolysis system to produce electricity, compost and other recovered materials (Tena 2023).

This hybrid model is the first of such projects to be established in Ghana (Tena 2023). Over the past two decades, the installed capacity of Ghana's energy system has been dominated by fossils and biomass. In 2021, 61% was sourced from Natural Gas and oil with 34% from biomass and the remaining accounted for by solar and hydro (Energy Commission 2022). Although the generation share of renewables is expected to increase to 1% in 2032 mainly due to an expansion in installed capacity of solar plants, waste to energy still remains at 100 kW; mainly from the SAFISANA plant (Agyenim et al. 2020, Ghana Grid Company 2022). Out of the total 21,466.26 GWh electricity generated in 2021, the sources were composed of 35% hydro, 64.52% thermal, 0.25% renewables (solar and waste) and 0.2% from emergency imports (Ghana Grid Company 2022).

This study therefore seeks to develop a business case for the new hybrid waste to energy plant in Ghana by assessing the techno-economic feasibility of the three renewable energy systems. As with all renewable energy projects, the economic feasibility must be proven in order to establish the basis for future projects. The business case will support decision making for policy makers for the management of waste in Ghana. Furthermore, the business case will provide evidence to investors and project developers on critical direction to stimulate further investment. Since this is a pilot plant which is currently under construction, all referenced data for the assessment is related to the plant design and the preliminary operational phase. The scope of the project specifically covers a total installed capacity of 400kW with 50 tons of mixed municipal solid waste (MSW) being received from the city of Kumasi.

2 Methodology

Study approach

For the hybrid power project under study, the feasibility analysis is carried out in the RETScreen® Clean Energy Management software focusing on the technical and financial aspects which are detailed in sections "Data collection" and "Financial analysis" respectively. The RETScreen Expert software enables decision making for planning, analysis and monitoring of energy projects. The platform analyses the benchmark, feasibility and/or performance of energy projects or a portfolio of facilities and provides results based on technical, economic and environmental aspects. It also allows the assessment of several technologies, project designs as well as different scenarios. The software is workbook based with each sheet containing specific formulas to produce output results based on the case specific data put in (Minister of Natural Resources Canada 2005).

Figure 1 provides a representation of the entire study approach which involves three stages: data collection at the plant and technical as well as financial analysis in RETScreen. Finally, based on the results of the financial analysis, a business case is presented which provides guidance regarding the conditions under which a viable investment can be made feasible.

2.1 Data Collection

Study area

The hybrid waste to energy pilot plant is situated specifically at Gyankobaa which is a small rural community located within Atwima Nwabiagya Municipality of the Ashanti region which is 25 km from the location of the plant. It is specifically located on Longitude −1.8 and Latitude 6.6. The Municipality has an estimated population of 103,698 with an annual growth rate 2.6%. The main economic activity

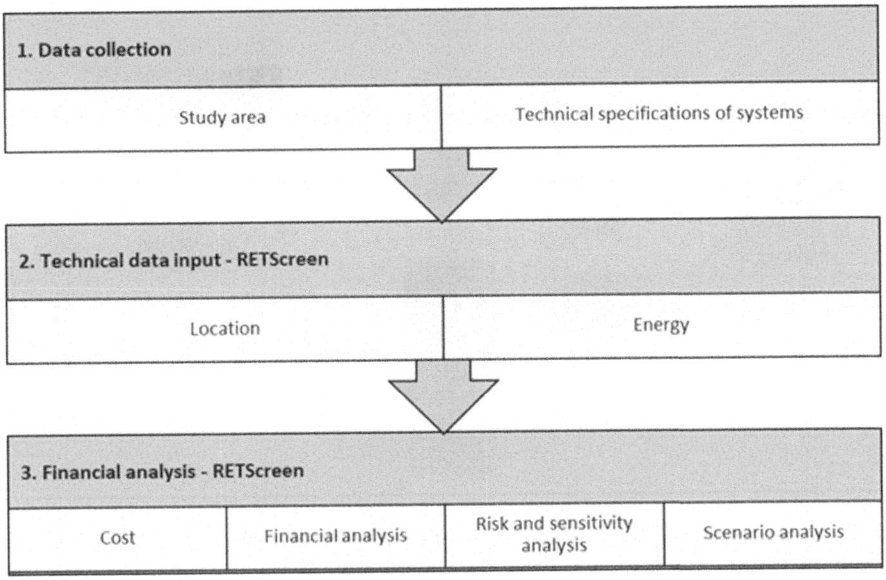

Fig. 1 Schematic of study approach

is farming and livestock rearing. The electricity coverage is about 70% with 24 communities without access to electricity (Ministry of Finance 2019). The general waste management situation in Ghana is characterized by unsegregated waste which is mainly disposed of in landfills which is the same within the Atwima Nwabiagya Municipality although the rural population is expected to contain more organics as compared to the urban municipalities (Miezah et al. 2015; Ofori 2016).

Technical design

The data input in RETScreen began with the input of the location data. The weather data was automatically generated from the NASA database incorporated into the software. In the energy worksheet of RETScreen, three tabs are created for each of the three systems with a total installed capacity of 400 kW: 200 kW solar PV system, 1000 m^3 biogas system linked to a 100 kW Combined Heat and Power (CHP) plant and a plastic pyrolysis system also linked to a 100 kW CHP plant (Fig. 2). Since most of the facilities within the plant are already installed, the scope of this study focuses on the technoeconomic study rather than the design of the system. Thus, all technical details from the plant are adopted without further calculations.

Solar PV system

The solar photovoltaic system consists of roof mounted solar panels, battery, inverter and cables. In order to calculate the output of the solar system, RETScreen uses location specific data from the closest climate data location from the NASA database.

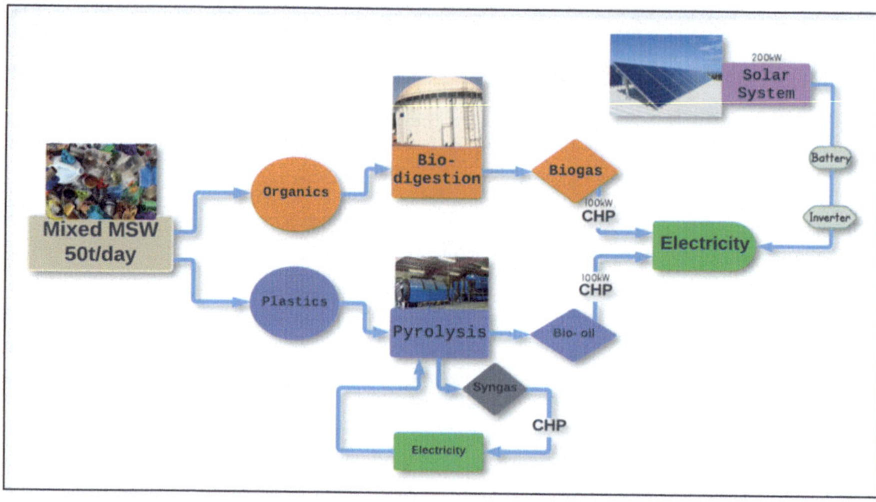

Fig. 2 A representation of the technical design for the hybrid power plant

Based on the selected location, the closest data location is 25 km away from the plant site.

The PV modules are roof mounted with a fixed solar tracking mode. Based on the solar irradiance data received at the selected location, maximum energy output from the modules is received at a tilt angle of 5 °C. Talesun TP6L72M mono-Si module is already installed at the site. This is because it is more locally accessible, and the technical specifications are adequate. Table 1 presents the technical specifications of the module. The miscellaneous losses from the module is assumed to be 10% since it is roof mounted. Also, the area is relatively rural and surrounded by thick forest and farmlands with low atmospheric dust although it is expected to be slightly higher during the dry season months from November to March each year. The module lifespan is assumed to be 25 years.

Anaerobic digestion (AD) system

The plant is expected to receive 50 tons of mixed MSW per day. The waste is sorted using both manual and mechanical procedures. A recent study on three plants in Ghana using similar sorting processes indicate that about 48% of organics is expected from the waste received at the plant (Sarquah et al. 2023). The organic waste is transferred to a 1000 m^3 biodigester which is expected to produce biogas at a rate of 45 m^3/h. The biogas is then transferred to a cleaning chamber and finally, the methane is directed to a 100 kW Combined Heat and Power (CHP) engine with an efficiency of 35% to be converted into electrical energy for the grid. A transformer is located onsite to convert the electrical energy for the grid. The plant is expected to operate 335 days/year with 30 days for both planned and unscheduled repair and maintenance works. The energy output from the AD system is calculated as in Eq. (1):

Table 1 Specifications of the components of solar PV system

Parameter	Specification
• Manufacturer and module	• Talesun (TP6L72M)
• Type	• Mono-Si
• Solar tracking	• Fixed (Roof mounted)
• Tilting	• 5%
• Power capacity	• 200 kW
• Efficiency	• 20.1%
• Nominal operating cell temperature	• 45 °C
• Solar collection area	• 995 m^2
• Inverter capacity	• 150 kW
• Inverter efficiency	• 95.5%
• Capacity factor	• 14.1%

$$E_{gen} = P \times T \times E_{ff} \quad (1)$$

where, E_{gen} is the energy generated from the system, P is the Power capacity of the CHP (kW), T is the operational hours per year (hrs/yr), Eff is the conversion efficiency of the CHP (%)

Pyrolysis system

The pyrolysis system is expected to receive 1 ton of mixed plastic waste per day. The pyrolysis process typically produces bio-oil, syngas and tar (Bassey et al. 2023). The pyrolysis system in this study is designed to be self-sustaining in terms of energy demand. Thus, the syngas is directed into a 50 kW CHP to power the pyrolysis process. The bio-oil is directed into a 100 kW CHP to produce electricity for the grid. The energy generation from the pyrolysis system is calculated as with Eq. (1).

Energy exported to grid

The energy exported to grid (E_{grid}) is calculated as a balance between the total energy generated (TE_{gen}) and the internal energy demand (E_d). E_d is mainly from the segregation system as well as minor onsite uses such as the office building. The national end-user tariff was applied to the internal energy consumed during the financial analysis. The energy output from the plant is calculated within the RETScreen software as in Eq. (2):

$$TE_{gen} = \sum E_{pv} + E_{AD} + E_{pyr} \quad (2)$$

where: E_{pv}, E_{AD}, E_{pyr} is the energy output from the Solar PV, AD and Pyrolysis systems respectively.

Since the plant was not under full scale operation during the period of the study, the complete log of energy consumption was not obtained. The total internal energy demand is therefore calculated using the Eq. (3).

$$E_d = P \times T \tag{3}$$

where P is the power rating of the equipment and T is the total operating hours per year.

Therefore, the total energy exported to grid is calculated as in Eq. (4):

$$E_{grid} = TE_{gen} - E_d \tag{4}$$

2.2 *Financial Analysis*

The financial analysis was carried out on the finance worksheet of the RETSCREEN software. The project costs values were input in the cost worksheet and automatically exported to the finance worksheet. The parameters are grouped into two based on the level of control from the project managers: Internally and externally controlled parameters. Reference is made to a Power Purchasing Agreement between the Government of Ghana and Jacobsen Jelco Ghana Limited as an Independent Power Producer which provides insight into typical Power Purchasing Agreements in the local context and gives an indication of the cost buildup for energy projects in Ghana (Parliament of Ghana 2014)

Internally controlled parameters

The main factors to be considered here are the Capital costs, Capital financing (Debt to Equity, Debt rate and Debt term) and Operation and Maintenance (O&M) Costs. All costs are escalated at the rate of inflation as specified in section "external economic parameters".

The equipment was procured as turnkey systems from local suppliers which includes consultancy design supply and installation. This ensures access to local personnel for repair and maintenance. The Capital Cost (CAPEX) is the sum of the costs of the individual systems and assumes that for each of the systems, a Turnkey installation is done. The capital costs are calculated as in Equation 5:

$$C_{pv} + C_{bio} + C_{py} + C_{seg} + land \tag{5}$$

Which are the costs of the individual systems as follow: solar PV, AD technology, Pyrolysis and segregation system respectively. The Capital Cost of Equipment was made up of 12% for the solar PV and 29% each for the AD, pyrolysis and segregation systems (Table 2).

The operation and maintenance costs are estimated to be 2% of the CAPEX which involves cleaning, servicing and periodic replacement of minor worn-out cables etc. One of the reasons for the low O&M is the rural nature of the area making cost of operations relatively low.

Table 2 Summary of the financial parameters used for the analysis

Parameter	Value
Initial Equipment costs	€1,710,000
Land	€150,000
Project lifetime	20 years
Discount rate	12%
Debt to equity ratio	85%
Debt interest rate	3%
Debt term	15%
Operation and Maintenance	2% of CAPEX
Electricity export escalation rate	25%

External economic parameters

The external factors considered for the financial analysis indicates influences from national and international factors which may affect the feasibility of the plant. These factors are usually beyond the control of the stakeholders but may have a high impact on operations. Global economic situations such as the COVID-19 pandemic and the Russia Ukraine conflict has hugely affected most economies and the energy sector including Ghana (International Energy Agency 2022). The details for the calculation of the parameters are provided in the following sections (Table 3).

Inflation

All costs are escalated at the local inflation rate. Based on the projected lifetime of the project, the historical inflation of the country is used to forecast potential future trends. Figure 3 shows the 25-year trend of Ghana's annual inflation rate based on data from The World Bank (2023). The average inflation rate for the period was 16.12%; the highest of within the period being 41% in 2001 while a low 7% is observed in 2019. This average figure is a reflection of all possible local economic conditions with a potential to impact the feasibility of the plant.

Revenues and Rate of escalation of revenue

The potential sources of revenue for the plant are the sale of products such as electricity, tipping fees and district heating (Asante et al. 2023). However, due to the distance of the plant from the regional capital and the rural nature of the surrounding

Table 3 Summary of external economic parameters

External economic parameters	Source of value	Value (%)
Escalation of cost	Average 25-year Inflation	16.12
Income tax	Corporate Income tax (with tax holidays)	17
Electricity export escalation rate	Average 25-year increase in BGC tariffs	25
Incentives and grants	Land and import duties (20% of Capex)	20

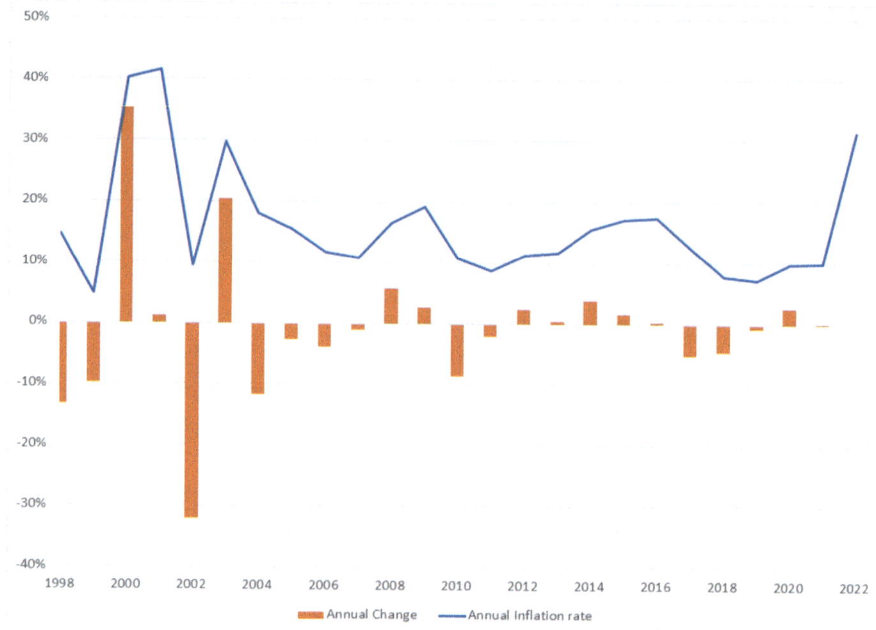

Fig. 3 The 20-year trend of inflation in Ghana

communities, tipping fees will become a major disincentive to attract waste management companies to dispose of waste at the plant. Also, district heating is not available in Kumasi.

The revenue estimations for the plant are based on the local tariff for Bulk Generation Charge (BGC) as published by the Public Utilities Regulatory Commission (PURC) in June 2023. The PURC is mandated to set realistic tariffs for public utilities including electricity to enhance competition and international standards. Bulk Generation Charge (BGC) which is the aggregated cost of power generated by the Independent Power Producers (IPPs) and the Volta River Authority (VRA). The End User Tariffs is obtained by a sum of the Bulk Generation Charge, Transmission Service Charge and the Distribution service charge (Acheampong et al. 2021). The proposed power plant is considered an IPP and thus revenues will be assumed to be based on the BGC tariffs. The most recent tariff released by the PURC for the second quarter of 2023 is GHp104.3646/kWh which translates into EUR0.088/kWh at the average interbank exchange rate for May 2023 from the Bank of Ghana (PURC 2023, Bank of Ghana 2023).

The rate of escalation of the revenues used in this study is 25%. This was obtained from the average of the year-on-year increments between 2018 and the second quarter of 2023. From Fig. 4, the cumulative change is greater than 17times the 15-year period although some years remained constant, the post COVID economic crises which is evident in most countries has caused a drastic increase in tariffs with 137% since

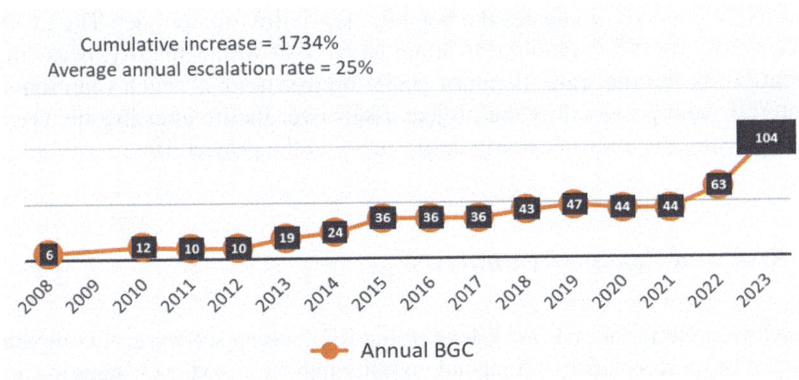

Fig. 4 The 25-year annual trend of escalation of tariffs for Bulk Generation Companies in Ghana (PURC 2020, 2023).

2021. This can be seen as the worst-case scenario whiles best-case scenarios occurred in 2015–2017 and also 2011–2012. The long-term average figures give a realistic representation of both worse case and best-case scenarios. Also, trends indicate that End User Tariffs are relatively stable around election years with no major reviews such as 2012, 2016, 2020 (PURC 2020).

The tariff pricing in Ghana is based on the Automatic Adjustment Formula (AAF) which was originally introduced in 2002 and reviewed in 2011. In applying the AAF, the PURC performs a comparative analysis of actual and projected Hydro-Thermal Generation Mix, Ghana Cedi-US Dollar Exchange Rate, Inflation Rate and Fuel Prices (such as natural gas) (Acheampong et al. 2021). Thus, the tariffs are a reflection of both local economic situation (inflation) as well as international market prices (fuel prices) and also various sources of energy. The use of the tariff in the escalation rate will therefore incorporate such internal and external influencing factors.

Tax–average for project lifespan

All corporate entities are required to pay Corporate Income Tax according to the Income Tax Act 2015 (Act 896). Companies within the waste management sector receive a tax holiday of seven years within which the applicable rate is 1%; and 25% from the 8th year onwards (Ghana Revenue Authority 2023). Over the 20 years expected lifespan of this project, the average annual tax rate is calculated as 17%.

Financial viability indicators

The viability of the proposed system will be assessed with Net present value (NPV), Levelised Cost of Energy (LCOE)/Energy Production cost and the Internal Rate of Return (IRR) based on the analysis results from provides results RETScreen. The RETScreen software calculates the NPV as the value of all future cashflows, discounted at the discount rate at present. The present value of all cash inflows is compared to the present value of all cash outflows over the investment period. A positive NPV generally indicates the potential feasibility of a project. The LCOE is the electricity export rate required to attain an NPV of whiles the IRR is calculated as the after-tax internal rate of return (IRR) on assets (%), which represents the true interest yield provided by the project assets over its life after income tax. It is calculated using the after-tax yearly cash flows and the project life.

2.3 *Risk and Sensitivity analysis*

This is carried out in the risk worksheet of the RETScreen software. A comparative analysis of the performance of financial viability indicators to the changes in various parameter enables the development of business case and provides basis for the future direction of the project.

Based on the results of the sensitivity analysis, two scenarios are assessed for the best case to attain economic viability. The technical design of the three technologies and the externally controlled parameters remains the same for both scenarios.

- Scenario 1—as described above
- Scenario 2—the internally controlled parameters are changed to improve viability of the project to develop the business case

3 Results and Discussion

3.1 *Technical Performance*

The expected daily solar irradiation data as against precipitation is represented in Fig. 5 with an average daily solar radiation of 4.34 kWh/m^2/d. The total energy potentially generated from the plant is 810 MWh/yr.

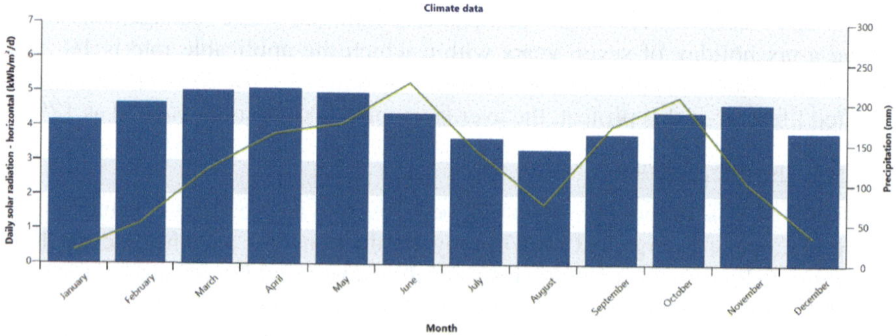

Fig. 5 The average monthly solar radiation as against the precipitation

3.2 *Financial Viability*

Based on the BGC tariff for May 2023, the total revenue expected in the first year is €70,457. The cumulative cashflows for the 20-year period indicate that positive cashflows begin in the 5th year although the debt financing period is 15 years. This is an indication of a good debt repayment potential with a low risk to the lender. The NPV is positive, IRR is 20.1% and LCOE is €0.331/kWh. The average residential and commercial End User Tariff (EUT) as announced in May 2023 is €0.127 (PURC 2023). Comparing the LCOE, the BGC tariff and the EUT, the cost of energy production is too high to sell the power and make the plant financially sustainable. The sensitivity analysis indicates that the most sensitive variables are the electricity exported to grid, debt financing (debt to equity, debt rate and debt term) and also the tariffs. However, the tariffs are beyond the control of the project. Thus, adjustments must be made to the internally controlled variables to provide a viable business case for the plant. Although the current externally controlled variables such as inflation are relatively high, the model provides us with a more optimistic approach to focus on the parameters which can be controlled by the project developer (see Fig. 6).

Scenario 2

In the Scenario 2, the following adjustments are made to the internally controlled variable.

- The source of the financing is assumed to be a grant with no debt.
- The CAPEX is reduced by removing the cost of the segregation system. The segregation system can be removed by sourcing already segregated organic and plastic waste, for instance from abattoirs and food processing industries. The Safisana system for instance receives only source segregated organic waste from market centers and industries (Agyenim et al. 2020). Also, source segregated plastic is gradually becoming very common in Ghana (Fiati 2021). With greater engagement with the municipality, the high cost of the segregation system can be removed for source segregated waste to be directed to the plant.

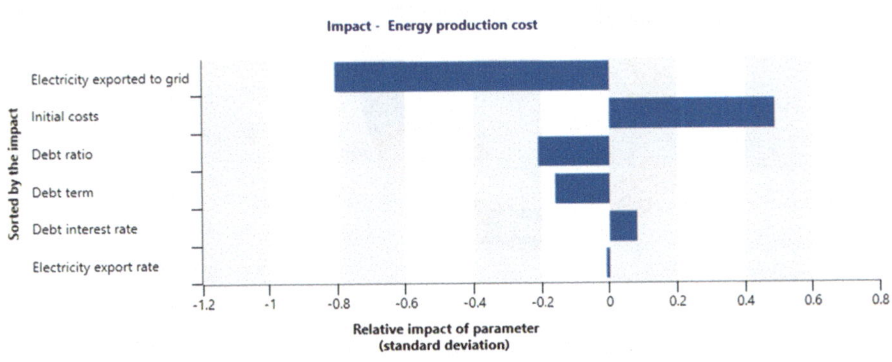

Fig. 6 The risk analysis indicating the impact of various variabilities on the LCOE

With the adjustments, the LCOE reduces to EUR 0.1447/kWh; which is still above the current selling price of the power (BGC Tariff). However, sensitivity analysis provides details into the various scenarios under which the plant can have an energy production cost matching or below the applicable tariffs. The most viable option is to increase the CAPEX by 16.7% whiles also increasing the energy exported to grid by 100% (see Fig. 7).

This may be a feasible option considering the cost of the various systems in the cost buildup (Fig. 8). Comparing the relative the cost buildup and the expected energy output, solar PV is the viable technology to expand at a relatively lower investment cost.

The cost of solar PV systems is reducing at a high rate each year and the LCOE for solar PV projects reduced from USD0.445/kWh in 2010 to 0.049/kWh in 2022 (International Renewable Energy Agency 2023). Since the procurement for the plant was done in 2019, the rate of reduction in the installed cost of Solar PV systems globally provides an opportunity for such systems (IEA et al. 2021). Programs such

- Remove analysis

Initial costs		Electricity exported to grid								MWh
		0.00	202.53	405.06	607.59	810.13	1,012.66	1,215.19	1,417.72	1,620.25
€		-100.0%	-75.0%	-50.0%	-25.0%	0.0%	25.0%	50.0%	75.0%	100.0%
0	-100.0%	8,697,050.00	8,697,050.00	8,697,050.00	8,697,050.00	8,697,050.00	8,697,050.00	8,697,050.00	8,697,050.00	8,697,050.00
201,667	-83.3%	8,697,050.00	8,697,050.00	8,697,050.00	8,697,050.00	8,697,050.00	8,697,050.00	8,697,050.00	8,697,050.00	8,697,050.00
403,333	-66.7%	8,697,050.00	96.64	48.32	32.21	24.16	19.33	16.11	13.81	12.08
605,000	-50.0%	8,697,050.00	209.74	104.87	69.91	52.44	41.95	34.96	29.96	26.22
806,667	-33.3%	8,697,050.00	329.34	164.67	109.78	82.34	65.87	54.89	47.05	41.17
1,008,333	-16.7%	8,697,050.00	452.94	226.47	150.98	113.23	90.59	75.49	64.71	56.62
1,210,000	0.0%	8,697,050.00	579.06	289.53	193.02	**144.77**	115.81	96.51	82.72	72.38
1,411,667	16.7%	8,697,050.00	708.77	354.38	236.26	177.19	141.75	118.13	101.25	88.60
1,613,333	33.3%	8,697,050.00	842.06	421.53	281.02	210.76	168.61	140.51	120.44	105.38
1,815,000	50.0%	8,697,050.00	979.60	489.80	326.53	244.90	195.92	163.27	139.94	122.45
2,016,667	66.7%	8,697,050.00	1,117.52	558.76	372.51	279.38	223.50	186.25	159.65	139.69
2,218,333	83.3%	8,697,050.00	1,256.70	628.35	418.90	314.17	251.34	209.45	179.53	157.09
2,420,000	100.0%	8,697,050.00	1,396.54	698.27	465.51	349.13	279.31	232.76	199.51	174.57

Fig. 7 The sensitivity analysis for scenario 2: comparing the LCOE (€/MWh) with changing CAPEX (initial cost in €) and electricity exported to grid (MWh)

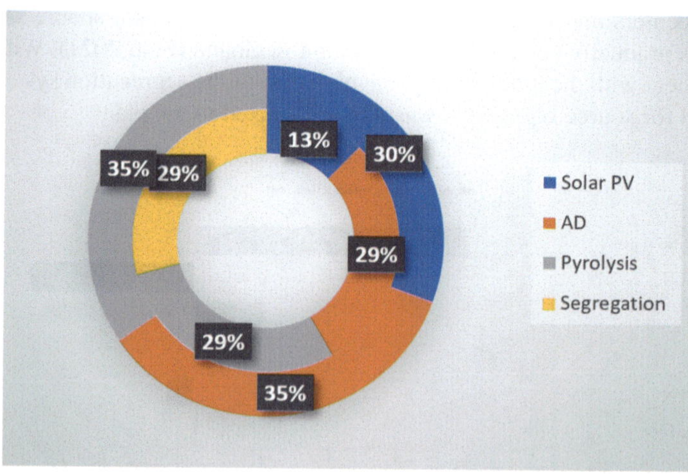

Fig. 8 The relative contribution of the various systems to the cost buildup (inner band) and the energy production (outer band)

as the Scaling Solar Program is targeted at increasing private investment in Africa by removing major barriers for large scale investors such as documentation, cost uncertainties and risks. The program has led to the establishment of projects in Zambia (100 MW), Ethiopia (250 MW) and Senegal (60 MW) with Togo, Madagascar, Cote d'Ivoire, and Niger set to receive similar benefits through grants. The large capacities of these projects, coupled with enhanced support from the program to investors ensured that the solar PV projects were awarded at relatively lower prices of about USD 60 and 78 MWh at the time (2015–2016) which was far below average prices in Africa. In 2018, the program supported lower prices of USD25/MWh (2019) in Ethiopia and USD 40/MWh in Senegal (2018) (IRENA and AfDB 2022).

This indicates that in order to improve the economic feasibility of the hybrid model, which relies only on revenues from the sale of electricity, the capacity must be expanded to produce at least 1.6 MWh/yr. Also, this expansion must be in favor of solar PV system (Gigure 1.34). However, the major limitation with this scenario is the dependence on grant support. The market for off-grid renewable energy in West Africa for instance depends largely on debt and equity rather than grants. However, most capital for energy project in Africa are financed through either debt and grant (IRENA and AfDB 2022). Public policy and financing play a critical role in stimulating investments and reducing the risks associated with financing renewable energy markets in Africa. Public concessional financing provides lower debt rates and more flexible debt terms as compared to independent investors. Private investments through Independent Power Producers (IPP) can be improved through public policy, public investments and risk sharing instruments. Thus, public financing is encouraged to drive such technological advancements.

4 Conclusion

The study assessed the technoeconomic feasibility of a hybrid waste to energy plant located in the Ashanti region of Ghana utilizing a combination of three technologies: solar PV, anaerobic digestion and pyrolysis. It is a novel model in Ghana in terms of the technical design. The plant receives 50 tons of unsegregated Municipal Solid Waste per day from the surrounding municipalities. A business case is developed to assess the economic feasibility of the current plant and the best scenario under which feasibility can be obtained. The RETScreen expert software was used. The results indicated that the plant is economically unsustainable whiles producing only 820 MWh of electricity to the grid per year. The most viable business case can be obtained when the capacity of the system is expanded to export at least 100% more electricity to the grid per annum at only 16.7% increase in the installed cost. Based on the installed cost for the three technologies, the recommended expansion should be done in favor of solar PV which is more likely to be done at such a low investment cost. Also, the although the country's economy has been strongly affected by recent global situations such as COVID-19, the model is still feasible within the current unfavorable externally controlled variables such as inflation and taxes. Thus, it can be concluded that the hybrid design is a feasible model which must be adopted in the developing country perspective although the capacity considerations are the capacity consideration must be carefully considered in the technology combination.

Nomenclatures

α Explanation
β Explanation

References

Acheampong, T.; Menyeh, B.O.; Agbevivi, D.E. (2021). Ghana 's Changing Electricity Supply Mix and Tariff Pricing Regime: Implications for the Energy Trilemma. Oil, Gas Energy Law Intelligence, 19, 1–28.

Adaramola, M.S.; Quansah, D.A.; Agelin-Chaab, M.; Paul, S.S. (2017) Multipurpose Renewable Energy Resources Based Hybrid Energy System for Remote Community in Northern Ghana. Sustain. Energy Technol. Assessments, 22, 161–170, https://doi.org/10.1016/j.seta.2017.02.011.

Agyenim, F.B.; Dzamboe, P.D.; Mohammed, M.; Bawakyillenuo, S.; Okrofu, R.; Decker, E.; Agyemang, V.K.; Nyarko, E.H. (2020) Powering Communities Using Hybrid Solar–Biogas in Ghana, a Feasibility Study. Environ. Technol. Innov., 19, 100837, https://doi.org/10.1016/j.eti.2020.100837.

Anyaoha, K.E.; Zhang, D.L. (2022). Transition from Fossil-Fuel to Renewable-Energy-Based Smallholder Bioeconomy: Techno-Economic Analyses of Two Oil Palm Production Systems. Chem. Eng. J. Adv., 10, 100270, https://doi.org/10.1016/J.CEJA.2022.100270.

Asante, K.; Gyamfi, S.; Amo-boateng, M. (2023) Techno-Economic Analysis of Waste-to-Energy with Solar Hybrid : A Case Study from Kumasi, Ghana. Sol. Compass, 6, 100041, https://doi.org/10.1016/j.solcom.2023.100041.

Bank of Ghana (2023) Interbank Exchange Rate: May 2023 Available online: https://www.bog.gov.gh/economic-data/exchange-rate/ (accessed on 4 July 2023)

Bassey, U.; Sarquah, K.; Hartmann, M.; Tom, A.; Beck, G.; Antwi, E.; Narra, S.; Nelles, M. (2023). Thermal Treatment Options for Single-Use, Multilayered and Composite Waste Plastics in Africa. Energy, 270, 126872, https://doi.org/10.1016/j.energy.2023.126872.

Behzadi, A.; Arabkoohsar, A.; Peri, V.S. (2021) Innovative Hybrid Solar-Waste Designs for Cogeneration of Heat and Power, an Effort for Achieving Maximum Efficiency and Renewable Integration. Appl. Therm. Eng., 190, https://doi.org/10.1016/j.applthermaleng.2021.116824.

Buragohain, S.; Mohanty, K.; Mahanta, P. (2021) Hybridization of Solar Photovoltaic and Biogas System: Experimental, Economic and Environmental Analysis. Sustain. Energy Technol. Assessments, 45, 101050, https://doi.org/10.1016/j.seta.2021.101050.

Van Den Berg, K.; Duong, T.C. (2018) Solid and Industrial Hazardous Waste Management Assessment - Options and Action Area to Implement the National Strategy. World Bank 2018, 107.

Energy Commission (2022) 2022 National Energy Statistics; Accra, Ghana

Esfilar, R.; Bagheri, M.; Golestani, B. (2021) Technoeconomic Feasibility Review of Hybrid Waste to Energy System in the Campus : A Case Study for the University of Victoria. Renew. Sustain. Energy Rev., 146, 111190, https://doi.org/10.1016/j.rser.2021.111190.

Fiati, K.E. (2021). The Plastic Waste Value Chain in Greater Accra Metropolitan Area: Assessing Flows and Spatial Patterns of Recycling.

Ghana Grid Company Limited (2022). 2022 Electricity Supply Plan for the Ghana Power System; Accra, Ghana.

Ghana Revenue Authority (2023) Corporate Income Tax (CIT) Available online: https://gra.gov.gh/domestic-tax/tax-types/corporate-income-tax/ (accessed on 4 June 2023).

Hadidi, L.A.; Omer, M.M. (2017) A Financial Feasibility Model of Gasification and Anaerobic Digestion Waste-to-Energy (WTE) Plants in Saudi Arabia. Waste Manag., 59, 90–101, https://doi.org/10.1016/j.wasman.2016.09.030.

Harvey, L.D.D. (2020) Clari Fi Cations of and Improvements to the Equations Used to Calculate the Levelized Cost of Electricity (LCOE), and Comments on the Weighted Average Cost of Capital (WACC). Energy, 207, 118340, https://doi.org/10.1016/j.energy.2020.118340.

Hoang, N.H.; Fogarassy, C. (2020). Sustainability Evaluation of Municipal Solid Waste Management System for Hanoi (Vietnam)-Why to Choose the "waste-to-Energy" Concept. Sustain., 12, 1–20, https://doi.org/10.3390/su12031085.

El Ibrahimi, M.; Khay, I.; El Maakoul, A.; Bakhouya, M. (2021) Techno-Economic and Environmental Assessment of Anaerobic Co-Digestion Plants under Different Energy Scenarios: A Case Study in Morocco. Energy Convers. Manag., 245, 114553, https://doi.org/10.1016/j.enconman.2021.114553.

El Ibrahimi, M.; Khay, I.; El Maakoul, A.; Bakhouya, M. (2022) Techno-Economic and Carbon Footprint Evaluation of Anaerobic Digestion Plants Treating Agro-Industrial and Municipal Wastes in North African Countries. Waste Manag., 154, 84–95, https://doi.org/10.1016/j.wasman.2022.09.019.

IEA, IRENA, UNSD, World Bank, (2021) Tracking SDG 7: The Energy Progress Report; Washington DC.

International Energy Agency (2022). World Energy Outlook Special Report: Africa Energy Outlook 2022

International Renewable Energy Agency (2023) Renewable Power Generation Costs in 2022; Abu Dhabi

IRENA and AfDB (2022). Renewable Energy Market Analysis : Africa and Its Regions; Abu Dhabi and Abidjan.

Mabalane, P.N.; Oboirien, B.O.; Sadiku, E.R.; Masukume, M. (2021). A Techno-Economic Analysis of Anaerobic Digestion and Gasification Hybrid System: Energy Recovery from Municipal Solid Waste in South Africa. Waste and Biomass Valorization, 12, 1167–1184, https://doi.org/10.1007/s12649-020-01043-z.

Miezah, K., Obiri-Danso, K., Kadar, Z., Fei-Baffoe, B., Mensah, M.Y. (2015). Municipal solid waste characterization and quantification as a measure towards effective waste management in Ghana. J. Waste Manag. 46, 15-27. https://doi.org/10.1016/j.wasman.2015.09.009

Minister of Natural Resources Canada (2005). 2001–2005 RETScreen International. Clean Energy Project Analysis: RETScreen Engineering and Cases Textbook; Third Edit.; Natural Resources Canada,; ISBN 0662391918.

Ministry of Finance (2019). Composite Budget for 2019–2022; Program Based Budget Estimates for 2019: Atwima Nwabiagya Municipal Assembly

Mutz, D., Hengevoss, D., Hugi, C., Gross, T. (2017). Waste-to-energy options in municipal solid waste management: A guide for decision makers in developing and emerging countries, Deutsche Gesellschaft für Internationale Zusammenarbeit (GIZ) GmbH. Eschborn, Germany. https://www.giz.de/en/downloads/GIZ_WasteToEnergy_Guidelines_2017.pdf. Accessed 15 March 2021

Njoku, P.O.; Edokpayi, J.N.; Odiyo, J.O. (2020) Modeling Landfill Gas Potential and Potential Energy Recovery from Thohoyandou Landfill Site, South Africa. J. Air Waste Manag. Assoc., 70, 820–833, https://doi.org/10.1080/10962247.2020.1778137.

Odoi-Yorke, F.; Adu, T.F.; Ampimah, B.C.; Atepor, L. (2023) Techno-Economic Assessment of a Utility-Scale Wind Power Plant in Ghana. Energy Convers. Manag. X, 18, 100375, https://doi.org/10.1016/j.ecmx.2023.100375.

Ofori, G.M.-L. (2016). Waste to Energy: An Alternative Energy Source for Ghana, Lund University.

Parliament of Ghana (2014). C. on M. and E. Power Purchasing Agreement and Energy Conversion Agreement between ECG and Jacobsen Jelco Ghana Limited; Accra.

Petrollese, M.; Cocco, D. (2020) Techno-Economic Assessment of Hybrid CSP-Biogas Power Plants. Renew. Energy, 155, 420–431, https://doi.org/10.1016/j.renene.2020.03.106.

PURC: Public Utilities Regulatory Commission (2020). A Study on the Trends in Electricity Tariffs in Ghana Between 2010 and 2020;

PURC: Public Utilities Regulatory Commission (2023). Publication of Electricity, Natural Gas and Water Tariffs.

Pueyo, A.; Bawakyillenuo, S.; Osiolo, H. (2016) Cost and Returns of Renweable Energy in Sub Saharan Africa: A Comparison of Kenya and Ghana

Sarquah, K.; Narra, S.; Beck, G.; Bassey, U.; Antwi, E.; Hartmann, M.; Sarfo, N.; Derkyi, A.; Awafo, E.A.; Nelles, M. (2023). Characterization of Municipal Solid Waste and Assessment of Its Potential for Refuse-Derived Fuel (RDF) Valorization. Energies.

Tena, N. (2023) 400kW Hybrid Waste to Energy Power Plant a First for Ghana Available online: https://www.esi-africa.com/industry-sectors/energy-efficiency/400kw-hybrid-waste-to-energy-power-plant-a-first-for-ghana/#:~:text=The 400-kilowatt production facility,organic manure for their farms. (accessed on 10 March 2023).

Tozlu, A.; Özahi, E.; Abuşoılu, A. (2016). Waste to Energy Technologies for Municipal Solid Waste Management in Gaziantep. Renew. Sustain. Energy Rev., 54, 809–815, https://doi.org/10.1016/j.rser.2015.10.097.

UNEP. (2017). Ethiopia's waste-to-energy plant is a first in Africa. Chemicals & Pollution Action. Available online: https://www.unep.org/news-and-stories/story/ethiopias-wasteenergy-plant-first-africa. Accessed on 10.03.2023

Weinand, J.M.; Hoffmann, M.; Göpfert, J.; Terlouw, T.; Schönau, J.; Kuckertz, P.; McKenna, R.; Kotzur, L.; Linßen, J.; Stolten, D. (2023) Global LCOEs of Decentralized Off-Grid Renewable Energy Systems. Renew. Sustain. Energy Rev. 2023, 183, 113478, https://doi.org/10.1016/j.rser.2023.113478.

World Bank (2023) Country Data Available online: https://data.worldbank.org/country/ghana?view=chart (accessed on 4 September 2023).

Xiong, J.; Ng, T.S.A.; Wang, S. (2016). An Optimization Model for Economic Feasibility Analysis and Design of Decentralized Waste-to-Energy Systems. Energy, 101, 239–251, https://doi.org/10.1016/j.energy.2016.01.080.

Uncited Reference

PURC: Public Utilities Regulatory Commission (2022). 2022-2025 Multi-Year Major Tariff Review for Natural Gas, Electricity Abd Water Services. Press RELEASE May, 2022.

Developments in Biogas Production and Substrate Use

A Life Cycle Assessment of Electricity Generation from the Anaerobic Digestion of Municipal Solid Waste in Ghana

Theophilus Baidoo, Patrick Boakye, Mizpah Ama Dziedzorm Rockson, Francis Kemausuor, and Satyanarayana Narra

Abstract Ghana relies heavily on fossil fuels for energy generation. This dependence on fossil fuels not only burdens the economy but also prevents the exploitation of the country's indigenous renewable energy resources. Anaerobic digestion (AD) is one of the promising waste-to-energy technologies that has great potential for generating clean energy while solving waste management problems in Ghana. A life cycle assessment (LCA) is needed to understand the overall environmental impact of waste-to-energy technologies and to implement future alternatives sustainably. This study applied an LCA to assess the inventory results of a hybrid waste-to-energy plant in the Atwima Nwabiagya Municipality in Ghana and its impact on the environment, including greenhouse gas emissions and energy generation. The study revealed that the anaerobic digestion plant will generate 805,574 kWh/year of electricity, equivalent to 352,589 Nm3/year of biogas, from 8,615 tons of municipal solid waste. The life cycle inventory further indicated that greenhouse gas emissions, including CH4 (3,786.8 kg), NOx (1,614.48 kg), and N2O (12.79 kg), are produced, which contribute to the overall carbon footprint. Environmental benefits were observed in 10 out of 11 impact categories in the CML-IA baseline method, except for abiotic depletion, which recorded an environmental burden. Contribution analysis revealed that the transport of waste to the plant site was the highest contributor to the environmental impact of the anaerobic digestion system. Scenario analysis suggested that replacing solar PV electricity with Ghana's electricity production mix (2018) leads to a substantial reduction in the abiotic depletion category (-99.61%) but results in increased burdens, notably in acidification (725%). The uncertainty analysis showed that data reliability and temporal correlation play significant roles in quantifying uncertainty

T. Baidoo · P. Boakye (✉) · M. A. D. Rockson
Department of Chemical Engineering, Kwame Nkrumah University of Science and Technology, Kumasi, Ghana
e-mail: patrickboakye@knust.edu.gh

F. Kemausuor
Department of Agricultural and Biosystems Engineering, Kwame Nkrumah University of Science and Technology, Kumasi, Ghana

S. Narra
Professorship of Material and Energy Valorisation of Biogenous Residues, Professorship of Waste and Resource Management, University of Rostock, 18059 Rostock, Germany

M. Narra and S. Narra (eds.), *Innovations in Circular Economy and Renewable Energy in Africa*, World Sustainability Series, https://doi.org/10.1007/978-3-031-68330-5_6

The study concluded that anaerobic digestion is a viable waste-to-energy technology in Ghana that can reduce greenhouse gas emissions and generate renewable energy, but further improvements are needed to optimize its environmental performance and reduce its resource consumption.

1 Introduction

The generation of solid waste is increasing rapidly worldwide due to population density, urbanization, economic growth, and industrialization (Lissah et al. 2021; Misganaw and Teffera 2021). Every day, Ghana generates about 22,500 tons of waste. Organic waste accounts for roughly 60% of the total (Debrah et al. 2022). In Ghana, the per capita waste generation is 0.47 kg/person/day (Miezah et al. 2015). Waste generated in 2016 contributed 3.2 million metric tons (MT) of CO_2 (8%) of the total greenhouse gas (GHG) emissions (Environmental Protection Agency 2019; World Bank 2020). Therefore, waste management practices can help to reduce greenhouse gas emissions and combat climate change (Kanhai et al. 2021). Several green technologies have been recommended for organic waste management, including composting (aerobic digestion), incineration, and anaerobic digestion (AD) (Mahajan et al. 2019; Adanu et al. 2020). These waste management techniques are well-established and practiced in most developed countries (Mahajan et al. 2019). In contrast, Ghana, among other developing nations, has a poor waste management strategy (Adanu et al. 2020). Landfills are rarely engineered where available and other techniques are not properly managed (Adanu et al. 2020). Although some waste management facilities do exist in the country, an assessment of the environmental impact of their operations has not been conducted (Afrane et al. 2021). To address this gap, this study employed a life cycle assessment (LCA) to evaluate the environmental impacts of an anaerobic digestion plant in the Atwima Nwabiagya Municipality in Ghana. Anaerobic digestion as a waste management technique has proven to be useful in converting organic waste into products that have a positive impact on society with fewer environmental effects (Kumar and Samadder 2020). The life cycle assessment (LCA) tools are well-built to holistically assess the environmental impacts of products and processes and to propose solutions to meet ecological needs (Nabavi-Pelesaraei et al. 2019).

Anaerobic digestion is one of the promising waste-to-energy technologies in Ghana that can generate clean energy while solving waste management problems in the country (Afrane et al. 2021; Cudjoe and Mensah 2021). Several small-scale biogas digesters are currently in operation in Ghana, making it the most widely used technology for managing organic waste, particularly sewage (Ofori-Boateng et al. 2013). However, the life cycle assessment of anaerobic digestion is least studied in Africa, especially in Ghana (Maepa et al. 2017). Recent critical reviews on LCA applications in the environmental assessment of waste management systems and waste-to-energy technologies worldwide (Khandelwal et al. 2019; Iqbal et al. 2020; Zhang et al. 2021) highlighted the inadequacy of LCA studies in waste management, especially in low-income regions like African countries. Most of these studies have primarily concentrated on countries with high-income and medium-income levels, thereby rendering their findings unsuitable for direct application to the unique context of Ghana's waste management system. This discrepancy arises due to the variations observed in waste quantities, characteristics, and treatment techniques.

Galgani et al. (2014) contributed to climate change mitigation through LCA of anaerobic digestion, in Northern Ghana. However, the study did not address several impact categories such as eutrophication, acidification, and human toxicity potentials. Hence, this study aims to conduct a comprehensive life cycle assessment on the anaerobic digestion of municipal solid waste (MSW) as a waste-to-energy technology in Ghana. The study aimed to fill this gap by quantifying and compiling the inventory for the input and output products of the anaerobic digestion process and estimating the environmental impact of the electricity generation via the anaerobic digestion process using the CML-IA baseline method. The study findings will also provide valuable insights into the environmental performance of anaerobic digestion technology by identifying environmental hotspots and informing decision-making processes to drive future innovations for improving its use.

2 Methodology

Experimental site selection and plant description

The environmental impact of electricity generation from the anaerobic digestion of MSW was assessed in this study using a hybrid waste-to-energy plant located in the Atwima Nwabiagya Municipality (6° 32′ 6.75″ N, 1° 36′ 2.00″W), Ghana. The hybrid waste-to-energy plant consists of a pyrolysis plant, solar PV, and a biogas plant. The focus of this study, however, was on the biogas plant (AD). The AD plant was designed to produce electricity and manage waste by processing about 8,615,000 kg of municipal solid waste per year. The AD plant comprises a 1000 m^3 CSTR closed-dome anaerobic digester, a digestate storage tank, and a 100-kW combined heat and power plant (CHP). The plant performs anaerobic digestion with a retention time of 21 days. It feeds the electricity produced to the national grid and uses the heat for heating the digesters. The fan-coolers dissipate the excess heat. A 2000 m^2 roof-mounted solar PV system provides energy for the operation of the facility and reduces the carbon footprint of the process.

Goal and scope definition

Defining the goal and scope is a crucial step in the LCA process. The study defined the primary functions of the AD facility as waste management and electricity production. It chose a functional unit of "generation of 805,574 of Electricity per year" for the study. The system boundary as shown in Fig. 1 includes the transportation of municipal solid waste (MSW) to the facility, the segregation of MSW, the anaerobic digestion of OFMSW, and energy generation from the biogas. Generation and collection of waste, the digestate storage and its application on the field, the distribution of electricity, and the construction of the facility were excluded in the analysis (Bruno et al. 2023). The impact of building the AD plant and CHP unit is excluded in the study because it has relatively low emissions compared to the use phase (Sandanayake et al. 2018).

Life cycle inventory

The main activity in the LCA's life cycle inventory phase is data collection. The data for the foreground system were obtained the plant owners, academic publications, and technical documents. The EcoInvent Database version 3.8 provided the background data (Ciroth et al. 2021).

Transportation of waste

The transportation of waste is the first stage of the system boundary. Diesel-refused skip trucks with a capacity of 21–23 tons collect the MSW from households and commercial areas and deliver it to the waste treatment site. The study obtained data on the transportation of the MSW from Zoomlion Ghana Limited, the largest private waste management company in Ghana. The data collection methods included questionnaires and direct interviews. The data included the quantity of waste transported, the distance covered, and the fuel consumed by vehicles. The study chose an estimated average distance of 10 km from the waste sources to the waste treatment plant. This distance covered neighboring municipalities that contributed to feedstock supply within a 10 km radius.

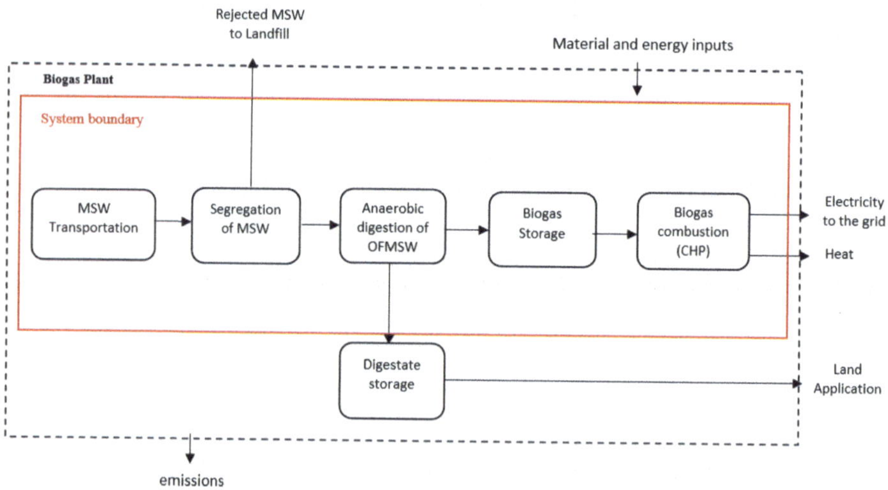

Fig. 1 The system boundary of the biogas plant

Segregation of municipal solid waste

The waste treatment facility has a sorting unit that segregates the mixed waste into organic waste and non-degradable fractions. MSW in the chosen region was not source-separated. The study performed material and energy balance calculation on the sorting unit based on data from the plant owner. The calculation determined the input and output parameters.

Anaerobic digestion of the organic fraction of municipal solid waste

The AD plant comprises a 1000 m^3 CSTR closed-dome anaerobic digester, a digestate storage tank, and a 100-kW combined heat and power plant. The reactor receives organic fraction of the municipal solid waste and mixes it with fresh water. The digester operates at a mesophilic temperature of 38 °C, which the study assumed as the temperature for modelling. The study determined the thermal requirement of the anaerobic digester using energy balance across the unit.

Emission estimations

The study performed emission estimations on the anaerobic digester and the combined heat and power plant units. The emissions included greenhouse gases such as methane (CH4), nitrous oxides (NOx), and dinitrogen monoxide (N_2O), as commonly reported in literature (Ayodele et al. 2017). The study assumed 60% methane content in the biogas and 2.5% methane loss for the AD following Scheutz and Fredenslund (2019). The emission factors for Danish decentralized CHP plants by Nielsen et al. (2010) are used to estimate the emissions from the CHP plant. According to these factors, combustion of 1 GJ of biogas produces 1.6 g of N_2O and 202 g of NOx.

Life Cycle Impact Assessment (LCIA)

The study evaluated the environmental impacts of the system by analyzing the results from the inventory analysis, specifically the inventory table, within the context of the goals and scope of the study. It performed the LCIA following the guidelines of ISO 14040/14044 and using OpenLCA 1.11.0 as the LCA assessment tool to set up the model for the anaerobic digestion process. The LCIA involved three steps: 1) selecting and classifying impact categories, 2) characterization, and 3) normalization. The study used the CML-IA baseline characterization method (CML - Department of Industrial Ecology 2016) of the Institute of Environmental Sciences (CML) and World (2010) normalization method for the LCIA because it is commonly used in several studies (Grosso et al. 2012; Biancini et al. 2020). The impact method consists mainly of impact categories such as abiotic depletion potential (kg Sb eq), abiotic depletion potential (fossil fuels) (MJ), acidification potential (kg SO_2 eq), eutrophication potential (kg $PO_4{}^{3-}$ eq), global warming potential 100a (kg CO_2 eq), human toxicity potential (kg 1,4-DCB eq), marine ecotoxicity potential (kg 1,4-DCB eq), ozone depletion potential (kg CFC-11 eq), photochemical oxidant creation potential (kg C_2H_4 eq), and terrestrial ecotoxicity potential (kg 1,4-DCB eq).

2.1 Interpretation

Scenario analysis

Scenario analysis per ISO 14040/44 (2006a, 2006b) is a common form of sensitivity. This involves the assessment of potential variations in LCA results resulting from discrete choices (Hauschild et al. 2018; Igos et al. 2019). It is employed to assess the uncertainty arising from various choices such as functional units, allocation procedures, assumptions in life cycle inventory, life cycle impact assessment methodologies, temporal variability, and spatial variability.

Three scenarios are considered in this analysis:

The first, which was the baseline, was modelled using an open-ground solar PV plant as the electricity source for the anaerobic digestion system.

The second scenario labelled as 'scenario 1' was modelled using an electricity production mix for Ghana in 2018 as provided by the EcoInvent Database in OpenLCA with the following consumption composition: Solar PV (0.2%), Natural Gas (34.4%), Oil (25%) and Hydro (40.4%) as a substitute for the Solar PV in the baseline scenario.

The third scenario labelled 'Scenario 2' was modelled using electricity produced from the biogas engine (combined heat and power plant) as substitute for the Solar PV in the baseline scenario.

Uncertainty analysis

The level of uncertainty in the results is contingent on the uncertainty in the model parameters used. To quantify the uncertainty in the scenarios and the sensitivity analysis, error propagation was employed through Monte Carlo simulations using OpenLCA 1.11.0. Monte Carlo simulation is the prevalent sampling technique in LCA and is supported by major LCA software platforms. (Hauschild et al. 2018; Igos et al. 2019). The fundamental principle of this sampling technique involves iteratively performing model calculations by sampling the values from predetermined probability distributions for each input parameter. Consequently, the model output can be represented as a probability distribution (Hauschild et al. 2018; Igos et al. 2019).

Standard deviations were assigned to individual parameters in the process inventories by utilizing data quality indicators within the pedigree matrices. The incorporation of pedigree matrices allows the assessment of data reliability and completeness as well as geographical, technological, and temporal correlations (Ciroth et al. 2016). For each scenario, the characteristic results were computed through 1000 iterations with dependent sampling.

3 Results and Discussion

Inventory analysis

Table 1 lists the unit and value of each input and output flow for each process stage. It also indicates the data source for each flow, which includes the plant owner, the waste management company, and calculations done based on these data.

The inventory analysis showed that the anaerobic digestion system relies on various resources as inputs and produces notable outputs. The transportation stage requires a large amount of MSW (8,615,000 kg), to be collected and sorted, which involves fuel consumption and emissions. The anaerobic digestion stage depends on biowaste (3,800,000 kg) and freshwater (1,400,000 kg) as essential inputs for the digestion process, which may affect the availability and quality of these resources. The system also benefits from solar energy input from a solar PV plant (48,180

Table 1 Inventory data for anaerobic digestion of MSW

Process stage	Unit	Value	Data source
Transportation	kg-km	86,150,000	Own calculation based on Waste Management Company data
Segregation of MSW			
Municipal solid waste	kg	8,615,000	Plant Owner
Electricity from solar PV	kWh/year	48,180	Own calculations based on the Plant owner's data
Anaerobic digestion			
Biowaste	kg	3,800,000	Plant Owner
Freshwater	kg	1,400,000	Plant Owner
Electricity from solar PV	kWh/year	314,484	Own calculation based on Plant Owner's data
Net biogas production	Nm3/year	352, 589	Plant Owner
Emissions associated with AD			
Methane loss	kg/year	3,786.8	Own calculations based on the plant Owner's data
CHP			
Electricity generated	kWh/year	805,574.0	Plant Owner
Net Thermal energy generated	kWh/year	868,621	Own calculation based on Plant Owner's data
Emissions associated with CHP			
N2O from CHP	kg/year	12.79	Own calculation based on Nielsen et al. (2010)
NOx from CHP	kg/year	1614.48	Own calculation based on Nielsen et al. (2010)

kWh/year), which reduces its reliance on fossil fuels. Moreover, the system showcases its renewable energy generation capability by producing biogas and electricity through a solar tower and a CHP unit. The system successfully converts biowaste into valuable energy resources, yielding a substantial net energy generation. The study suggests that optimizing the transportation stage and increasing the efficiency of the anaerobic digestion process can enhance the environmental performance of the system. The results of this study are consistent with other empirical studies on anaerobic digestion in Ghana and other countries. For instance, Cudjoe and Mensah (2021) assessed the electricity generation potential of biogas from food waste using anaerobic digestion technologies in Accra and Kumasi, Ghana. They found that with an estimated waste generation ranging from 899,000 to 3,359,000 tons per year in Accra and 915,000 to 3,159,000 tons per year in Kumasi, there was a potential for anaerobic digestion system to generate power between 80.43 and 300.49 GWh per year for Accra and 60.63 and 209.31 GWh per year for Kumasi.

Impact assessment

Life cycle inventory data were used to model and compute the environmental impacts of the anaerobic digestion system using the CML-IA baseline method, which included characterization and normalization steps. During the characterization phase, the relative contribution of each impact category was adjusted to a scale of 100%. The contribution analysis as shown in Fig. 2 revealed that the transportation stage was the highest contributor to most of the impact categories, followed by the anaerobic digestion stage.

The impact assessment results as shown in Table 2 indicate that the anaerobic digestion system has both environmental benefits and burdens on different impact categories. The system has a positive impact on global warming, acidification, eutrophication, ozone depletion, photochemical oxidation, and various types of ecotoxicity and human toxicity. These benefits are mainly attributed to the avoidance of emissions from conventional waste disposal methods and electricity generation from fossil fuels, which are the main sources of greenhouse gases and air pollutants in Ghana (Amo-Asamoah et al. 2020). However, the system has a negative impact on abiotic depletion due to its resource consumption, especially for the transportation stage, which requires a high amount of fuel. Therefore, the system has a trade-off between reducing emissions and consuming resources, which affects its environmental performance and sustainability.

To facilitate a direct comparison between the impact categories, a normalization process was performed to standardize all impact categories according to a reference value. Normalization provides a context for characterized results by relating them to the total environmental impacts of a certain region or population (Guinée et al. 2002). In this study, the normalization was based on the world average data for the year 2000 (Goedkoop et al. 2013).

According to the normalization results (Table 3), the aggregated value of the environmental impact amounted to -4.98×10^{-5}. The transportation stage contributes the most (4.83×10^{14}), followed by -4.54×10^{11} benefit from electricity generation, -7.42×10^{12} benefit from the segregation, and -4.84×10^{13} benefit from

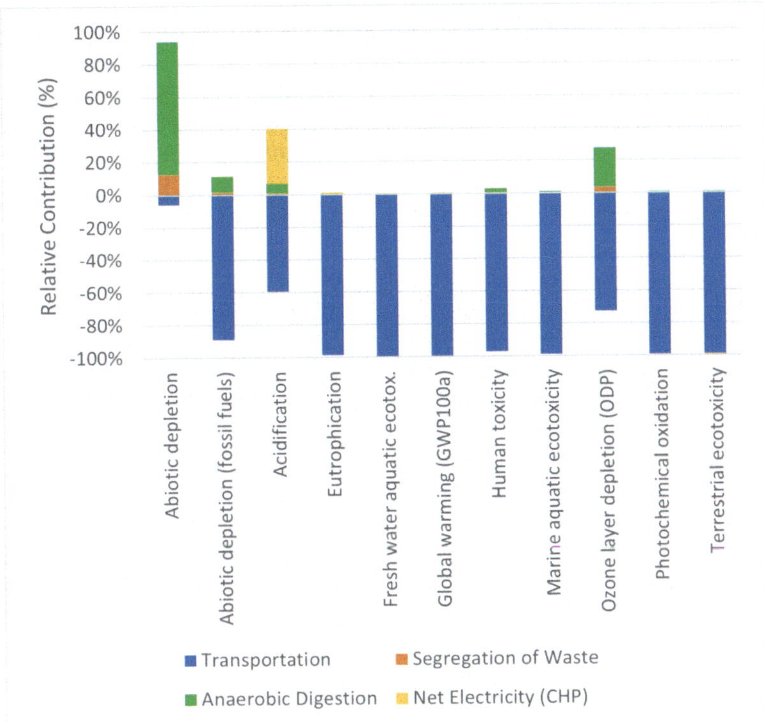

Fig. 2 Characterization results of the generation of 805,574 kWh of electricity per year using anaerobic digestion of MSW

the anaerobic digestion stage. Therefore, the transportation stage was recognized as the primary contributor to the environmental impacts including 7.12×10^{-9} impacts from abiotic depletion.

This finding corroborates the study by Poeschl et al. (2012), which identified the collection and transportation of municipal solid waste (MSW) and chemical fertilizer application as the major contributors to greenhouse gas emissions from anaerobic digestion systems, due to their high demand for fuel consumption. However, this study differs from Poeschl et al. (2012) in terms of the impact categories considered and the functional unit used. Poeschl et al. (2012) focused only on global warming potential and used ton of MSW as the functional unit, while this study considered multiple impact categories and used kWh of electricity as the functional unit. These differences may affect the comparability and relevance of the results.

The results of this study also relate to the local or regional context of Ghana, where waste management is a major challenge and electricity supply is unreliable (Amo-Asamoah et al. 2020). Anaerobic digestion can provide a viable solution for managing organic waste and generating renewable energy in Ghanaian municipalities, as demonstrated by Amo-asamoah et al. (2020), who evaluated the feasibility of using anaerobic digestion for managing organic waste in Ghanaian municipalities.

Table 2 Contribution of different life cycle stages to the characterized impacts of the biogas plant (baseline)

Impact category	Total	Transportation	Segregation	Anaerobic digestion	Electricity generation
Abiotic depletion	3.39E+00	−1.01E-01	2.11E+00	1.38E+00	0.00E+00
Abiotic depletion (fossil fuels)	−1.86E+06	−2.56E+06	4.20E+05	2.74E+05	4.90E+00
Acidification	−2.71E+02	−1.42E+03	2.06E+02	1.35E+02	8.07E+02
Eutrophication	−2.17E+04	−2.21E+04	7.99E+01	5.21E+01	2.13E+02
Freshwater aquatic ecotoxicity	−1.66E+07	−1.67E+07	6.12E+04	3.99E+04	6.72E+01
Global warming (GWP100a)	−5.34E+06	−5.41E+06	3.69E+04	2.41E+04	3.39E+03
Human toxicity	−1.55E+06	−1.65E+06	6.45E+04	4.21E+04	1.94E+03
Marine aquatic ecotoxicity	−8.01E+09	−8.18E+09	1.04E+08	6.78E+07	2.10E+03
Ozone layer depletion (ODP)	−1.49E-03	−8.63E-03	4.30E-03	2.82E-03	0.00E+00
Photochemical oxidation	−1.16E+03	−1.17E+03	1.10E+01	7.20E+00	0.00E+00
Terrestrial ecotoxicity	−1.21E+04	−1.22E+04	1.04E+02	6.81E+01	-4.67E-02

They found that anaerobic digestion can reduce greenhouse gas emissions by up to 90% compared to landfilling. They also suggested that anaerobic digestion should be central to the waste management system of municipalities.

3.1 Interpretation

The scenario and uncertainty analysis results of the model are presented by comparing different scenarios and using Monte Carlo simulations.

Scenario analysis

Two scenarios are considered in this analysis: scenario 1 uses the Ghana electricity grid in 2018 as the electricity source for the anaerobic digestion process, while scenario 2 uses electricity from biogas engine. The choice of these scenarios reflects the different options for electricity supply in Ghana, which vary in terms of availability, cost, and environmental impacts.

Table 3 Normalized environmental impact of the anaerobic digestion system

Impact category	Total	Transportation	Segregation	Anaerobic digestion	Electricity generation
Abiotic depletion	1.62E-08	−6.23E+06	1.30E+08	8.51E+07	0.00E+00
Abiotic depletion (fossil fuels)	−4.90E-09	5.22E+14	−8.57E+13	−5.59E+13	−1.00E+09
Acidification	−1.14E-09	1.25E+12	−1.81E+11	−1.19E+11	−7.10E+11
Eutrophication	−1.37E-07	1.61E+11	−5.82E+08	−3.80E+08	−1.55E+09
Freshwater aquatic ecotoxicity	−7.04E-06	2.37E+12	−8.70E+09	−5.67E+09	−9.55E+06
Global warming (GWP100a)	−1.28E-07	4.24E+13	−2.89E+11	−1.89E+11	−2.65E+10
Human toxicity	−6.00E-07	2.75E+12	−1.08E+11	−7.02E+10	−3.23E+09
Marine aquatic ecotoxicity	−4.13E-05	1.98E+14	−2.52E+12	−1.64E+12	−5.08E+07
Ozone layer depletion (ODP)	−6.59E-12	−1.31E+09	−6.53E+08	−4.28E+08	0.00E+00
Photochemical oxidation	−3.15E-08	3.72E+10	−3.50E+08	−2.29E+08	0.00E+00
Terrestrial ecotoxicity	−1.10E-08	1.11E+12	−9.42E+09	−6.17E+09	4.23E+06

From Table 4, scenario 1, emission savings in abiotic depletion was significantly lower compared to the baseline, and slightly lower in the abiotic depletion (fossil fuels) category. However, the remaining impact categories had a significant increase in emissions compared to the baseline. The scenario had the highest impact on the acidification impact category (725%).

In scenario 2, where electricity from a biogas was used, the scenario did not have a significant impact on the human toxicity category, freshwater aquatic ecotoxicity, and marine aquatic ecotoxicity but was slightly lower in eutrophication, abiotic depletion (fossil fuels), global warming, photochemical oxidation, and terrestrial ecotoxicity as compared to the baseline. Emissions in the remaining impact categories were significantly lower than the baseline. The lowest emission reduction was recorded in the ozone layer depletion category (2637%). The main reason for this difference is that scenario 1 relies on an electricity production mix that has a high share of fossil fuel-based energy sources that have higher emissions and resource consumption than biogas, which is a renewable and carbon–neutral energy source (Amo-Asamoah et al. 2020).

The environmental impacts of the scenarios are assessed using four impact categories: abiotic depletion, marine aquatic ecotoxicity, acidification potential, and global warming potential.

Table 6.4 Comparison of environmental impacts for all scenarios per impact categories

Impact category	Base	Scenario 1	Sensitivity analysis of S1 (%)	Scenario 2	Sensitivity analysis of S2 (%)
Abiotic depletion	3.39E+00	1.33E-02	−99.61%	1.36E+00	−59.74%
Abiotic depletion (fossil fuels)	−1.86E+06	−1.64E+06	−11.73%	−1.54E+06	−17.21%
Acidification	−2.71E+02	−2.24E+03	724.95%	5.79E+02	−313.69%
Eutrophication	−2.17E+04	−4.97E+04	129.20%	−2.02E+04	−6.71%
Freshwater aquatic ecotoxicity	−1.66E+07	−3.79E+07	128.54%	−1.65E+07	−0.52%
Global warming (GWP100a)	−5.34E+06	−1.19E+07	123.15%	−5.15E+06	−3.56%
Human toxicity	−1.55E+06	−3.73E+06	140.93%	−1.56E+06	0.55%
Marine aquatic ecotoxicity	−8.01E+09	−1.85E+10	131.27%	−8.01E+09	−0.03%
Ozone layer depletion (ODP)	−1.49E-03	−7.04E-03	372.24%	3.78E-02	−2637%
Photochemical oxidation	−1.16E+03	−2.62E+03	125.94%	−1.07E+03	−7.63%
Terrestrial ecotoxicity	−1.21E+04	−2.77E+04	128.88%	−9.83E+03	−18.76%

According to results in Fig. 3a, all scenarios show an equal and significant benefit with the Net Electricity contributing 807.24 kg SO_2 eq. to the acidification potential in the baseline scenario. The primary cause of acidification is the emission of large quantities of sulfur dioxide (SO_2) and nitrogen oxides (NOx) into the atmosphere. The electricity consumed represents the generation of electricity from renewable or low-impact sources, which have significantly lower emissions of acidifying substances than fossil fuel-based energy sources. Transportation consistently emerged as a hotspot in all the scenarios. This is because of the combustion of fossil fuels, which releases pollutants into the atmosphere. The Photochemical oxidant creation potential, eutrophication potential, and global warming potential all showed similar results from Fig. 3a, b, c, and d respectively. Transportation of the waste contributed the most to the impact results followed by the anaerobic digestion stage in all scenarios. Emissions from the anaerobic digestion stage could be due to the leakage of biogas from pipelines and flaring of excess methane.

The results of this study are consistent with previous studies that found that the electricity source is a key factor affecting the environmental performance of anaerobic digestion systems (Tian et al. 2021; Barahmand and Samarakoon 2022). The

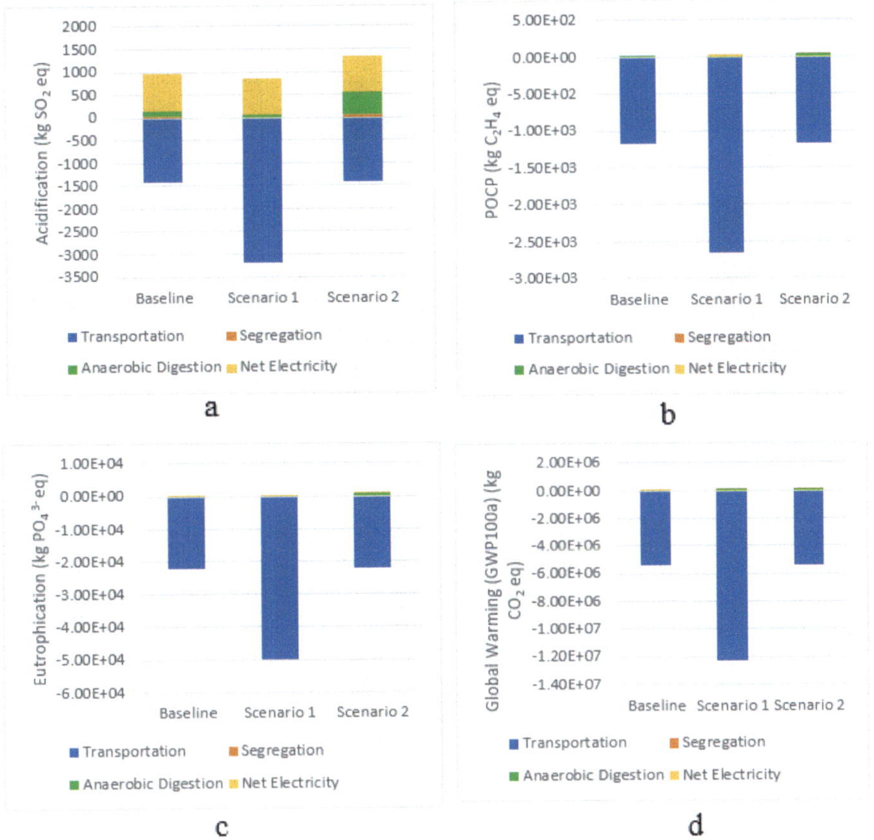

Fig. 3 Sensitivity analysis results for all scenarios in the **a** acidification potential, **b** photochemical oxidant creation potential, **c** eutrophication potential, and **d** global warming potential categories

sensitivity results also suggest some trade-offs and synergies among the impact categories. For example, scenario 1 has a trade-off between abiotic depletion and global warming potential, while scenario 2 has a synergy between these two impact categories. The results also imply some improvement measures for reducing the hotspots of transportation and biogas leakage, which contribute to most of the impacts in all scenarios. Transportation is responsible for the emissions of pollutants from fossil fuel combustion, while biogas leakage is responsible for the emissions of methane from pipelines and flaring systems. These hotspots can be reduced by optimizing the transportation routes and distances, and by implementing regular maintenance and inspection of digester systems and flaring systems.

Uncertainty analysis

Figure 4 shows the results of the uncertainty analysis performed with Monte Carlo simulations in all scenarios for each impact category.

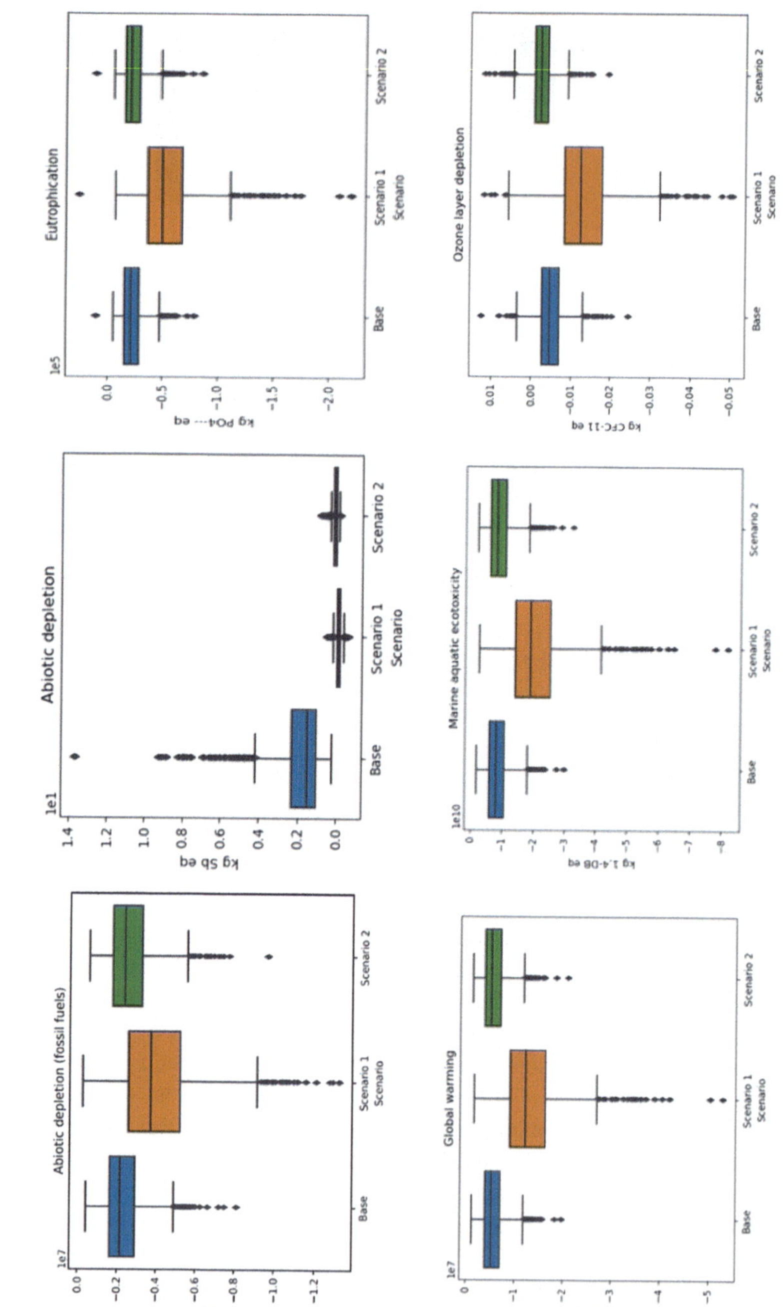

Fig. 4 Monte Carlo Simulation results for the uncertainties of the various scenarios for each category

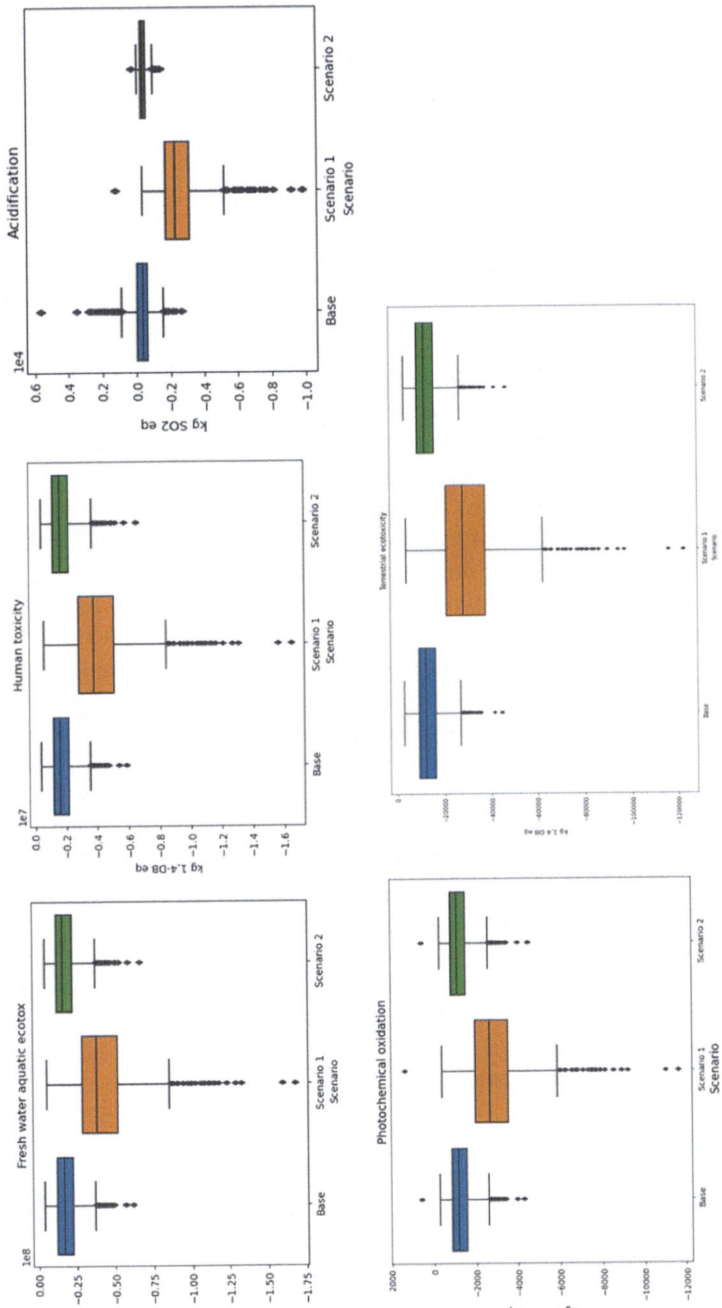

Fig. 4 (continued)

The analysis indicates that the quantity uncertainty of the LCA results is mainly influenced by the data reliability and temporal correlation of the LCI data. The background data, which are often based on qualified estimates from outdated datasets, introduce a large amount of uncertainty in the assessment. This finding is consistent with Bamber et al. (2020), who emphasized the need for comprehensive uncertainty analysis in LCA studies to enhance the robustness and validity of the results. The uncertainty can be reduced by using more recent and reliable data, updating, and validating the data regularly, and integrating and verifying different data sources.

4 Conclusion

This study has demonstrated the viability of anaerobic digestion as a waste-to-energy technology in Ghana that can reduce greenhouse gas emissions and generate renewable energy. However, it needs further improvements to optimize its environmental performance and reduce its resource consumption. The study has also provided valuable insights into the inventory results and impact assessment of the anaerobic digestion system using the CML-IA baseline method, as well as the main contributors to the environmental impact and possible improvements for the system. It has contributed to the literature on anaerobic digestion and life cycle assessment in Ghana and other developing countries, and has provided useful information for policy makers, stakeholders, and practitioners who are interested in implementing anaerobic digestion systems as a sustainable energy solution in Ghana. However, the study has some limitations that affect the generalizability and applicability of the results, such as the lack of data in databases for the Ghana scenario and the absence of a national life cycle inventory database, which add uncertainty to the study. Future studies can use a more comprehensive database to expand the scope of impact assessments.

Moreover, this study has a narrow system boundary that excludes some important aspects of the anaerobic digestion system, such as the use of digestate as fertilizer and the end-of-life of the biogas plant. These aspects may have significant environmental and social impacts that future research should consider. It also does not assess the economic viability and cost-effectiveness of the anaerobic digestion system, which future research can address using life cycle costing or cost–benefit analysis. Moreover, other studies can explore the socioeconomic impacts of using an anaerobic digestion system as a renewable energy source in Ghana.

References

Adanu, S.K., Gbedemah, S.F. and Attah, M.K. (2020) 'Challenges of adopting sustainable technologies in e-waste management at Agbogbloshie, Ghana', Heliyon, 6(8). Available at: https://doi.org/10.1016/j.heliyon.2020.e04548.

Afrane, S. *et al.* (2021) 'Techno-economic feasibility of waste-to-energy technologies for investment in Ghana : A multicriteria assessment based on fuzzy TOPSIS approach', *Journal of Cleaner Production*, 318(April), p. 128515. Available at: https://doi.org/10.1016/j.jclepro.2021.128515.

Amo-Asamoah, E., Asumadu, G. and Edwards, D.J. (2020) 'Potential for waste to energy generation of municipal solid waste (MSW) in the Kumasi metropolis of Ghana', 14(6), pp. 1315–1331. Available at: https://doi.org/10.1108/IJESM-12-2019-0005.

Ayodele, T.R., Ogunjuyigbe, A.S.O. and Alao, M.A. (2017) 'Life cycle assessment of waste-to-energy (WtE) technologies for electricity generation using municipal solid waste in Nigeria', *Applied Energy*, 201, pp. 200–218. Available at: https://doi.org/10.1016/j.apenergy.2017.05.097.

Bamber, N. *et al.* (2020) 'Comparing sources and analysis of uncertainty in consequential and attributional life cycle assessment: review of current practice and recommendations', *International Journal of Life Cycle Assessment*, 25(1), pp. 168–180. Available at: https://doi.org/10.1007/s11367-019-01663-1.

Barahmand, Z. and Samarakoon, G. (2022) 'Sensitivity Analysis and Anaerobic Digestion Modeling: A Scoping Review', *Fermentation*, 8(11). Available at: https://doi.org/10.3390/fermentation8110624.

Biancini, G. *et al.* (2020) 'Comprehensive Life Cycle Assessment Analysis of an Italian Composting Facility concerning Environmental Footprint Minimization and Renewable Energy Integration Comprehensive Life Cycle Assessment Analysis of an Italian', pp. 0–21.

Bruno, M. *et al.* (2023) 'Life Cycle Assessment and Cost–Benefit Analysis as Combined Economic–Environmental Assessment Tools: Application to an Anaerobic Digestion Plant', *Energies*, 16(9), pp. 0–19. Available at: https://doi.org/10.3390/en16093686.

Ciroth, A., Barreiros, T. and Jürgens, M. (2021) 'ecoinvent v.3.8 in openLCA', (November). Available at: https://ecoinvent.org/wp-content/uploads/2021/09/Change-Report-v3.8.pdf.

Ciroth, A. *et al.* (2016) 'Empirically based uncertainty factors for the pedigree matrix in ecoinvent', *International Journal of Life Cycle Assessment*, 21(9), pp. 1338–1348. Available at: https://doi.org/10.1007/s11367-013-0670-5.

CML - Department of Industrial Ecology (2016) 'CML-IA_aug_2016'. Available at: https://www.universiteitleiden.nl/en/research/research-output/science/cml-ia-characterisation-factors#downloads.

Cudjoe, D. and Mensah, P. (2021) 'Chemosphere Environmental impact analysis of municipal solid waste incineration in African countries', *Chemosphere*, 265, p. 129186. Available at: https://doi.org/10.1016/j.chemosphere.2020.129186.

Debrah, J.K. *et al.* (2022) 'Barriers and Challenges to Waste Management Hindering the Circular Economy in Sub-Saharan Africa'.

Environmental Protection Agency (2019) 'Ghana ' s Fourth National Greenhouse Gas Inventory Report', *National Greenhouse Gas Inventory to the United Nations Framework Convention on Climate Change*, (February), pp. 1–318. Available at: www.epa.gov.gh.

Galgani, P., van der Voet, E. and Korevaar, G. (2014) 'Composting, anaerobic digestion and biochar production in Ghana. Environmental-economic assessment in the context of voluntary carbon markets', *Waste Management*, 34(12), pp. 2454–2465. Available at: https://doi.org/10.1016/j.wasman.2014.07.027.

Goedkoop, M.J. *et al.* (2013) 'Category indicators at the midpoint and the endpoint level ReCiPe 2008', *ResearchGate*, (June 2016), p. 126.

Grosso, M. *et al.* (2012) 'The implementation of anaerobic digestion of food waste in a highly populated urban area : an LCA evaluation'. Available at: https://doi.org/10.1177/0734242X12453611.

Guinée, J.B. *et al.* (2002) 'Handbook on Life Cycle Assessment Operational Guide to the ISO Standards', *Environmental Impact Assessment Review*, 23(1), pp. 129–130. Available at: https://doi.org/10.1016/s0195-9255(02)00101-4.

Hauschild, M. Z. Rosenbaum, R.K. and Olsen, S.K. (2018) *Life Cycle Assessment: Theory and Practice, Life Cycle Assessment: Theory and Practice*. Cham: Springer International Publishing. Available at: https://doi.org/10.1007/978-3-319-56475-3_33.

Igos, E. *et al.* (2019) 'How to treat uncertainties in life cycle assessment studies?', *International Journal of Life Cycle Assessment*, 24(4), pp. 794–807. Available at: https://doi.org/10.1007/s11367-018-1477-1.

Iqbal, A., Liu, X. and Chen, G.H. (2020) 'Municipal solid waste: Review of best practices in application of life cycle assessment and sustainable management techniques', *Science of the Total Environment*, 729, p. 138622. Available at: https://doi.org/10.1016/j.scitotenv.2020.138622.

Kanhai, G. *et al.* (2021) 'Urban Municipal Solid Waste management: Modeling air pollution scenarios and health impacts in the case of Accra, Ghana', *Waste Management*, 123, pp. 15–22. Available at: https://doi.org/10.1016/j.wasman.2021.01.005.

Khandelwal, H. *et al.* (2019) 'Application of life cycle assessment in municipal solid waste management: A worldwide critical review', *Journal of Cleaner Production*. Elsevier Ltd, pp. 630–654. Available at: https://doi.org/10.1016/j.jclepro.2018.10.233.

Kumar, A. and Samadder, S.R. (2020) 'Performance evaluation of anaerobic digestion technology for energy recovery from organic fraction of municipal solid waste: A review', *Energy*, 197, p. 117253. Available at: https://doi.org/10.1016/j.energy.2020.117253.

Lissah, S.Y. *et al.* (2021) 'Managing urban solid waste in Ghana : Perspectives and experiences of municipal waste company managers and supervisors in an urban municipality', pp. 1–18. Available at: https://doi.org/10.1371/journal.pone.0248392.

Maepa, M. *et al.* (2017) 'Review: life cycle assessments in Nigeria, Ghana, and Ivory Coast', *International Journal of Life Cycle Assessment*, 22(7), pp. 1159–1164. Available at: https://doi.org/10.1007/s11367-017-1292-0.

Mahajan, R. *et al.* (2019) 'Evaluating anaerobic and aerobic digestion strategies for degradation of pretreated pine needle litter', *International Journal of Environmental Science and Technology*, 16(1), pp. 191–200. Available at: https://doi.org/10.1007/s13762-017-1601-y.

Miezah, K. *et al.* (2015) 'Municipal solid waste characterization and quantification as a measure towards effective waste management in Ghana', *Waste Management*, 46, pp. 15–27. Available at: https://doi.org/10.1016/j.wasman.2015.09.009.

Misganaw, A. and Teffera, B. (2021) 'Development of DPSIRO framework indicators for municipal solid waste management : a case of Bahir Dar city, Ethiopia', *Journal of Material Cycles and Waste Management* [Preprint], (0123456789). Available at: https://doi.org/10.1007/s10163-021-01266-9.

Nabavi-Pelesaraei, A. *et al.* (2019) *Life Cycle Assessment (LCA) Approach to Evaluate Different Waste Management Opportunities, Advances in Waste-to-Energy Technologies*. Available at: https://doi.org/10.1201/9780429423376-12.

Nielsen, M., Nielsen, O.-K. and Thomsen, M. (2010) *Emissions from decentralised CHP plants 2007 - Energinet.dk Environmental project no. 07/1882. Project report 5 – Emission factors and emission inventory for decentralised CHP production*. Available at: http://www.dmu.dk/Pub/FR786.pdf.

Ofori-Boateng, C., Teong, K. and Mensah, M. (2013) 'The prospects of electricity generation from municipal solid waste (MSW) in Ghana : A better waste management option', *Fuel Processing Technology*, 110, pp. 94–102. Available at: https://doi.org/10.1016/j.fuproc.2012.11.008.

Poeschl, M., Ward, S. and Owende, P. (2012) 'Environmental impacts of biogas deployment - Part I: Life Cycle Inventory for evaluation of production process emissions to air', *Journal of Cleaner Production*, 24, pp. 168–183. Available at: https://doi.org/10.1016/j.jclepro.2011.10.039.

Sandanayake, M., Zhang, G. and Setunge, S. (2018) 'Author ' s Accepted Manuscript for contractors Reference :', *Journal of Building Engineering* [Preprint]. Available at: https://doi.org/10.1016/j.jobe.2018.10.023.

Scheutz, C. and Fredenslund, A.M. (2019) 'Total methane emission rates and losses from 23 biogas plants', *Waste Management*, 97, pp. 38–46. Available at: https://doi.org/10.1016/j.wasman.2019.07.029.

Tian, H. *et al.* (2021) 'Life cycle assessment of food waste to energy and resources: Centralized and decentralized anaerobic digestion with different downstream biogas utilization', *Renewable and Sustainable Energy Reviews*, 150(May), p. 111489. Available at: https://doi.org/10.1016/j.rser.2021.111489.

World Bank (2020) 'Country Environmental Analysis: Ghana', *World Bank, Washington, DC.* ©, (April), p. 202. Available at: https://www.sprep.org/att/IRC/eCOPIES/Countries/Vanuatu/38.pdf.

Zhang, J. *et al.* (2021) 'Sustainable municipal waste management strategies through life cycle assessment method: A review', *Journal of Environmental Management*, 287(February), p. 112238. Available at: https://doi.org/10.1016/j.jenvman.2021.112238.

Anaerobic Co-digestion of Cassava Peel Residue: A Review of Potential Co-substrates Pre-treatment and Operating Factors for Improved Methane Yield

Zipporah Kyerewaa Asiedu and Satyanarayana Narra

Abstract Researchers are increasingly turning to alternative energy sources due to global concerns over issues such as global warming, ozone depletion, and rising crude oil prices. However many countries are still facing limitations when it comes to clean energy sources. Fortunately biogas is a renewable and economically feasible alternative to fossil fuels for generating energy and heat. Biochemical processes such as anaerobic digestion can be used to generate biogas from organic waste. However low nutrient levels can hinder this process making co-digestion a viable option. This review summarizes recent studies on anaerobic co-digestion (ACD) of cassava peel residue (CPR) with other substrates to identify the most effective combination for achieving high-quality biogas yields. Additionally it outlines the factors that influence the co-digestion process. This review aims to identify the best potential co-substrate for cassava peel residues' anaerobic co-digestion and provide a recommendation for future reference.

Keywords Anaerobic co-digestion · Biogas generation · Cassava peel residue · Substrates · Pre-treatment

1 Introduction

Global concerns about global warming, ozone depletion, limitation of fossil fuel, and the rise of crude oil prices have led researchers to pursue alternative clean and renewable energy sources. Meanwhile, many countries are suffering from the limitation of clean energy sources (Ismail et al. 2022). As of 2010 world biofuel production has mainly been intensive on first-generation biofuels, ethanol and biodiesel produced from starch, sugars and vegetable oils (Singh et al. 2016). To date good research on

Z. K. Asiedu (✉) · S. Narra
Professorship of Material and Energy Valorisation of Biogenous Residues, Professorship of Waste and Resource Management, University of Rostock, 18059 Rostock, Germany
e-mail: zipporah.asiedu@uni-rostock.de

M. Narra and S. Narra (eds.), *Innovations in Circular Economy and Renewable Energy in Africa*, World Sustainability Series, https://doi.org/10.1007/978-3-031-68330-5_7

making clean renewable energy such as biofuel has concentrated on solid biomass which is one of the most proficient compared to other energy sources. Meanwhile, developing countries that are not self-sufficient in energy supply and consumption turn to dump tremendous amounts of biomass wastes in landfills which can be a source of energy production (Ismail et al. 2022; Onokwai et al. 2019; Godfrey et al. 2019; Dimitriadis and Bezergianni 2017). The biofuel production process occurs when biomass is converted into a proper high-energy form. Biomass to biofuel conversion is significant because biomass is suitable for energy production. Even though biomass is a prominent renewable and sustainable energy option, the reliable and efficient conversion method is a big challenge. As a result, scientists continue to study chemical processes to advance the transformation of biomass into biofuels. Biomass sources include energy crops, wood wastes, aquatic plants, crops, agricultural waste products and municipal and animal feces. Biomass conversion technologies are broadly categorized into two: biochemical and thermochemical conversion technologies.

Biochemical conversion technologies refer to biomass conversion into complementary products through certain physical, chemical, and biological pre-treatments (Zhang et al. 2016). It involves using enzymes of bacteria and other microorganisms that helps in the decomposition of biomass waste into energy. Biochemical conversion technologies of biomass are more moderate environment-friendly and sustainable than other conversion technologies; in addition, the biochemical process requires low processing temperatures and high selectivity of products generated (Gupta and Mondal 2020). Conversely, the biochemical process involves lengthy stages such as pre-treatment fermentation downstream separation and upstream gas purification (Chen and Wang 2017). Typically, the pre-treatment process is used to degrade lignocellulosic biomass and convert cellulose and hemicellulose into sugars aiming to help reach ideal conversion effects not to produce final products. Presently, there are three advanced biochemical conversion methods; anaerobic digestion (AD) anaerobic fermentation and aerobic degradation (composting).

The anaerobic digestion (AD) process operates without oxygen to break down glucose in the organic materials. It converts very sugar with the help of different microorganisms into biogas (CH_4 and CO_2) and liquid by-products (Singh et al. 2016). There are four stages in the AD process: hydrolysis acidogenesis acetogenesis and methanogenesis. A group of different microorganisms carries out every step. Biogas is a clean and economically feasible renewable energy resource and potential replacement for fossil fuels in energy and heat generation (Maurus et al. 2020). Biogas is made up of various gases such as methane (CH_4) water vapor (H_2O) carbon dioxide (CO_2) hydrogen sulfide (H_2S) hydrogen (H_2) and siloxanes (Ismail et al. 2022; Mirmohamadsadeghi et al. 2019). Improving the efficiency of biogas production via anaerobic digestion requires enhancing the characteristics of the substrates and the operating conditions of the digester. Different subtracts materials can vary widely in physical and chemical features that affect digestibility in AD reactors thus biogas production. The ideal nutrient balance that results in increased digestion performance and biogas yields by mixing two or more substrates and digesting them simultaneously is known as anaerobic co-digestion (ACD) (Xie et al.

2016). The co-digestion process influences the substrates to complement each other concerning texture carbohydrate content, moisture, nutrient balance, and pH. This is especially evident when carbohydrate-rich subtracts co-digested. Some significant benefits of anaerobic co-digestion over mono-digestion include increased biogas production and methane concentration (Sawyer et al. 2017). Although co-digestion carries numerous benefits, many researchers have difficulty performing anaerobic co-digestion leading to a system upset mainly due to inappropriate substrate ratios and operating conditions.

Cassava is a significant food crop in Africa and the fast and low-cost growing staple food in the world which has been extensively employed as raw materials for various agri-industries to produce starch bioethanol and other bio-based products/chemicals (Ismail et al. 2022; Dias 2012). Cassava processing wastes are also noted with great potential as feedstock for bioconversion process or bio-fuel production from the 1st Generation 2nd Generation and integrated processes as they have a complex biochemical composition with high organic content thereby providing economic and environmental sustainability to cassava industries (Aso et al. 2018). Since more than 65% of global annual cassava output is processed for human consumption a massive quantity of cassava peeling residue (CPR) is generated and counters as a potential substrate for the anaerobic digestion to generate biogas (Lansche et al. 2020; Chavadej et al. 2019; Latif and Müller 2015; Dias 2012). These cassava wastes are often left to rot in an open field even though reasonably high starch contents are obtained from the process; this eventually becomes an environmental problem. The primary composition of cassava wastes CH_4 (55–65%), CO_2 (35–45%,) and small amounts of other gases (Kohmuean et al. 2020; Onokwai et al. 2019). However, cassava peel residues have a high value of organic carbon which results in an exceptionally very high C/N ratio. According to (Ozoegwu et al. 2017) a high C/N ratio is indicative that the material is not appropriate for biogas production and will not significantly yield biogas except when the cassava peels residue is pre-treated or/and blended with other materials with low C/N ratio.

Although there are some potential drawbacks to producing biogas from cassava peel residue (CPR), these are likely due to a lack of knowledge about pre-treatment options and finding the right mixing proportion of different substrates. In this review, the results of numerous laboratory studies that have explored the feasibility and potential of using different pre-treatment methods and co-substrates for anaerobic co-digestion (ACD) of CPR as the primary substrate will be summarized. Analyzed co-substrate for CPR by researchers who have achieved high-quality biogas yield in the ACD process were elaborated on. Finally, the significant factors that influence the aerobic co-digestion process and provide recommendations for future reference were outlined.

2 Methodology

2.1 State of Cassava Production and Peel Residue

Statistics indicate that cassava is the world's fourth most important staple crop in developing nations coming only after rice wheat and maize and is an essential component in the diet of over one billion people (FAO 2020; Spencer and Ezedinma 2017). In addition, cassava is also the world's third most significant source of food carbohydrates and the top food energy supplier for sub-tropical tropical regions. Cassava is an essential source of farm income in many Sub-Saharan Africa (SSA) countries; it provides economic and subsistence value for 800–1000 million people in more than 90 countries including Angola, Barbados, Brazil, Cambodia, China, Cook Islands, Democratic Republic of Congo, Dominica, Ghana, Democratic Republic Mozambique, Nigeria, Suriname, Uganda, and United Republic of Tanzania (Dias 2012). Cassava production in Sub-Saharan Africa has increased almost five times since 1961 from 31.5 million mt to 157.8 million mt in 2013; furthermore, cassava is substituting other root crops specifically yam in the humid zone maize in the non-humid zone and other food crops in the sub-humid zone at significant rates (Spencer and Ezedinma 2017; Kleih et al. 2013). Over 90% of cassava production takes place on small farms which results in numerous cassava root-based food products like akpakpuru attire casabe, chickwangue, farina (farinha de mandioca), fufu, fuku, gaplek, gari, ijapu, konkonte, lafun, landang, peujeum and thundam (Koyama et al. 2015). Studies indicated that in sub-Saharan Africa cassava cash incomes were the highest in farms with access to mechanized cassava processing equipment for the preparation of gari (Kemausuor et al. 2015). Cassava can be planted most time of the year and is available all year long with more than 2 years' harvest window (Ozoegwu et al. 2017).

Cassava is harvested by uprooting the stem to pull the roots from the ground. In general, there are four kinds of wastes generated from cassava processing: peels from initial processing (cassava peeling residue (CPR), fibrous by-products from crushing and sieving starch residues after starch settling, and wastewater effluents (Hansupalak et al., 2016). The cassava peeling residue is composed of cellulose hemicelluloses and lignin, but the structures of these materials are complex with recalcitrant and heterogeneous characteristics and are highly resistant to enzymatic hydrolysis; thus, there is a need to pre-treat the CPR by a delignification process. However, cassava peel residues have a high value of organic carbon and a low value of total nitrogen which results in a particularly high C/N ratio. According to Ozoegwu et al. (2017) high C/N ratio is indicative that the material is not appropriate for biogas production and will not significantly yield biogas except when the CPR is pre-treated/ pre-process and/or co-digest with nitrogen-rich substrates such as urea and animal and food wastes to adjusts the C/N ratio to the optimum value. Adelekan and Bamgboye (2009) reported the analysis of mineral content of cassava residue peels: 48.7 wt.% C; 1 wt.% N; 1.1 wt.% K; 1.6 wt.% P; 0.16 wt.% NO3; 0.15 wt.% Na; 0.9 wt.% Ca; 125 mg/kg Zn; 15 mg/kg Cu; 180 mg/kg Mn; 16.7 mg/ kg Pb; 48.7 wt.% C/N; and 52.6 wt.% ash). This result classifies CPR as a lignocellulose residue. And in 2020,

Aisien and Aisien (2020) experiment confirmed the lignin content of CPR with the compositions of 20–31 wt.% hemicelluloses 16–42 wt.% cellulose, and 6–8 wt.% lignin 81.9–93.9 wt.% organic matter and 4.1–6.5 wt.% crude protein.

2.2 *Pre-treatment of Cassava Peel Residue*

Proper pre-treatment processes turn to make cellulose more accessible to the action of the cellulase enzymes by the ability to alter biomass lignin and hydrolyze the hemicelluloses. In addition, pre-treatments in the digestion of biomass target to assist ideal coin reaching effects in the process and not to produce final products (Hallac et al. 2010; Chen and Wang 2017). Pre-treatments can be classified as either mechanical thermal chemical or biological (Jia et al. 2019). Bayitse (2014) research review studies advise that cassava peels require pre-treatment before use as a substrate; hence a multi-stage and high-rate digestion system might be adopted for difficulties in the digestion of the cassava peels. Other researchers have appreciated the unique nature of cassava peels residue and have also confirmed that for sufficient biogas production from cassava peels there must be an appropriate pre-treatment method in place (Cuzin et al. 1992; Suhaida et al. 2012; Shah et al. 2015; Gopinathan and Vivek 2016; Archibong 2016; Ben and Michael 2018; Mulaji 2020).

Ben and Michael (2018) and Igbum et al. (2019) experimental studies focused on chemical pre-treatment methods such as the use of chemical alkalis solutions; (NaOH, KOH, NaHCO3, $Ca(OH)_2$) of different concentrations in the maintenance of the CPR substrate pH for ideal biogas production. Nonetheless in their studies the application of a buffer solution was not considered which is believed to resist more change in CPR substrate compared with alkalis solutions. Regarding these Aisien and Aisien (2020) further investigated the buffer chemical pre-treatment on an anaerobic co-digestion of CPR with cow dung slurry. They pre-treated the CPR on two alkaline chemicals: sodium hydroxide (NaOH), calcium hydroxide (Ca (OH), and a buffer ammonium chloride (NH_4Cl) solution. The research experiment focused on fresh and stale cassava peel residue; the substrate digestion was done in six batch anaerobic biodigesters of 10-L capacity; each was used for 40 days. They discovered that the average amount of biogas generated was 2540 cm^3/day, and the cumulative volume of biogas and methane produced were 51,666 cm^3, 33,700 cm^3;51,29 cm3, 32,310 cm^3, and 95,592 cm^3, 58,890 cm^3 for NaOH, $Ca(OH)_2$ and NH_4Cl chemical pre-treatment methods respectively. The methane yield for NaOH, $Ca(OH)_2$, and NH_4Cl chemical pre-treatment methods were 53.3%, 58.7%, and 62%, respectively. The average pH and temperature were 6.71, 32 °C; 6.69, 31.4 °C; and 7.00, 30.6 °C for NaOH, $Ca(OH)_2$, and NH4Cl chemical pre-treatment methods respectively. Moreover, the stale cassava peels and cow dung slurry and the use of NH_4Cl pre-treatment chemical gave the best biogas production and methane yield of 104,961 cm^3 and 62.3% respectively. They reported that the NH_4Cl buffer pre-treatment chemical showed a substantial increase compared to the alkaline pre-treatment chemicals mentioned. This indicates the buffer's contribution to stabilizing the pH of the mixed substrate firmly resisting the frequent change in pH resulting from the acidic nature of the composite substrates. According to Aisien and Aisien (2020) conclusion it is advisable always to apply buffer pre-treatment chemicals rather than alkaline pre-treatment chemicals for a complex substrate such as cassava peels to produce a substantially large volume of biogas and a high yield of methane.

A gap identified from existing studies on CPR pre-treatment methods for sufficient biogas production is the investigation of the organic solvent such as Oraganosolv pre-treatment. The organosolvation process (OP) can occur in various organic or aqueous-organic solvent mixtures with or without a catalyst such as HCL or H_2SO_4 (Zhao et al. 2009). Typical solvents used for the process include ethanol methanol acetone and ethylene glycol; organic acids such as oxalic, acetylsalicylic and salicylic acids can be used as catalysts in the organosolvation process. OP is one of the few pre-treatment processes th+±/at does not require a significant size reduction of feedstock to achieve satisfactory cellulose conversion making the process less energy intensive (Olanbiwoninu and Odunfa 2015). Additionally, OP can be merged with other pre-treatment methods to complete a clean and effective biomass fractionation process or improve pre-treatment yield. The critical advantage of organosolv over different chemical pre-treatments is that relatively pure and low-molecular-weight lignin is retrieved as a by-product. The solvents can be recovered and recycled to reduce operation costs (Huijgen et al. 2011). Generally, all pre-treatment processes should be effective on a wide range and loading of lignocellulosic material and should result in the recovery of most the lignocellulosic components in a useable form in separate fractions; it should have low capital and operational cost.

2.3 *Co-digestion of Cassava Peel Residue with Various Substrates*

According to the literature and earlier studies done on energy recovery of cassava peel residue through anaerobic digestion it is well established that the co-digestion of cassava peels with two or more substrates with high nitrogen content will establish positive synergisms in the digestion medium that result in higher methane yield than a mono-digestion of a single substrate. To develop positive synergisms with an optimal carbon-to-nitrogen (C/N) ratio for anaerobic digestion around 20–30, CPR with a high C/N ratio of 40–50 can be co-digested with the co-substrate with a low C/N ratio. As shown in Table 1 the potential co-substrates used in anaerobic co-digestion of cassava peel residue are identified as the organic fraction of animal wastes (poultry piggery swine and cattle) and food waste (vegetable and fruits).

Table 1 Carbon-to-nitrogen (C/N) ratio of the primary and co-substrates reviewed in this study

Common Co-substrates	C/N Ratio	Refs.
Poultry droppings	5–10	Achebe et al. (2018), Etta et al. (2017), Bayitse (2014), Ofoefule and Uzodinma (2009)
Pig manure	18	Bayitse (2014), Ren et al. (2014), Panichnumsin et al. (2010)
Cow dung	18–24	Ismail et al. (2022), Konan et al. (2018), Mama and Agunwamba (2017), Bayitse (2014), Ofoefule and Uzodinma (2009)
Swine manure	8–10	Villa et al. (2020), Ofoefule and Uzodinma (2009)
Vegetables and fruits	11–15	Sawyerr et al. (2019), Muller (2018), Gopinathan and Vivek (2016)
Urea	0.4	Mulaji (2020), Nkodi et al. (2016)

2.4 *Animal Wastes (Poultry, Piggery, and Cattle)*

Manure is a mixture of animal feces and urine with nutrient richness necessary for optimal bacterial growth. Manure gives the balance of nutrients at an appropriate carbon–nitrogen (C/ N) ratio and high buffering capacity with a stable pH needed to increase methane production hence an excellent co-substrate in digestion (Etta et al. 2017; Bayitse 2014). Examples of manure are swine dung, poultry droppings, and cow dung. Cow dung is the most suitable material for high-yield biogas production because it has high nitrogen content. Pre-fermentation of the cow dung occurs in the ruminants' stomachs and there are micro-organisms in cow dung that help in the bio-digestion process (Bayitse 2014). Sanael-Moghadam et al. (2014) experimental study observed a higher efficiency and biogas production for a ratio of 20:80 (cassava peel residue: CD; C/N 15.5) which displayed an increase of 137% in the production of biogas as compared to the value of the mono-digestion of substrates. Likewise. Sawyerr et al. (2017) study was designed to perform co-digestion of cassava peel (CPR) with cattle manure (CM) at different ratios through a batch anaerobic digestion under mesophilic temperatures of 35 ± one °C. The results showed that the final cumulative methane yields by the co-digestion of CM: CPR mixing ratios of 80:20 and 20:80 were 738.76 mL and 838.70 mL respectively. Both studies indicate that the proper proportion of co-substrates influenced biogas production and methane yield. And it shows that the proportion of 20:80 (CPR:CD) is the optimal ratio to attain an ideal C/N ration for high production of biogas.

Additionally Bayitse (2014) investigated the critical volatile solids as well as the C/N ratio needed for optimum biogas production of cassava peel residue with cow dung poultry droppings and swine dung manure in proportions of 20:80, 40:60, 50:50, 60:40 to 80:20 (w/w) and concluded that the optimum substrate ratio for cassava peel residue and manure is found to be 20:80. Furthermore Etta et al. (2017) compared the use of different co-substrates with a 5–10 vol% ratio using CPR water hyacinth (Eichhornia crassipes) poultry droppings and cow dung under mesophilic conditions. Comparatively, their results showed that the combination of all four substrates generated the highest optimal total and percentage biogas at the shortest retention time as to the single double and triple mixture treatments. The same study observed a positive correlation in biogas production with and without inoculum where the four combined substrates yielded biogas of 815 MLS, 875 ml and 1340 MLS from respective weights of 1 kg, 2 kg, and 3 kg within 15 days without inoculum while with inoculum biogas yield within 15 days was 840 MLS 1170 MLS and 1690 MLS from the 1 kg, 2 kg and 3 kg weights respectively; indicating that the amount of biogas generated from the combination of the four substrates (WH+CPR+PD+CD) could be facilitated in the presence or absence (with or without) of inoculum. Hence, these two studies indicated that there is no significant difference in gas yields for different sources of animal manure co-digested with cassava peel residue. In Etta et al. (2017) study poultry droppings recorded the lowest biogas volume as expected and this is probably because of the impurities (such as saw dust and feeds) found in poultry droppings. There is need to use pure poultry dung free from other impurities for anaerobic digestion to the optimal biogas yield.

Achebe et al. (2018) conducted a study to determine and confirm other research findings on the kinetics for optimal biogas production of CPR with poultry droppings (PD) in a batch anaerobic co-digestion. Their experiment was carried out in five substrate ratios of (100% PD and 0% CPR), (75% PD and 25% CPR), (50% PD and 50% CPR), (25% PD and 75% CPR) and (0% PD and 100% CPR) respectively were incubated for forty-four (44) days at an ambient mesophilic temperature of 37 ± 1 °C at a pH of 6.7. They observed that the gas production of the co-substrates with a high amount of CPR did not begin on the first day but subsequent days and then increased steadily until it reached a maximum biogas yield between the 18th and 22nd day of anaerobic digestion. And this is due to the high cellulose content in the CPR substrate. Their results showed that the digester with the co-substrates of 75%:25% (PD: CPR) of 24.6L/g volatile solid (VS) generated the highest amount of biogas. Their result confirms that biogas production could be enhanced effectively through the co-digestion process of substrates and the better synergy of microorganisms in the digester due to C/N nutrient balance due to the limited quantity of cassava peel residue. All the studies reported that a lower biogas production volume was obtained from substrate slurry with CPR singly and in combination with other manure substrates probably due to the presence of cyanogenic acids which tend to inhibit bacteria activities. Ahou et al. (2019) reported that cyanogenic glucosides in CPR release cyanide acid which is highly toxic to methanogenic arches reducing biogas production.

2.5 *Food Waste (Vegetables and Fruits)*

Typically, food wastes are generated from households (55–80%), followed by commercial or market areas (10–30%) and food waste represents 68% of biodegradable organics of municipal solid wastes (MSW). The C/N ratio of food waste ranges from 11–15 and has a moisture content of 82–86% (Chow et al. 2020). Food wastes are potential co-substrates for anaerobic digestion because of their high biodegradability and suitable C/N ratio. Sawyerr et al. (2019) demonstrated an improvement in methane biogas yield from co-digestion of vegetable & fruits and cassava biomass (which consist of cassava peel, stem residues and other solid waste) as compared to mono digestion of substrates in a mesophilic control environment using batch reactor. The experiment was carried out in a 600 ml digester of four (4) CB: VF ratios explicitly 100:0, 60:40, 40:60 and 50:50 for 31 days under mesophilic conditions (37 ºC). The cassava biomass was physically pre-treatment by milling the feedstocks into small pieces before anaerobic digestion to enhance the enzymatic digestibility of the lignocellulosic structure of the cassava residue. The bio-digesters were inoculated with cow dung.

Sawyerr et al. (2019) achieved a general bio-methane potential ranging from 1124.26 to 1641.82 mL CH_4/g VS. Hence co-digestion of CB and V&F with inoculum at a ratio of 40:60 achieved the maximum methane yield indicating a methane yield enhancement of 23.08% higher than that of the mono-digestion feedstock. This result confirms that anaerobic co-digestion of the suitable substrate improves biogas yield. However, the ratio mixes of the substrate to balance the C/N ratio is of significant importance as this could alter the digestion process in the digester thereby changing the biogas yield and the rate. During the experiment Sawyerr et al. (2019) observed a significant decrease in methane yield of all the co-digestion feedstock after day one. And they explained that this could be attributed to the acidification in the batch reactors which confirms that hydrolysis and the alcoholic fermentation of the vegetable waste are rapid compared to other organic substrates. Furthermore impurities in food waste should be properly removed before the feeding of the digester to avoid blockage of the pump system and hydrodynamic problems (Sawyerr et al., 2019; Muller, 2018).

2.6 *Urea*

Annuntipul (2004) obtained a better biogas production yield for co-digested cassava tuber with urea (0.03%) in an experimental study. The study concluded that the amount of biogas produced highly depends on the ratio of urea to cassava tubers. Nkodi et al. (2016) also investigated the effect of different urea concentrations in co-digestion with cassava peel residue under mesophilic conditions. Their result showed a 24.33% enhancement of biogas yield in the co-digestion with urea compared to digestion without urea. They also observed that cumulative biogas yield decreased in the order with a decrease in the urea concentration of 80.79, 50.01, 49.186, 25.71, and 1.23L/Kg TS, respectively. Both studies confirmed that adding urea to various concentrations can be a good supplement for enhancing biogas production. Additionally, the application of urea concentration to co-digestion of CPR can be applied without livestock manure.

2.7 Significant Factors Influencing Pre-treatment and Anerobic Co-digestion

The optimum performance and practical function of anaerobic digesters depend on several factors. The feasibility of co-digestion is where enough control is required to prevent any failure. Some of the significant influences that need sufficient power to avoid digester failure in co-digestion are co-substrate mixing ratio and nutrient balance (C/N ratio) operating temperature organic loading rates (OLR) and hydraulic retention time (HRT) in the digester (Chow et al. 2020; Siddique and Wahid 2018; Xie et al. 2016).

2.8 Mixing

In AD, mixing can be done through mechanical mixing biogas recirculation or slurry recirculation. Improving the ability of the microorganism population to obtain nutrients providing good contact between microbes and substrates and promoting homogenous feed to the ACD digester is a result of the proper mixing of substrates. Also, the forming of inhibitory intermediates such as scum is reduced by mixing. A mixing system must be installed practically in co-digestion to ensure sufficient homogeneity besides improving digester performance. Hence pH should be considered to get an appropriate condition within the digester during mixing for the most effective digester performance. Many researchers (Lemmer et al. 2017; Yao et al. 2017) have noted that maintaining an optimum pH level maximizes methane yield. However, the optimal pH for microorganisms is between 6.8 and 7.0.

2.9 Co-substrate Mixing Ratio and Nutrient Balance (C/N Ratio)

The C/N ratio of organic materials represents the correlation between the amount of carbon and nitrogen present. This ratio is the balance of food that a microbe needs to grow to perform digestion of organic matter. The carbon to nitrogen (C/N) ratio affects the ACD process because it is the balance of food/ nutrients supply that microorganisms need to grow to perform digestion of organic matter to maximize biogas production (Chow et al. 2020; Reilly et al. 2016). Lower C/N values lead to higher ammonia concentrations and are toxic to methanogenic bacteria. Meanwhile large VFAs are produced when the C/N ratio exceeds the optimal value. Therefore, to attain an optimal C/N balance in the co-digestion it is recommended to have a mixture of substrates with a low and high C/N ratio. The ideal C/N ratios in AD are usually between 20:1 and 30:1. Hence, the perfect mixing ratio for cassava peel residues to lower nitrogen sources (such as animal and food wastes) is recommended at 20:80 to achieve optimum nutrient balance in the digester. This is because CPR is rich in lignocellulose (Ismail et al. 2022; Sawyerr et al. 2019; Muller 2018; Bayitse 2014). Lower C/N values lead to higher ammonia concentrations and are toxic to methanogenic bacteria. Meanwhile large VFAs are produced when the C/N ratio exceeds the optimal value. Therefore, to attain an optimal C/N balance in the co-digestion it is recommended to have a mixture of substrates with a low and C/N ratio. The ideal C/N ratios in AD are usually between 20:1 and 30:1. Hence the perfect mixing ratio for cassava peel residues to lower nitrogen sources (such as animal and food wastes) is recommended to be 20:80 to achieve optimum nutrient balance in the digester (Ismail et al. 2022; Sawyerr et al. 2019; Muller 2018; Bayitse 2014).

2.10 Organic Loading Rate (OLR)

The organic loading rate is crucial in achieving an optimized ACD process and optimal microorganism activity (Chow et al. 2020; Neshat et al. 2017). OLR may be considered as the amount of dry organic solids or volatile solids to be loaded per unit time per unit volume of a digestion process. Likewise, the substrates determine the microorganism activity level of biodegradation in the digester. Therefore, the optimum OLR and HRT are influenced by the type of substrates fed into the digester. ACD operates at a much higher OLR value where a higher ORL enhances different microbial species for optimum digestion performance. However, there is a considerable risk of overloading the digester which can cause damage to the circulating pump. Therefore, there is a necessity to use a mathematical model to inform a maximum OLR for a stable ACD process (Siddique and Wahid 2018). The inefficiency of AD technology can be a result of a lower OLR.

2.11 Hydraulic Retention Time (HRT)

Hydraulic retention time (HRT) is needed for any microorganism to consume and synthesize the substrate actively. The microbial community and biogas yield are significantly affected by HRT and during AD, HRT can accommodate varying biochemical reaction kinetics. Several factors such as substrate properties and composition, operating conditions and reactor configurations determine the optimum HRT value of a digestion process. The HRT needs to be long enough but longer than enough HRTs will typically lead to the death of microorganisms because of the scarcity of nutrients. However, the optimal HRT may vary for different co- digestion processes depending upon the desired temperature limit and nature of microbes in the substrates. Both Sawyerr et al. (2019) and Bayitse (2014) reported that an HRT of 40–45 days could provide satisfactory co-digestion and stable biogas production for cassava peel residue with organic animal food wastes. This is because CPR's high carbon content allows sufficient time for introducing microbes to break down organic matter and generate biogas.

2.12 *Temperature*

Several researchers have highlighted the significant effects of temperature on the microorganism community, process kinetics, stability and methane yield; therefore, selecting and controlling the temperature is critical. Depending on the type of substrates the co-digestion process is under either mesophilic (approximately 35 °C) or thermophilic (about 55 °C) conditions. Due to a wider variety of microorganisms in the mesophilic temperatures to thermophilic temperatures the mesophilic process is more stable compared to the thermophilic process (Yang et al. 2018). A mesophilic condition is recommended as favorable for CPR co-digestion. Several researchers (Ismail et al. 2022; Sawyerr et al. 2019; Konan et al. 2018; Bayitse 2014) who performed ACD of cassava peel residues with food wastes and animal manure under a mesophilic condition observed and confirmed a better performance than the thermophilic condition for biogas production and digestion capability. This is because a mesophilic condition is a better option for ACD to overcome ammonia inhibition in the substrates.

3 Conclusion

Cassava peel residues are readily available as cheap biowaste resources for anaerobic co-digestion with manure to produce biogas and bio-fertilizer. This review paper summarizes the recent research on the ACD of cassava peel residue with some potential co-substrates. The review shows that ACD of cassava peel residue involving co-substrates of organic fraction of animal wastes (poultry, piggery, swine, and cattle) and food waste (vegetable and fruits) show potential in boosting biogas production and methane yield. ACD is an efficient method for biogas production and environmental protection. The main challenges in biogas production using ACD technology are process instability mainly due to inappropriate substrate ratios, the dynamic behavior of microorganisms and operating conditions. Hence the anaerobic co-digestion system can be designed by considering the operating conditions such as a slow mixing of substrates with low and high C/N ratios at the correct mixing ratio. Out of these five potential co-substrates comparatively the combination of four animal manure substrates (CPR+PD+CD+SD) generates the highest optimal total and percentage biogas at the shortest retention time as to the single, double and double mixture treatments. Since various studies indicated that there is no significant difference in gas yields for different sources of animal manure co-digested with cassava peel residue therefore depending on availability cow, piggery, poultry or swine dung can be utilized. Moreover, biogas generated from the combination of the animal manure substrates could be facilitated in the presence or absence (with or without) of inoculum.

Future research is required focusing preferrable pre-treatments that will facilitate digestion in the digester and enhance biogas production since cassava peels contain a high concentration of cyanide and lignin; therefore, a multi-stage and high-rate efficient co-digestion system is recommended for cassava peels and manure.

References

Achebe, C., Onokpite, E., & Onokwai, A., 2018. Anaerobic Digestion and Co-digestion of Poultry Droppings (PD) and Cassava Peels (CP): Comparative Study of Optimal Biogas Production. Journal of Engineering and Applied Sciences, 12(July), 87–93.

Adelekan, B. A., & Bamgboye, A. I., 2009. Comparison of biogas productivity of cassava peels mixed in selected ratios with major livestock waste types. African Journal of Agricultural Research, 4(7), 571–577.

Ahou, Y. S., Angeli, J. B., Awad, S., Lamine Baba-Moussa and Andres, Y., 2019. Lab-scale anaerobic digestion of cassava peels: A first step of energy recovery from cassava waste and water hyacinth, Environmental Technology, https://doi.org/10.1080/09593330.2019.1670266.

Aisien, F. A., & Aisien, E. T., 2020. Biogas from cassava peels waste. Detritus, 10(June), 100–108. https://doi.org/10.31025/2611-4135/2020.13910.

Annuntipul A., 2004. Biogas production from cassava tubers. A Thesis Submitted in Partial Fulfillment of the Requirements for the Degree of Master of Environmental, Biology Suranaree University of Technology.

Archibong, E., 2016. Mycological Deterioration and Pathogenicity Studies of Post-harvest Cassava. Food Science and Technology, 4(2), 23–30. https://doi.org/10.13189/fst.2016.040202.

Aso, S. N., Teixeira, A. A., & Achinewhu, S. C., 2018. Cassava Residues Could Provide Sustainable Bioenergy for Cassava Producing Nations. Cassava. https://doi.org/10.5772/intechopen.72166.

Bayitse, R., 2014. Anaerobic Co-Digestion of Cassava Peels and Manure: a Technological Approach for Biogas ... February.

Ben, M. G., & Michael, E. I., 2018. Biogas Production from Farm Waste (Cassava Peels and Swine Dung): Co-Digestion and Prospect on Economic Growth. 6(2), 58–63.

Chavadej S, Wangmor T, Maitriwong K et. al., 2019. Separate production of hydrogen and methane from cassava wastewater with added cas- sava residue under a thermophilic temperature in relation to digest- ibility. J Biotechnol 291:61–71. https:// doi. org/ 10. 1016/j. jbiot ec. 2018. 11. 015.

Chen, H., & Wang, L., 2017. Pretreatment Strategies for Biochemical Conversion of Biomass. Technologies for Biochemical Conversion of Biomass, 21–64. https://doi.org/10.1016/b978-0-12-802417-1.00003-x.

Chow, W. L., Chong, S., Lim, J. W., Chan, Y. J., & Chong, M. F., 2020. Anaerobic Co-Digestion of Wastewater Sludge : A Review of Potential Co-Substrates and Operating. Processes, 8(1)(39), 1–21.

Cuzin, N., Farinet, J. L., Segretain, C., & Labat, M. (1992). Methanogenic fermentation of cassava peel using a pilot plug flow digester. Bioresource Technology, 41(3), 259–264. https://doi.org/10.1016/0960-8524(92)90011-L.

Dias, P., 2012. Analysis of Incentives and Disincentives for Cassava in Mozambique. FAO, MASAP (Rome).

Dimitriadis, A., & Bezergianni, S., 2017. Hydrothermal liquefaction of various biomass and waste feedstocks for biocrude production: A state of the art review. Renewable and Sustainable Energy Reviews, 68(May 2016), 113–125. Godfrey L, Tawfic Ahmed M.

Etta, A. E. B., James, E., Ben, A., & Tiku, D. R., 2017. Biogas generation from co-digestion of four substrates; water hyacinth, cassava peels, poultry droppings and cow dung. Annual Research and Review in Biology, 13(4), 1–9. https://doi.org/10.9734/ARRB/2017/29459.

FAO, IFAD, UNICEF, WFP & WHO., 2020 The state of food security and nutrition in the world 2020. Building resilience for peace and food security (FAO, 2020).

Godfrey, L. et al. (2019). Solid Waste Management in Africa: Governance Failure or Development Opportunity? In Regional Development in Africa. London: IntechOpen. https://doi.org/10.5772/intechopen.86974

Gopinathan, C., & Vivek, A. T., 2016. Bioprocess Optimization for Enhanced Biogas Production from Grape Pulp Waste. International Journal of Science and Research (IJSR), 5(6), 673–676. https://doi.org/10.21275/v5i6.nov164132.

Gupta, G. K., & Mondal, M. K., 2020. Bioenergy generation from agricultural wastes and enrichment of end products. In Refining Biomass Residues for Sustainable Energy and Bioproducts. Elsevier Inc. https://doi.org/10.1016/b978-0-12-818996-2.00015-6.

Hallac, B. B., Sannigrahi, P., Pu, Y., Ray, M., Murphy, R. J., & Ragauskas, A. J., 2010. Effect of ethanol organosolv pretreatment on enzymatic hydrolysis of Buddleja davidii stem biomass. Industrial and Engineering Chemistry Research, 49(4), 1467–1472. https://doi.org/10.1021/ie900683q.

Hansupalak, N., Piromkraipak, P., Tamthirat, P., Manitsorasak, A., Sriroth, K., & Tran, T., 2016. Biogas reduces the carbon footprint of cassava starch: a comparative assessment with fuel oil. Journal of Cleaner Production, 134(Part B), 539–546. https://doi.org/10.1016/j.jclepro.2015.06.138.

Huijgen, W. J. J., Smit, A. T., Reith, J. H., & Uil, H. Den., 2011. Catalytic organosolv fractionation of willow wood and wheat straw as pretreatment for enzymatic cellulose hydrolysis. Journal of Chemical Technology and Biotechnology, 86(11), 1428–1438. https://doi.org/10.1002/jctb.2654.

Igbum, O.G.; Eloka-Eboka, A.C. and Adoga, S. 2019. Feasibility study of biogas energy generation from refuse dump in a community – based distribution in Nigeria. International Journal of Low–Car- bon Technologies, vol. 14, issue 2, pp. 58–78.

Ismail, N., Fauzi, N. F., Salehabadi, A., Latif, S., Awiszus, S., Müller, J., Mama, C. N., Agunwamba, J. C., Onokwai, A. O., Okonkwo, U. C., Osueke, C. O., Ezugwu, C. A., Eze, N. C., Fagbenle, E. O., & Olukanni, D. O., 2022. Production and purification of biogas from cassava peel using cow dung as inoculum. International Journal of Civil Engineering and Technology (IJCIET), 19(1), 2032–2043. https://doi.org/10.1007/s13762-021-03222-4.

Jia, Z., Zheng, Y., & Zhou, J., 2019. Effects of different pretreatment methods on the enzymatic hydrolysis of cassava residue. BioResources, 14(3), 6060–6078.

Kemausuor, F., Addo, A., & Darkwah, L., 2015. Technical and Socioeconomic Potential of Biogas from Cassava Waste in Ghana. Biotechnology Research International, 2015, 1–10. https://doi.org/10.1155/2015/828576.

Kleih U, Phillips D, Wordey M, Komlaga G. 2013. Cassava market and value chain analysis: Ghana case study London, UK: Natural Resources Institute, University of Greenwich, and Accra, Ghana: Food Research Institute; https://cava.nri.org/images/documents/publications/GhanaCassavaMarketStudy-FinalFebruary2013_anonymised-version.pdf.

Kohmuean P, Boonrod N, Wongkoblap A, 2020 Biogas production from cassava waste: effect of concentration. In: IOP conference series: materials science and engineering, pp 012115–22.

Konan, E. N. K., Gnagne, T., Kouamé, F. Y., Konan, F. K., Kouamé, M. K., & Tano, K., 2018. Optimum amount of inoculum for anaerobic digestion of cassava waste. Livestock Research for Rural Development, 30(4).

Koyama, N., Kaiser, J., Ciugu, K., & Kabiru, J., 2015. Market Opportunities for Commercial Cassava in Ghana, Mozambique and Nigeria. IDH-The Sustainable Trade Initiative, 6, 1–77.

Lansche J., Awiszus S., Latif S., Müller J., 2020 @@@Potential of biogas production from processing residues to reduce environmental impacts from cassava starch and crisp production-a case study from Malaysia. Appl Sci 10:2975. https:// doi. org/ 10. 3390/ APP10 082975.

Latif S., Müller J., 2015. Potential of cassava leaves in human nutrition: a review. Trends Food Sci Technol 44:147–158.

Lemmer, A., Merkle, W., Baer, K., Graf, F., 2017. Effects of high-pressure anaerobic digestion up to 30 bar on pH- value, production kinetics and specific methane yield. Energy, 138, 659–667.Maurus K, Ahmed S, Kazda M, 2020 Beneficial effects of intermittent feedstock management on biogas and methane production. Biore-sour Technol 304:123004. https:// doi. org/ https://doi.org/10.1016/J.BIORTECH.2020.123004.

Mama, Cordelia Nnennaya and Agunwamba, Jonah Chukwuemeka. 2017. 'Biogas Production from Blends of Cassava Waste Water and Cow Dung Under Changing Meteorological Parameters'. 1 Jan. 2017: 89–98.

Maurus K., Ahmed S., Kazda M., 2020 Beneficial effects of intermittent feedstock management on biogas and methane production. Biore- sour Technol 304:123004. https:// doi. org/ 10. 1016/J. BIORT ECH. 2020. 123004.

Mirmohamadsadeghi S., Karimi K., Tabatabaei M., Aghbashlo M., 2019. Biogas production from food wastes: a review on recent develop-ments and future perspectives. Bioresour Technol Rep 7:100202– 100212. https:// doi. org/ 10. 1016/j. biteb. 2019. 100202.

Mulaji, C. K.,2020. Biogas Production by Co-Digestion of Cassava Peels with Urea Insecticide activity of essential oils extracted from stem bark and roots of some plant species used in Democratic Rep ... Biogas Production by Co-Digestion of Cassava Peels with Urea. May, 2–6. https://doi.org/10.17950/ijset/v5s3/303.

Muller, M., 2018. Complimentary Contributor Copy. Protests and Riots Past: Present and Future Perspectives, January, 47–75. https://doi.org/10.1016/j.arthro.2012.05.044.

Neshat, S.A., Mohammadi, M., Najafpour, G.D., Lahijani, P., 2017. Anaerobic co-digestion of animal manures and lignocellulosic residues as a potent approach for sustainable biogas production. Renew. Sust. Energ. Rev. 79, 308–322.

Nkodi, T. M., Taba, K. M., Kayembe, S., Mulaji, C., & Mihigo, S., 2016. Biogas Production by Co-Digestion of Cassava Peels with Urea. International Journal of Scientific Engineering and Technology, 55(3), 139–141. https://doi.org/10.17950/ijset/v5s3/303.

Ofoefule, A. U., & Uzodinma, E. O., 2009. Biogas production from blends of cassava (Manihot utilissima) peels with some animal wastes. International Journal of Physical Sciences, 4(7), 398–402.

Olanbiwoninu, A. A., & Odunfa, S. A., 2015. Production of Fermentable Sugars from Organosolv Pretreated Cassava Peels. Advances in Microbiology, 05(02), 117–122. https://doi.org/10.4236/aim.2015.52012.

Onokwai Anthony O, Ugochukwu C Okonkwo, Christian O Osueke, Chinedu A Ezugwu, Ndubuisi C Eze, Reuben S Diarah and O Olawale, Quantifying Cassava Waste Generation and Biogas Production in Eha-Alumona Grinding Mills, International Journal of Civil Engineering and Technology, 10(01), 2019, pp. 2032–2043.

Ozoegwu, C. G., Eze, C., Onwosi, C. O., Mgbemene, C. A., & Ozor, P. A., 2017. Biomass and bioenergy potential of cassava waste in Nigeria: Estimations based partly on rural-level garri processing case studies. Renewable and Sustainable Energy Reviews, 72(November 2016), 625–638. https://doi.org/10.1016/j.rser.2017.01.031.

Panichnumsin, P., Nopharatana, A., Ahring, B., & Chaiprasert, P., 2010. Production of methane by co-digestion of cassava pulp with various concentrations of pig manure. Biomass and Bioenergy, 34(8), 1117–1124. https://doi.org/10.1016/j.biombioe.2010.02.018.

Reilly, M., Dinsdale, R., Guwy, A., 2016. The impact of inocula carryover and inoculum dilution on the methane yields in batch methane potential tests. Bioresour. Technol. 208, 134–139.

Ren, J., Yuan, X., Li, J., Ma, X., Zhao, Y., Zhu, W., Wang, X., & Cui, Z., 2014. Performance and microbial community dynamics in a two-phase anaerobic co-digestion system using cassava dregs and pig manure. Bioresource Technology, 155, 342–351. https://doi.org/10.1016/j.biortech.2013.12.120.

Sawyerr, N., Trois, C., Workneh, T., & Okudoh, V., 2017. Co-Digestion of Animal Manure and Cassava Peel for Biogas Production in South Africa. 101–106. https://doi.org/10.17758/eares.eap1117059.

Sawyerr, Nathaniel & Trois, Cristina & Seyoum Workneh, Tilahun. (2019). Identification and Characterization of Potential Feedstock for Biogas Production in South Africa. Ecological Engineering. 20. 103–116. https://doi.org/10.12911/22998993/108652.

Shah, P. U., Rava, N. P., & Shah, N. K., 2015. Adsorption of Copper from an Aqueous Solution by Chemically Modified Cassava Starch. Journal of Materials and Environmental Science, 6(9), 2573–2582.

Siddique, M. N. I., & Wahid, Z. A., 2018. Achievements and perspectives of anaerobic co-digestion: A review. Journal of Cleaner Production, 194, 359–371. https://doi.org/10.1016/j.jclepro.2018.05.155.

Singh, A., Rawat, K. S., Nautiyal, O. P., & Chavdal, T. V., 2016. Biomass To Fuel: Conversion Techniques. Energy Resources: Development, Harvesting and Management, September, 155–194.

Spencer, D. S. C., & Ezedinma, C., 2017. Cassava cultivation in sub-Saharan Africa. July, 123–148. https://doi.org/10.19103/as.2016.0014.06.

Suhaida, A., Cheng, G., Mel, M., & Has, M., 2012. Single-Step Bioconversion of Unhydrolyzed Cassava Starch in the Production of Bioethanol and Its Value-Added Products. Bioethanol, February. https://doi.org/10.5772/20475.

Villa, L. M., Orrico, A. C. A., Akamine, L. A., Junior, J. de L., & Sunada, N. da S., 2020. Anaerobic co-digestion of swine manure with sweet potato or cassava in different c/n ratios. Ciencia Rural, 50(10), 1–9. https://doi.org/10.1590/0103-8478cr20190734.

Xie, S., Hai, F. I., Zhan, X., Guo, W., Ngo, H. H., Price, W. E., & Nghiem, L. D., 2016. Anaerobic co-digestion: A critical review of mathematical modelling for performance optimization. Bioresource Technology, 222, 498–512. https://doi.org/10.1016/j.biortech.2016.10.015.

Yang, Z., Wang, W., He, Y., Zhang, R., Liu, G., 2018. Effect of ammonia on methane production, methanogenesis pathway, microbial community and reactor performance under mesophilic and thermophilic conditions. Renew. Energy. 125, 915–925.

Yao, Y. Chen, S., Kafle, G.K., 2017. Importance of "weak-base" poplar wastes to process performance and methane yield in solid-state anaerobic digestion. J. Environ. Manage. 193, 423–429.

Zhang, Zhanying & Harrison, Mark & Rackemann, Darryn & Doherty, William & O'Hara, Ian. (2016). Organosolv Pretreatment of Plant Biomass for Enhanced Enzymatic Saccharification. Green Chemistry. 18. 360–381. https://doi.org/10.1039/C5GC02034D.

Zhao, Xuebing & Cheng, Keke & Liu, Dehua. (2009). Organosolv pretreatment of lignocellulosic biomass for enzymatic hydrolysis. Applied microbiology and biotechnology. 82. 815–27. https://doi.org/10.1007/s00253-009-1883-1.

Innovations in Solid Fuel Production from Organic Waste Materials

Circular Entrepreneurship: Production of Fuel Briquettes from Organic Resources in a Rural Area of Uganda

Pauline Namutebi and Haniyeh Jalalipour

Abstract The collection, transportation, and proper disposal of waste pose significant challenges in Uganda, particularly in impoverished urban and rural areas. The mismanagement of municipal solid waste (MSW), which contains a significant amount of organic waste, poses considerable risks to human health and the environment. This research aims to present a best practice example of producing fuel briquettes from organic resources as a means to enhance the sustainability of solid waste management (SWM) in Uganda. The paper outlines the concept of a circular economy business model, promoting sustainable practices in African rural areas, and includes an economic analysis. The study describes the design, production, and sale of eco-friendly energy sources derived from organic waste. The raw feedstocks consist of 30 tonnes of organic waste collected monthly from two markets in Mukono, as well as seasonal agricultural waste from 20 surrounding farmers after the harvesting season. The collected waste undergoes manual sorting to remove impurities such as plastic bags, followed by drying and carbonization in locally made metallic drums. The resulting biochar is then mixed with molasses to produce 10 tons of briquettes monthly, with a production capacity expected to double by 2024. The cost-benefit analysis reveals that briquette production in Mukono requires an annual investment of about 18,006 USD, with a return time of approximately 7 months. This research highlights the potential economic benefits of fuel briquette production from organic resources in rural areas of Uganda.

Keywords Waste management · Economic analysis · Biochar · Agricultural waste · Sustainable practices

---

P. Namutebi (✉)
Fireball Energy Uganda Limited, Kabembe Mukono District, Uganda
e-mail: paulynflav@gmail.com

H. Jalalipour
Professorship of Waste and Resource Management, University of Rostock,
Justus-Von-Liebig-Weg 6, 18059 Rostock, Germany

M. Narra and S. Narra (eds.), *Innovations in Circular Economy and Renewable Energy in Africa*, World Sustainability Series, https://doi.org/10.1007/978-3-031-68330-5_8

1 Introduction

Uganda, a country located in East Africa deals with countless challenges associated with waste management, deforestation, and energy poverty. With a rapidly growing population, the existing waste management infrastructure struggles to cope with rising waste volumes leading to insufficient collection, treatment, and disposal (Edodi 2023). This negatively impacts the environment and the quality of life of communities through the release of greenhouse gases, the spread of diseases, foul odour emissions, blockage of drainage channels, and flooding (Oates et al. 2019).

Over the last few years, various strategies to promote effective waste management have been explored. One such initiative gaining popularity is the transformation of organic waste onto fuel briquettes. Fuel briquettes are compressed biomass fuels with viability to substitute traditional biomass fuels (Oates et al. 2019). These are produced through a densification process where organic wastes like agricultural residues are compressed into solid biofuel briquettes (Miito and Banadda 2016). Besides reducing waste and related environmental pollution, this process also generates sustainable and affordable energy alternatives for heating and cooking.

The gain in momentum of the fuel briquettes production and consumption can be associated with the demonstrated abilities to address a few interconnected challenges. Firstly, briquette-making positively contributes to waste management. As a developing country, Uganda suffers from inadequate waste disposal and management infrastructure. Yet, organic waste such as agricultural remains contributes to over 80% of the daily generated waste (Komaketch et al. 2014). By utilizing these organic residues for briquette production, waste is diverted from dumpsites/ landfills and transformed into a valuable resource while minimizing environmental pollution. Secondly, briquette fuels provide an affordable and widely available alternative fuel to especially rural communities where modern energy access is significantly low (Miito and Banadda 2016). This improves the quality of life by improving cooking efficiency and minimizing household fuel costs. In addition, briquette production reduces deforestation and associated carbon emissions by minimizing the dependence on traditional biomass fuels. This decreases the charcoal and firewood demand. Similarly, the briquette combustion process lowers harmful pollutant emissions which improves indoor air quality (Abondio 2020).

Numerous enterprises and organizations have formed in Uganda to support and boost briquette production. One such initiative is Fireball Energy Uganda Limited. This is a social enterprise that provides access to affordable, eco-friendly energy solutions in the form of Briquettes manufactured from organic waste. The study describes the design, production, and sale of Briquettes from market waste. It aims to highlight the potential economic benefits of fuel briquette production in rural areas of Uganda.

1.1 *Study Area*

The study area is Kabembe village located in the Mukono district, 33.7km from Kampala's capital city. Mukono is part of the Greater Kampala Metropolitan area with a total population of approximately 701,400 people (Uganda Bureau of Statistics 2020). So far, there has been no research on the generation and composition of municipal solid waste in the Mukono district. However, the Kampala Waste Management Plan (KCCA 2020) estimates the generation rate per capita at 0.37 kg per day in the Mukono district which equals 250 tonnes per day of municipal solid waste generation. Only 40% of this waste is collected by the Municipality and Private collectors and transported to an open dumpsite with about 15 hectares of area located at Katikolo village. The 60% of uncollected waste or that collected through illegal waste collectors usually ends up in illegal open dumpsites close to households and communities where it is commonly burned to reduce the accumulated quantities.

2 Materials and Methods

The raw feedstocks for carbonization consist of 30 tonnes of organic waste collected monthly from two markets in Mukono, as well as seasonal agricultural waste from 20 surrounding farmers after the harvesting season. The collected waste undergoes manual sorting to remove impurities such as plastic bags, followed by drying. The case study focuses on a small-scale briquette manufacturing enterprise in Kabembe Mukono, Uganda. Founded in 2018, Fireball Energy Uganda Limited aims to promote sustainable energy solutions that address environmental challenges related to traditional biomass fuels like charcoal and firewood. The company specializes in the production and distribution of high-quality fuel briquettes made from locally sourced organic waste materials. The company operates a small production unit equipped with briquetting machines and drying facilities. The enterprise works with farmers across the Mukono district and market vendors to collect and process organic agricultural waste. Through a carefully designed briquette production process, the organic waste is transformed into compact and energy-dense briquettes. Fireball Briquettes have a high calorific value and low emissions, making them an efficient clean-burning alternative to traditional biomass fuels.

3 Results

3.1 *Carbonization Briquetting*

The carbonization is done in locally made metallic drums. These drums are loaded with dry organic material, sealed, and heated to a carefully regulated temperature. The resulting biochar is then mixed with molasses and placed in a briquette press to produce 10 tons of briquettes monthly.

3.2 Cost Analysis

The cost analysis of Fireball Energy Uganda Limited for producing briquettes from locally available raw feedstocks is displayed in Table 1. Capital costs majorly include processing equipment like carbonizing drums, crushers, mixers, briquetting press, and solar dying facility. A total investment of about USD 10,000 is required to establish a fully operational briquette-making medium-small sized plant. Given a 5-year lifespan for the apparatus, the capital investment required is $2,027 USD each year. To operate a briquetting plant, raw material costs, energy costs, and labour are key considerations with respectively 50%, 7%, and 20% of total operational investments. The maintenance of the briquetting machinery requires a weekly removal of accumulated residue and tightening of bolts and nuts. Monthly, the oil from the generator is changed, air filters cleaned, and change of belts. This costs USD 30 monthly on average. Briquettes are an effective source of heating for small businesses such as poultry farms and households. The production and sale of 10 tonnes of briquettes results in a monthly revenue generation of USD 2739, which equals a return time of approximately 7 months.

Table 8.1 Financial analysis of briquettes production by Fireball Energy Uganda Limited

Investment	Contribution (%)	Cost (USD)
Capital investment		
Modern carbonizing drum	41%	4,110
Briquette press machine	27%	2,740
Briquette lighter machine	19%	1,918
Solar drying facility	14%	1,370
Total		**10,137**
Operational costs		
Agricultural waste sourcing (10 tonnes)	50%	8,219
Fuel costs–generator	7%	1,184
Labour/production costs	20%	3,288
Packaging	4%	658
Branding	2%	329
Transportation	8%	1,315
Administration costs	6%	986
Maintenance	2%	360
Total		**16,338**
Income		
Sales		**32,877**

3.3 A Circular Concept to Promote Biochar Production in Mukono

Biochar/ Briquette production from agricultural waste in rural areas of Uganda offers several advantages that can positively impact both the local communities and the environment.

3.3.1 Waste Management

Agricultural waste, such as crop residues and other organic materials, is often left to decay or burned openly, leading to air pollution and environmental degradation. By converting this waste into briquettes, rural communities in Uganda can effectively manage and utilize the agricultural residues, reducing waste accumulation and minimizing the release of harmful pollutants (Abondio 2020). Additionally, co-composting food waste with biochar can further enhance waste management by diverting organic materials from landfills and creating nutrient-rich compost.

3.3.2 Soil Improvement

Biochar is an excellent soil amendment that enhances soil fertility and structure. When applied to the soil, biochar increases water retention, nutrient availability, and microbial activity, making it ideal for restoring degraded soils in rural areas (Yang et al. 2017). Improved soil fertility can lead to better crop yields and food security for local farmers. Co-composting food waste with biochar enriches the compost with additional nutrients, further enhancing its soil improvement properties.

3.3.3 Climate Change Mitigation

Biochar is a stable form of carbon that sequesters carbon dioxide for an extended period. By producing and incorporating biochar into the soil, rural communities in Uganda can contribute to mitigating climate change by locking away carbon and reducing greenhouse gas emissions (Lehmann et al. 2021). Co-composting food waste with biochar also promotes carbon sequestration in the compost, providing an additional benefit for climate change mitigation.

3.3.4 Renewable Energy Source

The produced briquettes provide the local communities with renewable sources of energy for cooking, heating, or powering small-scale machinery, providing an alternative to traditional fossil fuels and reducing reliance on non-renewable energy sources (Miito and Banadda 2016).

3.3.5 Economic Opportunities

Establishing carbonization facilities in rural areas can create new economic opportunities for local communities. It can generate income and employment through the production and sale of biochar and briquettes, contributing to poverty alleviation and promoting rural development (Oates et al. 2019). Moreover, the co-composting of food waste and biochar can create marketable compost products that can be sold locally or regionally.

Knowledge Transfer and Capacity Building: Implementing biochar/briquettes production and co-composting in rural Uganda can involve knowledge transfer and capacity-building initiatives, empowering local communities with new skills. This knowledge can then be shared and disseminated among neighbouring communities, fostering regional collaboration and sustainable development.

4 Conclusion

This study demonstrates a best practice of organic waste management in a rural area in Uganda. Fireball Energy Uganda Limited is a small enterprise that collects market waste and uses a simple and cost-effective carbonizing technology to manufacture 10 tonnes of briquettes each month. A total expenditure of around USD 10,000 is required to set up a fully working medium-small-sized briquette-making unit. The manufacturing and sale of briquettes generate a monthly income of USD 2739. Given the apparatus's 5-year lifetime, the return on investment is around 7 months.

Overall, producing briquettes from market waste in rural Uganda offers a range of significant benefits. Firstly, it presents an effective waste management solution, reducing waste accumulation and enhancing cleanliness in rural areas. Secondly, it provides a renewable energy alternative to firewood and charcoal, curbing deforestation and preserving local ecosystems. Additionally, briquette production creates income-generating opportunities and employment, empowering communities and contributing to economic growth. Moreover, the use of briquettes leads to reduced smoke and indoor air pollutants, improving health outcomes for residents. This eco-friendly approach fosters environmental conservation, mitigates climate change, and contributes to a more resilient and sustainable future for rural Uganda.

References

Abondio, R.B., Komakech, A.J., Kambugu, R.K., Kiggundu, N., Wanyama, J., Zziwa, A. and Kyamanywa, S. (2020) Assessment of Municipal Organic Solid Waste, as a Potential Feedstock for Briquette Production in Kampala, Uganda. Journal of Sustainable Bioenergy Systems, 10, 62–75. https://doi.org/10.4236/jsbs.2020.102006

Edodi (2023) Managing the environment: issues and priority actions for sustainable waste management in Uganda, African Geographical Review, 42:3, 342-356, https://doi.org/10.1080/19376812.2022.2054438.

Kampala Capital City Authority (KCCA). 2020. Kampala Waste Management Plan, Accessed from: https://www.kcca.go.ug/media/docs/IFC-KWM-PIM-12-MARCH.pdf.

Lehmann, J., Cowie, A., Masiello, C.A. 2021. Biochar in climate change mitigation. Nat. Geosci. 14, 883–892 (2021). https://doi.org/10.1038/s41561-021-00852-8.

Miito, G. J., and N. Banadda. 2016. Waste to energy technologies for solid waste management in developing countries: a case study of Uganda. Agricultural Engineering International: CIGR Journal, 18(3):136–146.

Oates, L., Gillard, R., Kasaija, P., Sudmant, A., Gouldson, A. 2019. Supporting decent livelihoods through sustainable service provision: Lessons on solid waste management from Kampala, Uganda. Coalition for Urban Transitions. London and Washington, DC.: http://newclimateeconomy.net/content/cities-working-papers.

Uganda Bureau Of Statistics (UBOS).2020. UBOS Statistical Abstract, Accessed from: http://library.health.go.ug/sites/default/files/resources/UBOS%20Statistical%20Abstract%202020.pdf.

Yang, Ding., Yunguo, Liu., Shaobo, Liu., Xixian, Huang., Zhongwu, Li., Xiaofei, Tan., Guangming Zeng., Lu, Zhou., 2017. Potential Benefits of Biochar in Agricultural Soils: A Review. Pedosphere. Volume 27, Issue 4, Pages 645-661, ISSN 1002-0160.https://doi.org/10.1016/S1002-0160(17)60375-8.

Characterization of Three Lignocellulosic Biomass Materials as Novel Binders in Briquette Production

S. U. Yunusa, E. Mensah, K. Preko, S. Narra, A. Saleh, Safietou Sanfo, and N. S. Sadiq

Abstract With the global increase in population, energy demand has outweighed the available energy resources and has compelled people to seek alternative energy sources. Based on this, coupled with the interest to curtail nonrenewable fuels, biomass briquetting has emerged as a sustainable means of harnessing energy from biomass. One of the most pertinent aspects of briquette production is selecting and using the appropriate binder. Hence, a good understanding of the properties of binders is important. This paper characterized three biomass materials (sweet potato peel, locust bean pulp, and locust bean husk) as novel binders in briquette production. The parameters characterized are gross calorific value, proximate (moisture content, volatile matter, ash content, and fixed carbon), and ultimate parameters (Carbon, Nitrogen, Oxygen, Hydrogen, and Sulfur). To ascertain the veracity of the characterization, briquettes were produced using the three binders. The findings from the characterization showed that the biomass materials have calorific value between 21.26

S. U. Yunusa (✉)
WASCAL Graduate Research Programme on Climate Change and Land Use, Department of Civil Engineering, KNUST-Kumasi, Kumasi, Ghana
e-mail: yunusa.s@edu.wascal.org; suyunusa@abu.edu.ng

S. U. Yunusa · A. Saleh · N. S. Sadiq
Department of Agricultural and Bio-Resources Engineering Ahmadu, Bello University, Zaria, Nigeria

E. Mensah
Department of Agricultural and Biosystems Engineering, KNUST-Kumasi, Kumasi, Ghana

S. Narra
Professorship of Material and Energy Valorisation of Biogenous Residues, Professorship of Waste and Resource Management, University of Rostock, 18059 Rostock, Germany

K. Preko
Department of Physics, KNUST-Kumasi, Kumasi, Ghana

S. Sanfo
West African Science Service Centre on Climate Change and Adapted Land Use (WASCAL), Ouagadougou, Burkina Faso

Laboratoire de Développement Agricole et Transformation de l'Agriculture (DATA), CEDRESS Université Thomas Sankara, Ouagadougou, Burkina Faso

M. Narra and S. Narra (eds.), *Innovations in Circular Economy and Renewable Energy in Africa*, World Sustainability Series, https://doi.org/10.1007/978-3-031-68330-5_9

and 34.48 MJ/kg, volatile matter between 58.6% and 84.2%, ash content between 5.5% and 16.3%, and fixed carbon between 3.5% and 15.1%. The major elemental constituents are carbon, between 35.0% and 40.5%, and oxygen between 36.97% and 43.74%. The findings from the briquette's quality metrics showed a maximum compressed density of 0.67 g/cm^3, impact resistance index of 50%, and compressive strength of 360 kN/m^2. Thus, the study recommends the use of gelatinized sweet potato peel, locust bean pulp, and locust bean husk as binders in briquette production.

1 Introduction

Energy sources are currently being overburdened as a result of the global rise in population. Karpov et al. (2022) reported a steady rise over the last two decades, except for periods of global crisis (2007–2009) and pandemic (2020) where there was a decline in industrial activities. This has been part of the focus of the United Nations through its Sustainable Development Goal 7, which targets the provision of clean, sustainable, and affordable energy for all by 2030 (United Nations 2022). On this premise, as part of augmenting the shortfall and burden of fossils, energy has over the years been synthesized from biomass materials. Biomass can be transformed into energy through direct combustion in their loosed or densified forms (briquettes/pellets) as elaborated in Yunusa et al. (2023a, b), or indirectly into bio-fuels (biogas, biodiesel, bioethanol, etc.) through anaerobic digestion and pyrolysis (Niño et al. 2020). However, when used as fuel in either raw or processed form, the use of improved biomass cookstove has been recommended to curtail toxic gas emissions and improve energy efficiency (Yunusa et al. 2023a, b). This is perceived to be cleaner and more sustainable than conventional fossil fuels, especially as biomass is considered a carbon–neutral source of energy.

Briquetting is one of the commonest forms of energy production from biomass (Yunusa et al. 2023a, b). It involves densifying loosed biomass materials into solid fuels with the aid of binding agents and pressure. Through this, loosed biomass materials are transformed into more compact and uniformly stable products called briquettes (Sugebo 2022). Attaining a well-densified and efficient briquette is a function of choosing the appropriate biomass and binding material and densifying under an appropriate compression condition (Kpalo et al. 2020). Binding materials are very essential in briquetting as they enhance bonding and contribute to both the physical and combustion performance of the briquette, especially in low to medium-pressured briquetting. Binders are materials that bind particles together and prevent them from disintegrating (Sanjika and Chipula 2021). Hence, the quality of the briquette depends largely on the quality of the binding material (Sanjika and Chipula 2021; Zhang et al. 2018). Thus, a careful selection of binding material is pertinent. Based on this, it is largely encouraged to consider the quality, availability, and cost of the binder before selection (Obi et al. 2022).

Over the years, food-based binders especially starch (cassava starch and flour, corn starch, maize flour, wheat flour, rice flour, wheat starch, potato starch, etc.)

are reportedly the most used binders in briquetting (Zhang et al. 2018). They are, however, expensive and have low water resistance (Zhang et al. 2018). Therefore, it becomes imperative to evaluate more biomass materials as binders to minimize the cost of production and improve efficiency and sustainability.

Against this background, this paper aims at evaluating three biomass materials as novel binders in briquette production.

2 Materials and Methods

2.1 *Materials*

Three lignocellulosic biomass materials (sweet potato peel, locust bean pulp, and locust bean pod) were selected for the study. The selection was based on availability within the study area, cost, and projected performance.

2.2 *Material Collection and Sample Preparation*

A sample of approximately 3 kg of each biomass was collected for the study. Sweet potato peel (SPP) was obtained from Samaru, Zaria, Kaduna State Nigeria (latitude 11°10′5.65824″ N and longitude 7°39′25.88832″ E), locust bean pulp (LBP) was purchased from Sabon gari market, Zaria, Kaduna State, Nigeria (latitude 11°6′37.1826″ N and longitude 7°43′40.48392″ E) and locust bean husk (LBH) was obtained from Sabon gari, Kaduna State Nigeria (latitude 11°6′28.54044″ N and longitude 7°43′47.84988″ E). The samples were screened for foreign materials and stored in an airtight plastic to avoid contamination and moisture absorption.

Sweet potato peel: Sweet potato (*Ipomoea batatas* L.) peel was obtained in fresh form and was washed and sun-dried at an average temperature and relative humidity of 29 °C and 65% for 5 days. The dried sample was then ground into powder using an electric milling machine (Model 9z–23, Zhengzhou Shuliy Machinery Company, Ltd., China) and packaged in a polyethylene bag for analysis.

Locust bean pulp: The African locust bean (*Parkia biglobosa*) tree is a perennial tree of leguminous origin that belongs to the sub-family mimosoideae and family leguminosae (now family fabaceae) (Akande et al. 2010). While the seed has over the years been processed into a local seasoning referred to as 'Dawa Dawa', the pulp and pod have still not gained a lot of interest in many parts of Nigeria (Olujobi 2012). Hence, they mostly end up as waste or used as animal fodder. After screening out foreign materials, the pulp was manually mixed to homogenize the lump particles and sieved through a 1mm sieve size to obtain particles $\leq$ 1mm.

Locust bean husk: The husk was sun-dried at an average temperature and relative humidity of 29 °C and 65% for 2 days and milled using a hammer mill fitted with a 2mm screen, and was sieved through a 1 mm sieve size to obtain particles $\leq$1 mm.

2.3 *Characterization*

The characterized parameters were selected based on the major properties required of briquette binders, which include good adhesion, less emissive, and good thermal and combustion performance (Zhang et al. 2018). Thus, the properties evaluated were calorific value, proximate (moisture content, volatile matter, ash content, and fixed carbon), and ultimate properties (Carbon, Nitrogen, Oxygen, Hydrogen, and Sulfur).

Adhesion includes the ability of the material to gelatinize when mixed in boiling water and its ability to bind particles together. While calorific value, proximate parameters, and major elemental constituents (O, C, H) are good measures for evaluating the potential thermal and combustion performances. The minor elemental constituents (N and S) on the other hand are measures of assessing the nitrogen and sulfuric-based emissions during combustion (Mitchual et al. 2014a, b).

2.3.1 Gross Calorific Value

The gross calorific value indicates the energy content of the biomass material per unit mass (cal/g) (Musabbikhah et al. 2019). Thus, the higher the value, the better the thermal performance of the biomass when densified into a briquette. The gross calorific value was determined using a bomb calorimeter (Model: 6100, Parr Instrument Company, USA) following the method described in ASTM D5865-10a (2010).

2.3.2 Proximate Analysis

Proximate analysis shows the content of volatile matter, ash content, and fixed carbon in the samples. These parameters are expected to be within a certain range for briquette production feedstocks. Prefarably, a low moisture content between 6 and 14%, moderate to high volatile matter content (50–90%), a very low ash content (<4%), and fixed carbon content between 9 and 25% are required (Nikiema et al. 2022).

(a) Moisture content

The moisture content was determined following the method described in ASTM D3173-87 (1996). A gram of each sample was oven dried at 105 °C for an hour and the moisture content was computed using Eq. 1.

$$MC(\%)wb = \frac{W_i - W_f}{W_i} \times 100 \quad (1)$$

where, MC (%) wb = moisture content (web basis), W_i = initial weight of sample, W_f = final weight of sample

(b) Volatile matter

The volatile matter is the portion of biomass that changes into vapor or gas when heated to a temperature of 400 °C to 500 °C or more (Sunnu et al. 2021). The sample's volatile matter was determined using the modified procedure described in ASTM D3175-87 (1996) for all sparking fuels, and the percentage of volatile matter in each sample was estimated using Eq. 2.

$$Percentage\ volatile\ matter\ (\%) = (\frac{A - B}{A} \times 100) - \%MC \quad (2)$$

where A is the weight of the oven-dried sample, B is the weight of the sample after heating in the furnace, and MC is the percentage moisture content

(c) Ash content

The content of ash shows the slagging nature of the material during combustion. Thus, the lower it is, the better the heating value (Thabuot et al. 2015). Overall, an ash content of less than 4% is generally preferred for briquetting as it reduces slagging potential (Nagarajan and Prakash 2021). The ash content was estimated following the method described in ASTM D3174-02 (2002), and the percentage ash content was computed using Eq. 3 (Mitchual et al. 2014a, b).

$$Ash\,content\ (\%) = \frac{M_{ash}}{M_{oven-dry}} \times 100 \quad (3)$$

where M_{ash} is the mass of the ash and $M_{oven-dry}$ is the mass of the oven-dried sample.

(d) Fixed carbon

The fixed carbon is the residue or remnant of fuel after the volatile matter, moisture, and ash content have been removed (Sarkar 2015). The higher the fixed carbon content, the better the energy value (Mitchual et al. 2014a, b). The fixed carbon (FC) was estimated by subtracting the value of the volatile matter (VM) and ash content from 100% as given in Eq. 4 (Mansaray and Ghaly 1997).

$$\%\,FC = [100 - (\%MC + \%Ash + \%VM)] \quad (4)$$

2.3.3 Ultimate Analysis

The ultimate or elemental analysis in this study was restricted to carbon, nitrogen, sulfur, hydrogen, and oxygen. The nitrogen content was determined using the Micro-Kjeldahl method with a Micro-Kjeldahl digestion distillation apparatus, and the sulfur content was determined using the aluminum block method in a Tecator (Model 40 digestor, Tecator AB, Sweden). However, the content of carbon, hydrogen and oxygen was estimated from the result of proximate analysis with an accuracy of ±2% using Eqs. 5 to 7 (TT et al, 2022).

$$Carbon\ (\%) = 0.635\text{FC} + 0.460\text{VM} - 0.095\text{AC} \tag{5}$$

$$Hydrogen\ (\%) = 0.059\text{FC} + 0.060\text{VM} + 0.010\text{AC} \tag{6}$$

$$Oxygen\ (\%) = 0.340\text{FC} + 0.469\text{VM} - 0.023\text{AC} \tag{7}$$

where, FC = Fixed Carbon, AC = Ash Matter, and VM = Volatile Matter

2.4 *Evaluation of Binders in Briquette Production*

2.4.1 Binder Preparation

(a) Binder SPP

SPP binder was prepared in two ways, by using water (Fig. 1i) and by using LBH solution (Fig. 1j). The first method was done by mixing the required quantity with 100 ml of water at room temperature until it dissolved without clogs. This was followed by boiling 500 ml of water using an electric stove where the dissolved SPP was poured and stirred continuously for about 5 min to gelatinize into paste. While the second method involves gelatinizing with LBH solution instead of water. Here, 300 g of LBH was mixed with 8 L of water and boiled continuously until the reddish surface was observed to have turned pale yielding a reddish-brown solution. The solution was then used to gelatinize SPP following the same procedure as in the water preparation method.

(b) Binder LBP

In preparing binder LBP (Fig. 1f), the sample was first dissolved into 100 ml of water at room temperature, and mixed with 500 ml of boiled water. This was followed by continuous mixing for about 5 min to gelatinize appropriately.

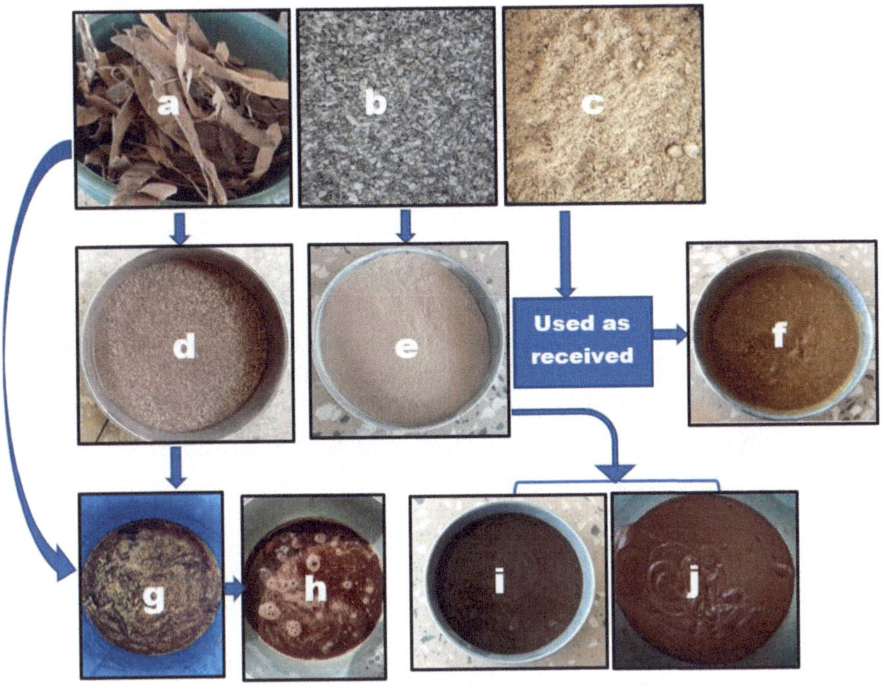

Fig. 1 Binder Preparation. **a** Raw LBH, **b** Raw SPP, **c** Raw LBP, **d** Milled LBH, **e** Milled SPP, **f** LBP binder, **g** Soaked LBH, **f** LBH binder, **i** SPP binder gelatinized with water, **j** SPP binder gelatinized with LBH solution

(c) Binder LBH

Binder LBH (Fig. 1h) was prepared using the cold method which involved soaking the sample in water at room temperature for 14 days with intermittent mixing to ensure it dissolved effectively. After the 14th day, the mixture was observed to have transformed into a relative gel form.

2.4.2 Briquette Production

As a way of validating the results of characterization, briquettes were produced using two-stage rice husks graded to 1.5 mm. However, in this study, only the binders gelatinized through the hot method (i.e., SPP prepared with LBH solution and LBP) were considered to enable comparison without bias. Three samples (1, 2, and 3) were prepared for each binder type at variable binder content (5%, 10%, and 15%) and dwell time (1.5, 1, and 0.5 min). The samples were densified at room temperature and a constant pressure of 4.5 MPa using a hydraulic piston press with a cylindrical mold of 8 cm diameter and height of 16 cm. Each binder was prepared accordingly as discussed in the previous section. Rice husk was mixed with the binder while it is hot, and 140 ml water at room temperature was added and manually mixed for about 5 min to fully homogenize. 200g of each sample mix was fed into the mold and densified with the aid of a 5 tons hydraulic jack. The densified briquettes were sun-dried for a week and allowed to cool in a ventilated room for an additional week before testing.

2.4.3 Briquette Testing

The briquettes were tested for compressed density, impact resistance, and compressive strength. The compressed density was determined immediately after ejection from the mold by measuring the diameter and length of the briquette in two perpendicular directions (±0.01 mm) using a digital vernier caliper and dividing by an estimate of the mass (±0.1 g) as given in Eq. 8 (Gendek et al. 2018).

$$\rho = \frac{m}{\frac{\pi}{4}d^2 l} \tag{8}$$

where ρ is the density (g/cm^3), m is the mass of briquette (g), d is the averaged diameter (cm) and l is the averaged length (cm)

The impact resistance test was carried out as per ASTM D440-86 (2002). The briquettes were dropped twice from a height of 2 m onto a steel plate. Then, the impact resistance index (IRI) was estimated following Eq. 3.8 (Richards 1990). However, only pieces that weighed 5% or more of the initial weight of the briquette were considered in the IRI estimation (Mitchual et al. 2014a, b).

$$IRI = \frac{N}{n} \times 100 \tag{9}$$

where N is the number of drops and n is the number of pieces that weighed 5% or more of the initial weight of the briquette after N drops.

The compressive strength was determined using a universal test machine (SM1000-TecQuipment Ltd., Nottingham, UK) with a load cell capacity of 100 kN. The sample briquette was placed in between the plates and was subjected to uniform loading until failure.

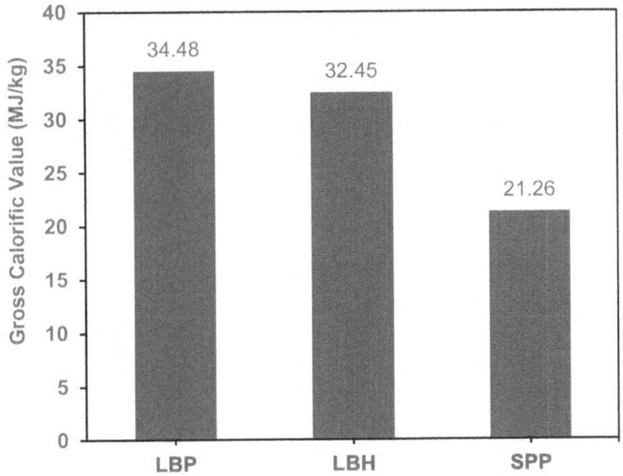

Fig. 2 Gross calorific value of the biomass samples

3 Results and Discussion

3.1 Biomass Characterization

3.1.1 Calorific Value

The result of the calorific value of the biomass samples is presented in Fig. 2. The obtained values range between 21.26 to 34.48 MJ/kg with binder LBP having the highest energy content (34.48 MJ/kg). The values obtained are higher than the values (15.06–20.09 MJ/kg) reported by Veereh and Narayana (2013), for six lignocellulosic biomass materials used in briquette production.

3.2 Proximate Analysis

The result of the proximate analysis of the three biomass samples is presented in Table 2. The volatile matter ranges between 58.6% and 84.2%, while the ash content and fixed carbon contents range between 5.5% to 16.3%, and 3.5% to 15.1%, respectively. Although the fixed carbon content in sample SPP and LBP are lower than the range (9–25%) reported by Nikiema et al. (2022), the volatile matter and ash contents which are major indicators in attaining thermally efficient briquettes conforms with the given range. In addition, the values of ash content agrees with the findings (0.82–9.87) of Tabakaev et al. (2017) for biomass materials intended for solid fuel production. Overall, the obtained values conform with the range of values required in feedstocks intended for briquette production as they possess low levels of moisture, relatively low content of ash and high content of volatile matter.

Table 2 Result of proximate analysis

Sample	Moisture content (%)	Volatile matter (%)	Ash content (%)	Fixed carbon (%)
LBP	6.7	83	6.8	3.5
LBH	10	58.6	16.3	15.1
SPP	6.7	84.2	5.5	3.7

Table 3 Result of ultimate analysis

Sample	C (%)	N (%)	O (%)	H (%)	S (%)
LBP	39.76	0.56	43.10	5.66	7.17
LBH	35.00	1.33	36.97	5.18	7.69
SPP	40.50	1.12	43.74	5.72	0.47

3.3 *Ultimate Analysis*

Table 3 presents the result of the ultimate analysis. The results revealed the major elemental constituents (i.e., C, O, and H) to be higher in sample SPP compared to other samples. This depicts that sample SPP has a better potential of yielding thermally efficient briquettes when used as binders than the other samples. Although sample LBP showed the lowest content of nitrogen, depicting the potential of being cleaner in terms of nitrogen-based gases, the sample had a high content of sulfur. It is important to note that the lower the content of nitrogen and sulfur, the lower the amount of nitrogen and sulfuric-based emissions during combustion (Mitchual et al. 2014a, b). Overall, with minor constituents (i.e., N and S) being low in sample SPP, it shows that when combusted, the sample will emit very low sulfuric and nitrogen emissions. Although SPP presents the best performance, the result of major elemental constituents is not significantly different from that of other sample, indicating that they also have good potential for use as binder. However, when using these samples, the ratio must be kept as low as possible to curtail sulfuric emissions.

3.4 *Briquette Quality*

The results of the quality test are presented in Figs. 3, 4, 5 and 6. As shown in Figs. 3 and 4, briquettes produced from 5 and 15% binder content presented the highest density and IRI values than those densified with 10% binder content. This was attributed to the fact that at a 5% binder ratio, the samples had the highest dwell time, thereby enhancing interparticle bonding. Although at 15% binder content, the dwell time was low, the presence of more binder content aided the densification process thereby yielding denser briquettes. With the values recorded at a 5% binder ratio better than those of 10%, it depicts that an increment in the content of the binder does not always translate to improved density and strength. However, other factors especially material type, pressure level, and dwell time are essential. Overall, there was no significant difference in the performance of binder SPP and LBP for compressed density and IRI. Figure 4 shows the image view of the developed briquettes.

Figure 6 shows the result of compressive strength. The briquettes densified with 15% content of binder LBP had the highest compressive strength (360 kN/m^2). Although density and impact resistance are good measures for evaluating briquette binding performance, Miao et al. (2019) believes that compressive strength is a better indicator for measuring the quality of the binders. Overall, it was observed that the compressive strength improves with an increase in the content of the binder.

Fig. 3 Pictorial view of the densified briquettes (authors own picture)

Fig. 4 Compressed density

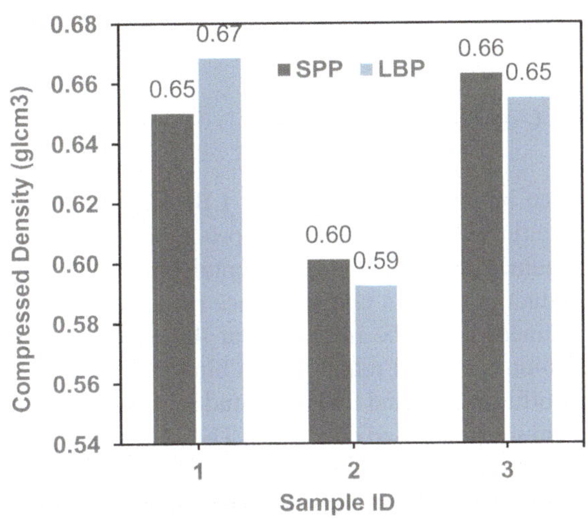

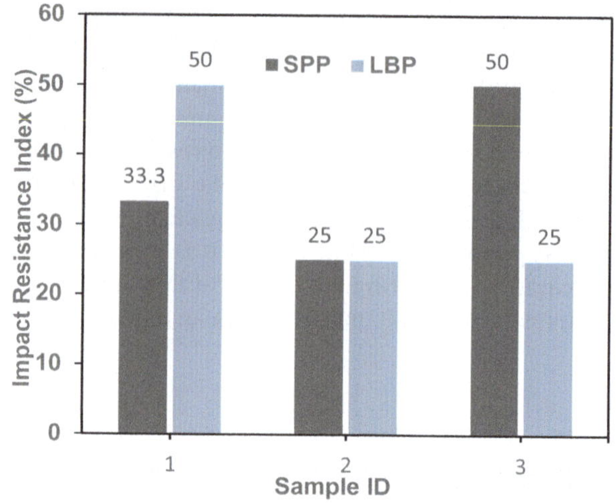

Fig. 5 Impact resistance

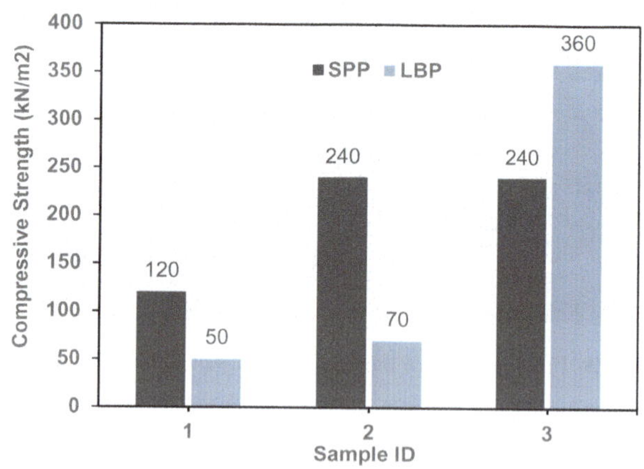

Fig. 6 Compressive strength

4 Conclusion

Three biomass materials (SPP, LBP, and LBH) were characterized to assess their potential as binders in briquette production. The characterization was based on energy content, proximate, and elemental composition. This was followed by briquette production using a two-stage rice husk as a way of validation. Based on the results obtained, it can be inferred that the experimented biomass materials possess the minimum content required of a binder. However, due to the relatively high content of sulfur in LBP and LBH, the study recommends the use of a very low percentage to avoid sulfuric-based emissions. This will be more suitable in high-pressured briquetting where binding content is kept very low. Thus, the study recommends the use of the three biomass materials as binders in briquette production.

References

TT AK, Mech N, Ramesh ST, Gandhimathi R (2022) Evaluation of composite briquettes from dry leaves in energy applications for agrarian communities in India. J. Clean. Prod. 350: 131312. https://doi.org/10.1016/j.jclepro.2022.131312.

Akande FB, Adejumo OA, Adamade CA, Bodunde J (2010) Processing of locust bean fruits: Challenges and prospects. African J. Agric. Res. 5: 2268–2271.

ASTM D3173–87 (1996) Standard Test Method for Moisture in the Analysis Sample of Coal and Coke, in: Annu. B. ASTM Stand., West Conshohocken, PA, USA.

ASTM D3174–02 (2002) ASTM: Standard test method for ash in the analysis sample of coal and coke from coal, in: Annu. B. ASTM Stand., West Conshohocken, PA, USA. www.astm.org.

ASTM D440 - 86 (2002) Standard Test Method of Drop Shatter Test for Coal, in: ASTM Int., West Conshohocken, PA, USA.

ASTM D5865–10a (2010) Standard Test Method for Gross Calorific Value of Coal and Coke, in: Annu. B. ASTM Stand. https://doi.org/10.1520/D5865-10A.2.

Gendek A, Aniszewska M, Malaťák J, Velebil J (2018) Evaluation of selected physical and mechanical properties of briquettes produced from cones of three coniferous tree species. Biomass and Bioenergy. 117: 173–179. https://doi.org/10.1016/j.biombioe.2018.07.025.

Karpov KA, Nesteruk LG, Shpakovskaya IA (2022) Features of changes in global energy consumption for the period 2000-2020. IOP Conf. Ser. Earth Environ. Sci. 990: 012004. https://doi.org/10.1088/1755-1315/990/1/012004.

Kpalo SY, Zainuddin MF, Manaf LA, Roslan AM (2020) A review of technical and economic aspects of biomass briquetting. Sustain. 12:. https://doi.org/10.3390/su12114609.

Mansaray KG, Ghaly AE (1997) Physical and thermochemical properties of rice husk. Energy Sources. 19: 989–1004. https://doi.org/10.1080/00908319708908904.

Miao Z, Zhang P, Li M, Wan Y, Meng X (2019) Briquette preparation with biomass binder. Energy Sources, Part A Recover. Util. Environ. Eff. https://doi.org/10.1080/15567036.2019.1682722.

Mitchual SJ, Frimpong-mensah K, Darkwa NA (2014) Evaluation of Fuel Properties of Six Tropical Hardwood Timber Species for Briquettes. J. Sustain. Bioenergy Syst. 4: 1–9.

Mitchual SJ, Frimpong-Mensah K, Darkwa NA (2014) Relationship between Physico-Mechanical Properties, Compacting Pressure and Mixing Proportion of Briquettes Produced from Maize Cobs and Sawdust. J. Sustain. Bioenergy Syst. 4: 50–60.

Musabbikhah, Saptoadi H, Subarmono, Wibisono MA (2019) Analysis and Selection of the Best Model of Biomass Briquette Based on Calorific Value. J. Phys. Conf. Ser. 1175:. https://doi.org/10.1088/1742-6596/1175/1/012270.

Nagarajan J, Prakash L (2021) Preparation and characterization of biomass briquettes using sugarcane bagasse, corncob and rice husk. Mater. Today Proc. 47: 4194–4198. https://doi.org/10.1016/j.matpr.2021.04.457.

Nikiema J, Asamoah B, Egblewogbe MNYH, Akomea-Agyin J, Cofie OO, Felix A, Gebreyesus G, Zipporah K, Njenga M (2022) Resources, Conservation & Recycling Advances Impact of material composition and food waste decomposition on characteristics of fuel briquettes. Resour. Conserv. Recycl. Adv. 15: 200095. https://doi.org/10.1016/j.rcradv.2022.200095.

Niño A, Arzola N, Araque O (2020) Experimental study on the mechanical properties of biomass briquettes from a mixture of rice husk and pine sawdust. Energies. 13:. https://doi.org/10.3390/en13051060.

Obi OF, Pecenka R, Clifford MJ (2022) A Review of Biomass Briquette Binders and Quality Parameters. Energies. 15:. https://doi.org/10.3390/en15072426.Sampson

Olujobi OJ (2012) Comparative Evaluation of Nutritional Composition of African Locust Bean (Parkia biglobosa) Fruits from Two Locations. Niger. J. Basic Appl. Sci. 20: 195–198. http://www.ajol.info/index.php/njbas/index.

Richards SR (1990) Physical Testing of Fuel Briquettes. Fuel Process. Technol. 25: 89–100.

Sanjika T, Chipula G (2021) Technical feasibility of producing binder-free water hyacinth briquettes for domestic energy use. African J. Sci. Technol. Innov. Dev. https://doi.org/10.1080/20421338.2021.1988417.

Sarkar DK (2015) Fuels and Combustion, in: Therm. Power Plant, pp. 91–137. https://doi.org/10.1016/B978-0-12-801575-9.00003-2.

Sugebo B (2022) A review on enhanced biofuel production from coffee by-products using different enhancement techniques. Mater. Renew. Sustain. Energy. https://doi.org/10.1007/s40243-022-00209-0.

Sunnu AK, Adu-Poku KA, Ayetor GK (2021) Production and Characterization of Charred Briquettes from Various Agricultural Waste. Combust. Sci. Technol. https://doi.org/10.1080/00102202.2021.1977803.

Tabakaev R, Shanenkov I, Kazakov A, Zavorin A (2017) Thermal processing of biomass into high-calorific solid composite fuel. J. Anal. Appl. Pyrolysis. 124: 94–102. https://doi.org/10.1016/j.jaap.2017.02.016.

Thabuot M, Pagketanang T, Panyacharoen K, Mongkut P, Wongwicha P (2015) Effect of Applied Pressure and Binder Proportion on the Fuel Properties of Holey Bio-Briquettes, Elsevier B.V. https://doi.org/10.1016/j.egypro.2015.11.583.

United Nations UN (2022) Progress Towards the Sustainable Development Goals. https://doi.org/10.1017/S0020818300010845.

Veeresh SJ, Narayana J (2013) Sustainable utilization of agro-waste for high calorific energy briquettes. Energy Sources, Part A Recover. Util. Environ. Eff. 35: 1375–1384. https://doi.org/10.1080/15567036.2010.525594.

Yunusa SU, Mensah E, Preko K, Narra S, Saleh A, Sanfo S (2023) A comprehensive review on the technical aspects of biomass briquetting. Biomass Convers. Biorefinery. https://doi.org/10.1007/s13399-023-04387-3.

Yunusa SU, Mensah E, Preko K, Narra S, Saleh A, Sanfo S, Isiaka M, Dalha IB, Abdulsalam M (2023) Biomass cookstoves: A review of technical aspects and recent advances. Energy Nexus. 100225. https://doi.org/10.1016/j.nexus.2023.100225.

Zhang G, Sun Y, Xu Y (2018) Review of briquette binders and briquetting mechanism. Renew. Sustain. Energy Rev. 82: 477–487. https://doi.org/10.1016/j.rser.2017.09.072.

Optimization of Solid Fuel Production by Co-hydrothermal Carbonization of Polyvinyl Chloride and *Pontideria Crassipes*: A Waste Valorization Study

Solomon Asare, Francis Kemausuor, Michael Kweku Commeh, Mizpah Ama Dziedzorm Rockson, Ahmad Addo, and Satyanarayana Narra

1 Introduction

Due to the world's rising population, coupled with the increasing energy demands and concerns about global warming, new alternative energy options with a low carbon footprint are being sought. Furthermore, the grave threat of global warming caused by greenhouse gases and the depletion of commercially available conventional fossil fuels restricts their use. To reduce our dependency on fossil fuels, there is the need to focus on converting waste into a more useful form such as energy, alternative fuel, and nutrient. The global waste generation reached 3.4×10^9 metric tons in 2020, with only a 20% recycling rate (World Bank data 2020). Municipal plastic waste (MPW) generation has been gradually increasing at a rate of 5% per year, but MPW recycling is only at 3% per year; the rest of MPW is either disposed of by landfill or incinerated (Zhao et al. 2020). Due to their versatility, plastics are widely used in automobiles, electronics, and construction. A popular plastic, Polyvinyl Chloride (PVC), shares a major percentage in construction, followed by Polyethylene (PE) and Polyethylene Terephthalate (PET). This has resulted in the increase of chlorine content in MSW.

S. Asare (✉) · F. Kemausuor · A. Addo
Department of Agricultural and Biosystems Engineering, College of Engineering, Kwame Nkrumah University of Science and Technology (KNUST), Kumasi, Ghana
e-mail: solomonasaresnr@gmail.com

M. K. Commeh
Technology Consultancy Centre, College of Engineering, KNUST, Kumasi, Ghana

M. A. D. Rockson
Department of Chemical Engineering, College of Engineering, KNUST, Kumasi, Ghana

S. Narra
Professorship of Material and Energy Valorisation of Biogenous Residues, Professorship of Waste and Resource Management, University of Rostock, Rostock, Germany

M. Narra and S. Narra (eds.), *Innovations in Circular Economy and Renewable Energy in Africa*, World Sustainability Series, https://doi.org/10.1007/978-3-031-68330-5_10

Hydrothermal carbonization can absorb the chlorine in its process water, thereby preventing its release into the atmosphere, polluting it as a result. Biomass and PVC meet all of the criteria for renewable and carbon–neutral energy sources, and their ubiquitous availability has made them a viable alternative to existing commercial fossil fuels (Yadav and Reddy 2020).

Hydrothermal carbonization (HTC) is the process of degrading organic matter by treating the reacting biomass components in water at a specific temperature and pressure to maintain the water in either a subcritical or supercritical state. HTC (also called wet torrefaction) is a technique for altering organic waste into a carbon-rich solid. Hydrothermal carbonization is accomplished at 180–260 °C, with feedstock inundated in water and heated in a restricted enclosure for 5–240 min under 2–6 MPa (Libra et al. 2011). HTC helps to prevail over the energy costs connected with drying the biomass since the process is aqueous based. HTC is currently commonly recommended as a method for converting a variety of wet biomasses. Typical HTC has been widely applied to convert biomass into energy. In co-hydrothermal carbonization (Co-HTC), two or more materials that have synergy is made to undergo HTC together. Typical feedstocks include watermelon and pineapple peels (Azaare et al. 2021), water hyacinth, and sewage sludge (Zhang et al., 2020a, b), biomass and plastics such as pomelo peels and PVC (Ning et al. 2022). Sewage sludge and biomass such as sewage sludge and coconut shells (Tu et al. 2021), sewage sludge and *chlorella vulgaris* (Du et al. 2021), and so on.

HTC simulates natural coal formation by transforming organic material in a primarily water-based feedstock into a coal-like substance one million times quicker than nature (Libra et al. 2011). HTCs are affected by reaction parameters such as temperature, type of feedstock, resident time, and mixing ratios. Gao et al. (2013) studied the impact of residence time on the chemical and structural properties of hydrochar produced by hydrothermal carbonization of water hyacinth and found that after four hours, the chemical and physical properties of WH were unaffected by resident time. However, more significant parameters are the temperature (Yao & Ma 2018; Zhao et al. 2020; Azaare et al. 2021) and the type of feedstock (Shen 2020). Mixing ratios significantly impact the yield of biochar where there exists a synergy between the feedstock used in a co-HTC. However, increasing pressure had little to no effect on the HTC process.

Hydrothermal carbonization (HTC) is a thermochemical procedure deployed to convert biomass with high moisture content to be used in various applications. The biomass is heated under pressure for 5–240 min while submerged in water at a temperature between 180 and 350 °C during HTC (Arellano et al. 2016). Other products are liberated during the process; however, the primary output of HTC is a substance called hydrochar. Additionally, it generates gas (mostly CO_2) and liquid (aqueous soluble) by-products (Fiori et al. 2014). The HTC can be made with or without a catalyst. Thus, the procedure can be classed as either Direct HTC or Catalytic HTC. When HTC is performed in the absence of a catalyst, the feedstock is only mixed with water and heated in an enclosed reactor with varying degrees of temperatures. This is the direct HTC process, but catalysts are utilized in a Catalytic HTC method (Nizamuddin et al. 2017). HTC yields solid, liquid, and gaseous products (Libra et al. 2011).

HTC process occurs in the subcritical zone. It is well established that the properties of water alter considerably under subcritical state. Increases in temperature below 374 °C result in a reduction in the dielectric constant, which weakens the hydrogen bonds in water and creates high ionization constants that speed up the dissociation of water into acidic hydronium ions (H_3O^+) and basic hydroxide ions (OH^-) (Marcus 1999). Additionally, subcritical water is a good medium for the acid-catalyzed reaction of organic compounds without the need for additional acid since it has an adequately higher H^+ concentration than liquid water (Ruiz et al. 2013). Water characteristics change substantially under different temperatures. Water provided to the process or found in the biomass acts as a good solvent and reaction media during HTC (Libra et al. 2011).

Several reactions occur during the action, and the exact kind of reaction pathways are yet undetermined. However, hydrolysis has a lower activation energy than most other processes, so hydrolysis may be claimed to govern the HTC (Heidari et al. 2019). Many bonds are broken during the reaction. In the hydrolysis process, the ester and ether bonds of the hemicellulose (at Temperature >180 °C), cellulose (at Temperature >200 °C), and lignin (at Temperature >220 °C) break into several fragments (Titirici & Antonietti 2010). Dehydration and decarboxylation proceed the pyrolysis reaction, which reduces H/C and O/C ratios (Matus et al. 2016). Water can enter the reaction medium due to dehydration. The dehydration of glucose is depicted in Eq. 1

$$C_6H_{12}O_6 - \rightarrow C_6H_4O_2 + 4H_2O \quad (1)$$

Decarboxylation reaction leads to the degradation of carboxyl and carbonyl groups and evolution of CO_2 and CO. Condensation polymerization immediately follows decarboxylation, which involves the participation of some of the fragments of the previous reactions. The combination of molecules occurs, resulting in the simultaneous formation of a larger molecule and the release of a small molecule.

Furthermore, aromatization processes produce aromatic polymer structures that are hydrothermally stable and are used to construct hydrochar (Tekin et al. 2014). It is worth mentioning that the central difficulty in knowing the interaction of the hydrothermal carbonization is that the reaction processes do not happen in sequence, and their order, difficulty, and relations are not revealed yet (Sevilla & Fuertes 2009).

An effective way to slow down the accumulation of waste and reduce its effect on the environment is to convert the waste into a more useful form such as energy. Due to their versatility, lightweight and low cost of production relative to other materials such as metals, concrete, and wood, the dependency on plastics has increased over the years (Al-Salem et al. 2017). PVC, a popular plastic, is widely used in packaging, building, construction, automotive, and agriculture. It is frequently used in construction compared to the usage of polystyrene (PS) and polyethylene (PE), increasing PVC-containing wastes.

Many countries use landfilling and incineration as a way of disposing of plastic waste to reduce expenses. Landfills are not sustainable due to their destruction of the soil structure, difficulty in decomposing, adverse effects on vegetation, and the exceeded occupation of land. Rainwater can penetrate landfill layers and contact the waste plastics containing different addictive, which will result in hydrophilic poisons dissolving in the leachate. As for burning, poisonous chemicals such as dioxins, furan, and heavy metals could be liberated into the atmosphere triggering a devasting effect on the environment, humans, and flora and fauna.

The hydrothermal carbonization (HTC) process can remove organic chlorine from PVC by dissolving it in the water thereby preventing the release of dioxins into the atmosphere (Poerschmann et al., 2015). Although it is efficient to eliminate chlorine from PVC waste by HTC, alkalis are usually needed to enhance the dechlorination efficiency (Huang et al., 2019). Lignocellulose biomass generally includes a high amount of alkali and alkaline earth metals (Shen 2020).

Pontederia crassipes (a common variety of water hyacinth (WH)), an invasive aquatic plant, has a high percentage of lignocellulose and can therefore be a useful source of biomass for producing energy through HTC. WH is a fast-growing aquatic plant, which is projected to cover four (4) million hectares of water bodies (Bote et al. 2020). WH poses a challenge to humans by disrupting transportation and tourism, blocking irrigation systems, impeding drainage, and choking hydroelectric dams. Co-hydrothermal carbonization of PVC and WH will not only ensure the safe disposal of PVC and removal of the nuisance plant but will also lead to the production of high-energy hydrocar, as an efficient waste energy product.

Research has been conducted on the co-HTC of plastics and biomass feedstock. However, this work is novel with the use of waste PVC and WH for the co-HTC process. In this work, high-energy hydrochar was produced from the co-HTC of PVC and WH, varying the temperature, resident time, and mixing ratio of feedstock. The thermal properties of the products were analyzed to ascertain the optimal operating parameters by using the response surface methodology.

2 Materials and Methods

2.1 Equipment Used

For the hydrothermal carbonization, a 100 ml stainless steel (Fig. 1) autoclave reactor was used. It consisted of the Teflon reacting chamber, the autoclave body, treaded cap, the upper and lower disk, and a rob for locking. The autoclave with its content is placed in a Muffle furnace (manufactured by Nabertherm Electronics Company, Germany).

Fig. 1 A 100 ml autoclave reactor used for the co-hydrothermal reaction (authors own pictures)

2.2 Experimental Design

The experiment was designed to determine how modifying the temperature, reaction time, and biomass ratio will affect the chemical and thermal properties of the hydrochar produced by co-HTC. The temperature ranged from 200 to 260 °C, the time from 60 to 120 min, and the biomass ratio from 1:3 to 3:1, respectively. When the boundaries of the parameters were selected, the Box Behnken Design method was used to generate the experimental points. The Design of Experiment software (version 13) was used to determine the number of runs. A total of 17 runs were obtained. Each operating parameter (factors) was set at three levels ($\alpha = \pm 1$), as shown in Table 1.

To achieve the desired output variables and optimize the operating parameters, the Response Surface Methodology (RSM) was used to model the output variables. For a comprehensive evaluation of the co-HTC process, the following output variables were chosen to respond to the changes in the co-HTC reaction conditions: The yield of hydrochar, dechlorination efficiency, higher heating value (HHV) (or calorific value), and energy densification ratio (EDR).

Table 1 Experimental design levels for HTC response surface experiment

	Levels			
Factors	−1	0	1	Unit
Temperature	60	120	260	°C
Residence time	60	90	120	Minute
Mixing ratio	1	2	3	

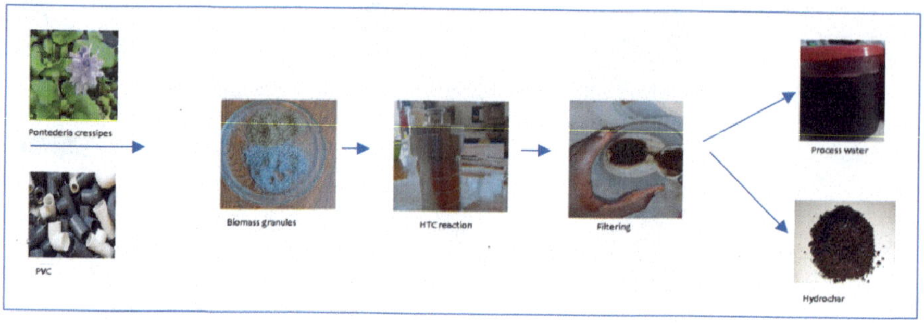

Fig. 2 The flow diagram of the co-hydrothermal carbonization experiment

2.3 Co-carbonization Process

For each treatment, 5 g of the feedstock was loaded into the reactor and 70 ml of distilled water was added. The autoclave reactor used has an inner Teflon volume of 100 ml. The PVC and water hyacinth were in the 1:3, 1:2, and 1:1 ratio. The inner Teflon container mixture was stirred for 5 min to obtain a slurry mixture. When the furnace was heated to the pre-set temperatures of 200, 220, and 260 °C, the reactor was placed in it to allow for the carbonization process for the pre-set time of 60, 90, and 120 min. Once the experiment has relapsed, the autoclave is removed from the furnace and cooled to room temperature using water. The autoclave is opened to allow the gasses to escape, and the mixture is filtered using 0.7 μm filter paper. The hydrochar was dried at 105 °C for 24 h before being grounded and sieved to a particle diameter range of 0–500 μm for further analysis. Each experiment was replicated three times. Figure 2 shows the flow diagram of the experiment.

The hydrochar yield from the Co-hydrothermal carbonization was calculated using Eq. 2.

$$\text{Hydrochar yield} = \frac{\text{Mass of hydrochar (g)}}{\text{Mass of feedstock (g)}} \times 100\% \tag{2}$$

2.4 Energy Content and Energy Densification Calculation

The modified Dulong formula (Eq. 3) was used to calculate the energy content (calorific value) of the feedstocks and hydrochar (Hosokai et al. 2016).

$$CV = 0.382C + 0.849\left(H - \frac{O}{8}\right) \tag{3}$$

where CV = Calorific value of bio-chars (MJ/kg), C, H, and O = mass percentages of Carbon, Hydrogen, and Oxygen, respectively.

Energy densification ratios (EDR) of the hydrochar were calculated from Eq. 12.

$$\mathrm{EDR} = \frac{\text{Energy content of hydrochar}}{\text{Energy content of raw feedstock}} \tag{4}$$

2.5 Dechlorination Efficiency

To investigate dechlorination efficiency, the chlorine concentration in the hydrochar was measured using ion chromatography following the procedure of EN 14,582:2007 (European standard). In an oxygen bomb filled with oxygen, the samples were ignited with ethanol. Organic chlorine was converted into inorganic chlorine during the combustion process and dissolved in a sodium carbonate and sodium bicarbonate absorbing solution. The chlorine in the liquid phase was then measured using ion chromatography. Equation 5 was used to calculate the dechlorination efficiency.

$$\mathrm{DE} = \left(1 - \frac{Mass\ of\ Chlorine\ in\ Hydrochar}{Mass\ of\ Chlorine\ in\ Feedstock}\right) \times 100\% \tag{5}$$

2.6 Analysis of Functional Group

Using a Fourier transform infrared (FT-IR) spectrometer, the functional groups of raw feedstocks and hydrochar were determined. The FT-IR spectrum was scanned using 24 scans at a spectral resolution of 4 cm^{-1} between 4000 cm^{-1} and 500 cm^{-1}.

2.7 Thermogravimetric Process

Microcomputer differential thermal (Hengjiu, Model HCT-4, Beijing China) balance was used to determine the combustibility of the various samples. The sample was dried for 24 h at 105 °C before being ground to less than 74 mm. In the experiment, 5 mg of the sample were precisely weighed, added to the sample crucible, and then placed on a differential thermobalance with a heating rate of 20 °C/min.

2.8 Statistical Analysis

The experimental data were subjected to analysis of variance (ANOVA) to analyze the statistically significant difference between test results using the Design Expert v13 software. The analysis performed were based on sum of squares due to error (SSE), R2, adjusted R2, and root mean squared error (RMSE) to validate the results. The level of significance was determined when the p value was less than 0.05 (i.e., $p < 0.05$). The normal plot of residuals and the residuals versus predicted responses as displayed in Fig. 3 were used for assessing the experimental errors.

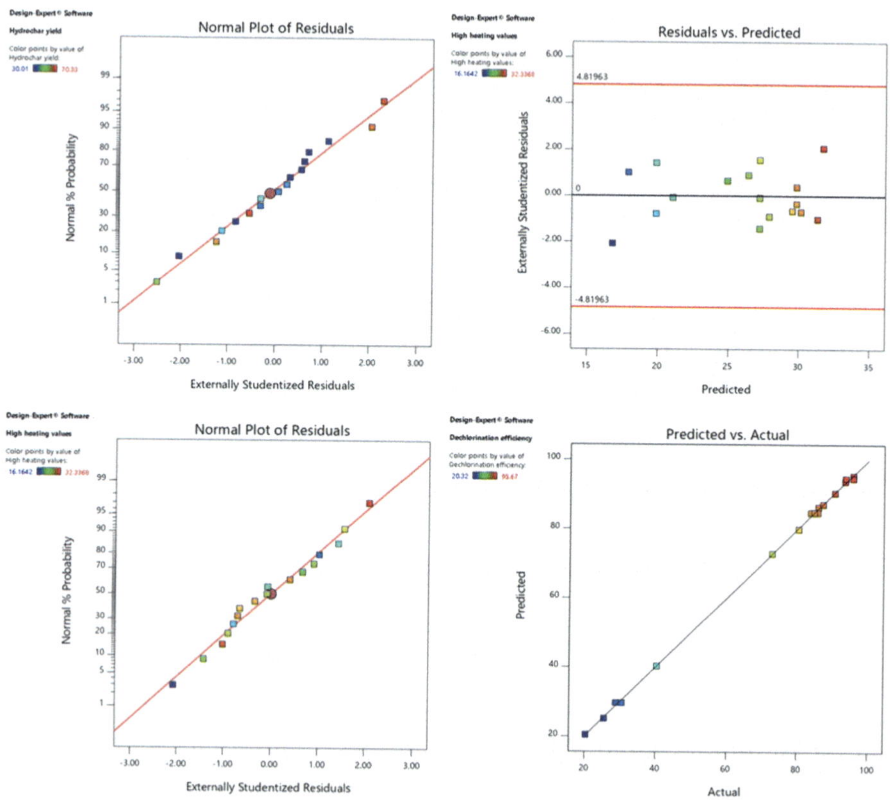

Fig. 3 **a** Normal plots of residuals, **b** Residuals against predicted values (HHV), **c** Normal plots of residuals (HHV), **d** Plot of predicted values against actual values (DE)

3 Results and Discussion

3.1 Optimization of Output Variables

The Design Expert software (Version 13) was used for the optimization process, model the output variable, and to undertake the analysis of variance, using the response surface method. The Box Behnken Method was used to generate the experimental runs used for iteration of the responses.

3.2 Effects of Process Parameters on Output Variables

Effect of Process Parameters on Dechlorination.

Table 2 shows the various varied combinations of the process parameters and their corresponding hydrochar that were obtained. Figure 4 shows the experimental errors of the data obtained which are depicted by the normal plot of residuals and the residuals against predicted responses. The normal plot of residuals of the variables are distributed in a straight line.

Table 2 Reaction parameters and output variables of the experiments

Sample	Hydrochar yield (%)	Dechlorination (%)	High heating value (MJ/Kg)
200-60-2	70.33	20.32	16.16
260-60-2	31.34	93.66	30.05
200-120-2	55.76	40.32	20.99
260-120-2	30.01	95.67	32.33
200-90-1	67.32	28.78	20.58
260-90-1	30.12	94.39	30.95
200-90-3	65.48	25.54	18.26
260-90-3	30.12	90.54	29.86
230-60-1	40.34	80.33	26.70
230-120-3	33.57	87.22	29.26
230-60-3	38.44	72.76	25.13
230-120-3	32.69	85.87	27.51
260-60-2	31.25	93.54	29.62
200-90-1	64.21	30.46	19.42
230-90-2	35.66	84.69	26.38
230-90-2	36.11	83.77	27.99
230-90-2	36.32	85.69	27.11

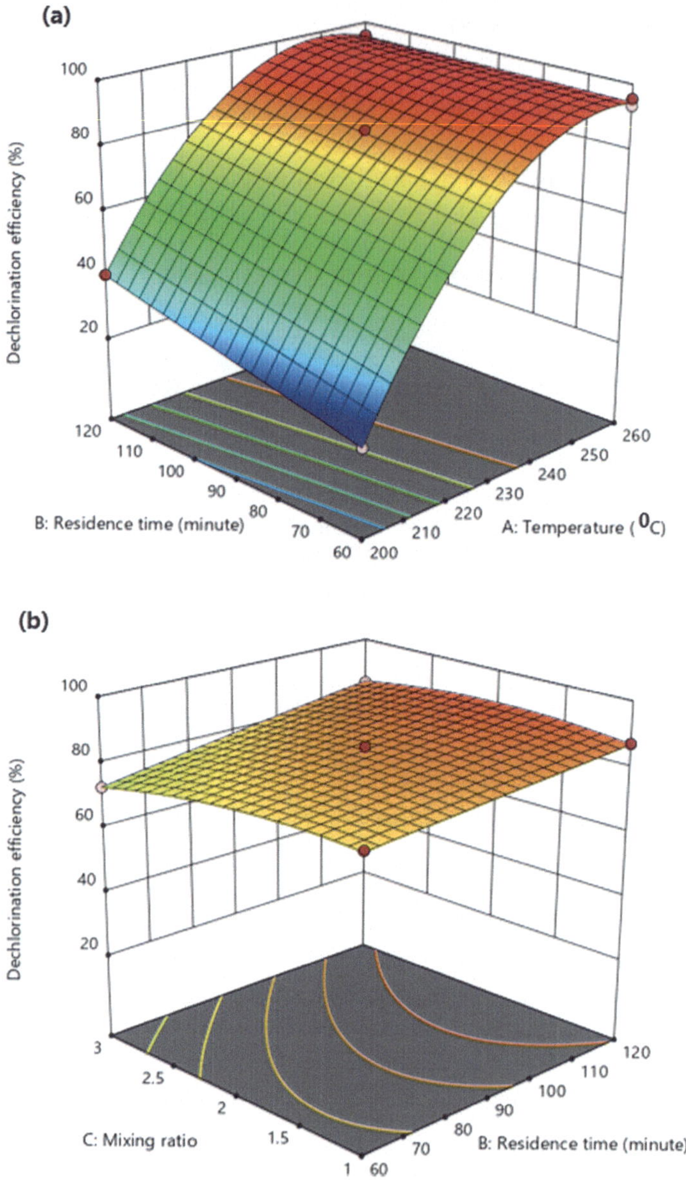

Fig. 4 Response surface plot of Dechlorination efficiency showing, **a** effect of reaction temperature and residence time; **b** effect of mixing ratio and residence time; **c** effect of mixing ratio and temperature

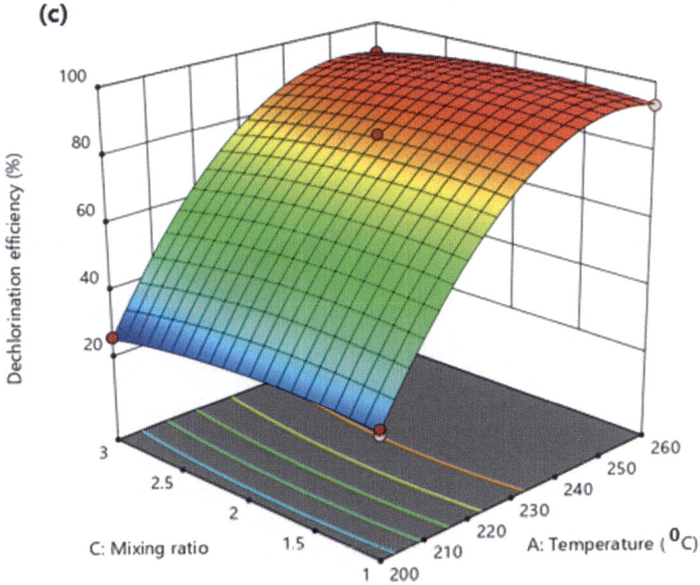

Fig. 4 (continued)

Figure 4 shows the effect of temperature, residence time, and mixing ratio on dechlorination efficiency. The plots of residuals against predicted responses were intermittently distributed over and beneath the x-axis as displayed in Fig. 5, signifying that the fitting of experimental results by the proposed models are within acceptable level (Kang et al. 2019).

The model that describes the effect of the parameters on DE is given as:

$$\begin{aligned} &\textit{Dechlorination efficiency} = \\ &\quad -1569.49427 + 12.72064\,\textit{Temperature} + 1.29576\,\textit{Residence time} \\ &\quad +3.44809\,\textit{Mixing ratio} - 0.005253\,\textit{Temperature} * \textit{Residence time} \\ &\quad +0.009565\,\textit{Temperature} * \textit{Mixing ratio} \\ &\quad +0.051833\,\textit{Residence time} * \textit{Mixing ratio} \\ &\quad -0.024326\,\textit{Temperature}^2 - 0.000112\,\textit{Residence time}^2 \\ &\quad -3.07100\,\textit{Mixing ratio}^2 \end{aligned}$$

The results of the analysis of variance (ANOVA) of the output variables are listed in Table 3. The analysis of the effect of operating parameters on dechlorination shows that temperature ($p < 0.0001$), residence time ($p < 0.0001$) and mixing ratio ($p < 0.0004$) has a significant effect on dichlorination. The response surface graph was visualized in Fig. 5. The temperature had a positive influence on removing chlorine from PVC. The lowest DE 20.32% was recorded at 200 60 2. As the temperature was increased to 230 °C, the DE was increased more than four times to 85.45%. The

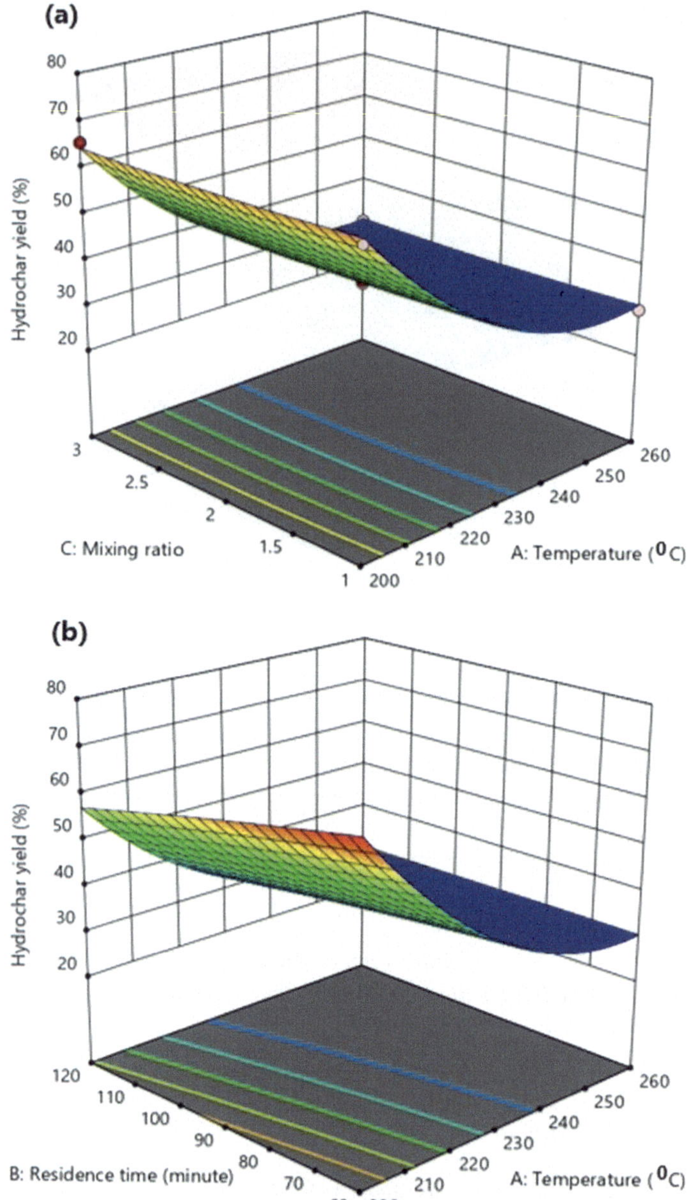

Fig. 5 Response surface plot of hydrochar yield showing, **a** effect of reaction temperature and residence time; **b** effect of mixing ratio and residence time; **c** effect of mixing ratio and temperature

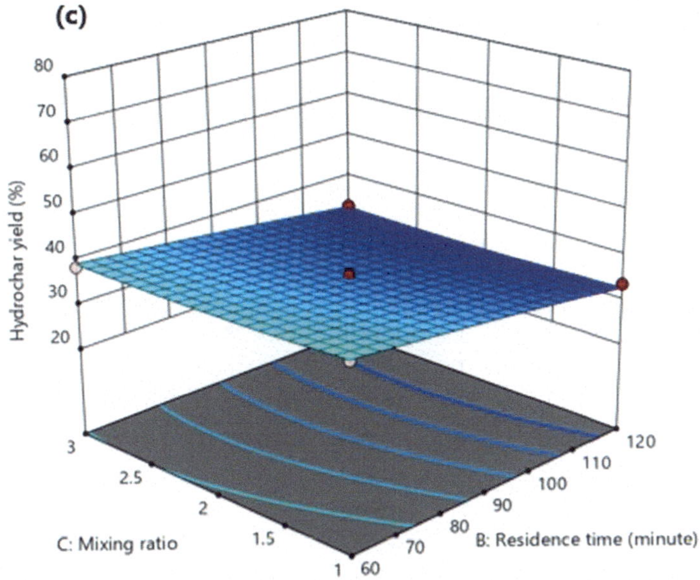

Fig. 5 (continued)

highest DE, 95.69% was recorded at 260 °C. There was, however, a slight increase in the DE as temperatures were increased from 230 to 260 °C. Dechlorination took place in two stages: elimination and substitution. Chlorine and hydrogen are removed from the two carbons in the vinyl structure by dehydrochlorination through the zipper mechanism to generate HCl. The hydroxyl (-OH) groups replaced the chlorine atoms on the polymer chain to produce C–OH groups (R2), which then reacted with oxygen atoms to form carboxylic acid (R^3). If two hydroxyl (-OH) groups are substituted, HCl can be produced by intramolecular and intermolecular dehydration between these unstable intermediates (Zhao et al. 2017).

As a result, both elimination and substitution could produce acid substances, creating an acidic environment for inorganics removal. Cellulosic materials were decomposed during HTC and hydrolyzed to produce 5-hydroxymethyl furfural or alcohol with a large free OH bond. Substances containing free OH will therefore improve the substitution route of the PVC reaction (Yao & Ma 2018).

Residence time also had a significant influence in degrading PVC during the co-hydrothermal carbonization process. The dichlorination efficiency almost doubled from 20.32% at 200 °C to 40.32% when the time was doubled from 60 to 120 min. When the time was increased to 90 min at the same temperature, DE increased to 30.46%. However, at higher temperatures, the effects of residence time on DE were minimal. At 260 °C, the DE increased from 93.66%, 94.39%, and 95.67% when the time was increased from 60, 90, and 120 min, respectively. This means increasing both temperature and resident time can improve the dechlorination process in the co-HTC.

Table 3 ANOVA table of output variables

Source	Sum of squares	df	Mean square	F-value	p-value	
High heating values						
Model	378.42	9	42.05	81.37	<0.0001	Significant
A-Temperature	324.39	1	324.39	627.77	<0.0001	
B-Residence time	20.25	1	20.25	39.19	0.0004	
C-Mixing ratio	5.33	1	5.33	10.31	0.0148	
AB	1.61	1	1.61	3.11	0.1211	
AC	0.1828	1	0.1828	0.3538	0.5707	
BC	0.0088	1	0.0088	0.0170	0.8999	
A^2	21.70	1	21.70	42.00	0.0003	
B^2	0.0004	1	0.0004	0.0007	0.9790	
C^2	0.0025	1	0.0025	0.0048	0.9465	
Residual	3.62	7	0.5167			
Lack of fit	1.54	3	0.5141	0.9910	0.4821	Not significant
Dechlorination efficiency						
Model	12,735.53	9	1415.06	1586.03	<0.0001	Significant
A-Temperature	9634.94	1	9634.94	10,799.08	<0.0001	
B-Residence time	227.38	1	227.38	254.86	<0.0001	
C-Mixing ratio	33.48	1	33.48	37.53	0.0005	
AB	104.38	1	104.38	116.99	<0.0001	
AC	0.3846	1	0.3846	0.4310	0.5325	
BC	9.67	1	9.67	10.84	0.0133	
A^2	1905.13	1	1905.13	2135.31	<0.0001	
B^2	0.0399	1	0.0399	0.0447	0.8386	
C^2	37.11	1	37.11	41.59	0.0004	
Residual	6.25	7	0.8922			
Lack of fit	0.7427	3	0.2476	0.1800	0.9048	Not significant
Hydrochar yield						
Model	3586.75	9	398.53	213.58	<0.0001	Significant
A-Temperature	2644.26	1	2644.26	1417.13	<0.0001	
B-Residence time	113.88	1	113.88	61.03	0.0001	
C-Mixing ratio	1.54	1	1.54	0.8252	0.3939	
AB	46.20	1	46.20	24.76	0.0016	
AC	0.1056	1	0.1056	0.0566	0.8188	
BC	0.2601	1	0.2601	0.1394	0.7199	
A^2	511.35	1	511.35	274.05	<0.0001	

(continued)

Table 3 (continued)

Source	Sum of squares	df	Mean square	F-value	p-value	
B^2	0.4756	1	0.4756	0.2549	0.6292	
C^2	1.31	1	1.31	0.7036	0.4293	
Residual	13.06	7	1.87			
Lack of fit	7.99	3	2.66	2.10	0.2426	Not significant

3.3 Effect of Process Parameters on Yield of Hydrochar

PVC and water hyacinth in suspension with water were converted to hydrochar by co-HTC. Table 1 shows the output of the hydrochar yield at different operating parameters. Before analyzing the exact outcomes, the normal residual plots (Appendix 1) and residuals against predicted responses (Appendix 2) were used to assess the experimental errors. They have a linear relationship with minor variation. These evaluations show that the model fits the experimental data well. As a result, the developed models can be considered adequate for predicting and optimizing the biochar fuel production process, signifying that the residuals are normally distributed and thus confirming their normality (Anupam et al. 2016). The model expressed in terms of actual factors for the yield of hydrochar is given as:

$$
\begin{aligned}
&\textit{Hydrochar yield} = \\
&+919.58984 - 6.68607\,\textit{Temperature} - 0.872349\,\textit{Residence time} \\
&-4.65110\,\textit{Mixing ratio} + 0.003494\,\textit{Temperature} * \textit{Residence time} \\
&+0.005012\,\textit{Temperature} * \textit{Mixing ratio} + 0.00800\,\textit{Residence time} * \textit{Mixing ratio} \\
&+0.012603\,\textit{Temperature}^2 - 0.000386\,\textit{Residence time}^2 + 0.0577677\,\textit{Mixing ratio}
\end{aligned}
$$

Considering the ANOVA analysis of the variables affecting the yield of hydrochar, temperature and residence time were the most prevailing factors that affected the yield of hydrochar, both with p values <0.0001. This is visualized in the response surface plot in Fig. 6. The temperature had a downward trend with the yield of hydrochar. The highest yield of hydrochar (70.33%) occurred at 200 °C 60 min and mixing ratio of 2:1. As the temperature was increased to 230, the yield of hydrochar decreased to 40.34%, 36.11%, and 32.69 at 60-, 90-, and 120-min residence times, respectively. It shows that lower temperatures favor high hydrochar yield. Temperature's primary function is to provide heat for disintegration for the fragmentation of the biomass bond (Nizamuddin et al., 2017). A more considerable percentage of the feedstock composition is putrefied when the carbonization temperature is increased, which results in a reduction in the mass yield of the hydrochar. Hemicellulose, cellulose, and lignin in WH, were carbonized at HTC temperatures 180, 220, and 250 °C, respectively. Hemicellulose is more unstable than cellulose and, as a result, degrades more easily into its monomers when heated. The hemicellulose monomers are xylose, mannose, glucose, and galactose. During the co-HTC process, xylose dissolves in the water as a pyranose ring, forming furfural (Tekin et al. 2014). In a hydrothermal environment, as the density of water increases, so does hydrolysis, causing an increase in the decomposition of lignin to its low molecular weight components. Eventually, with the continual elevation of the reaction temperature to 250 °C, the lignin component of the biomass began hydrolyzing, forming phenolic substances such as catechol and phenols.

3.4 Effect of Process Parameters on High Heating Value

For consideration as a fuel, the heating values of hydrochar are an important parameter. A high heating value indicates a high carbon content and the removal of volatile matter. The HHV was increased from 16.16 MJ/kg at the temperature of 200 °C, 60 min and a ratio of 2, to 32.34 MJ/Kg. This is an enhancement of the raw feedstocks which have HHV of 11.2 MJ/Kg and 13.4 MJ/Kg for PVC and WH, respectively. Table 3 shows that increasing temperatures and resident time had a positive influence on the HHV of the hydrochar produced with p-values of <0.0001. The model that describes the relationship between the actual parameters and the HHV is given as:

$$
\begin{aligned}
\textit{High heating values} = & -169.41704 + 1.43745\,\textit{Temperature} \\
& +0.202189\,\textit{Residence time} - 2.06156\,\textit{Mixing ratio} \\
& -0.000652\,\textit{Temperature} * \textit{Residence time} + 0.006594\,\textit{Temperature} * \textit{Mixing ratio} \\
& -0.001562\,\textit{Residence time} * \textit{Mixing ratio} - 0.002596\,\textit{Temperature}^2 \\
& +0.000011\,\textit{Residence time}^2 - 0.325188\,\textit{Mixing ratio}^2
\end{aligned}
$$

3.5 FTIR Analysis

Fourier Transform Infrared Radioscopy (FTIR) analysis was used to investigate the chemical functional groups present in the hydrochar at various temperatures as well as the feedstock (PVC and WH) to better comprehend the chemical changes that transpire during co-HTC. Figure 7 shows the FTIR spectrum of the hydrochar and feedstocks. As shown in Fig. 7, the FTIR spectra of untreated WH and PVC exhibit distinct features, designating that the two raw materials were characterized by different chemical compounds. The adsorption peak that occurred between 600 and 900 cm^{-1} for untreated PVC can be attributed to C–Cl stretching in alkyl halide groups of PVC chains (Wei et al. 2022). The other peaks happening at 931 cm^{-1}, 950–1200 cm^{-1}, 1252 cm^{-1}, 1400–1700 cm^{-1}, and 2800–3000 cm^{-1}, were assigned to the vibrations of trans-CH wiggling, C-H stretching, CH_2 deformation, C = O in aromatic groups, and sp^3 C-H stretching in CH_n groups, respectively. Idiosyncratic bands at

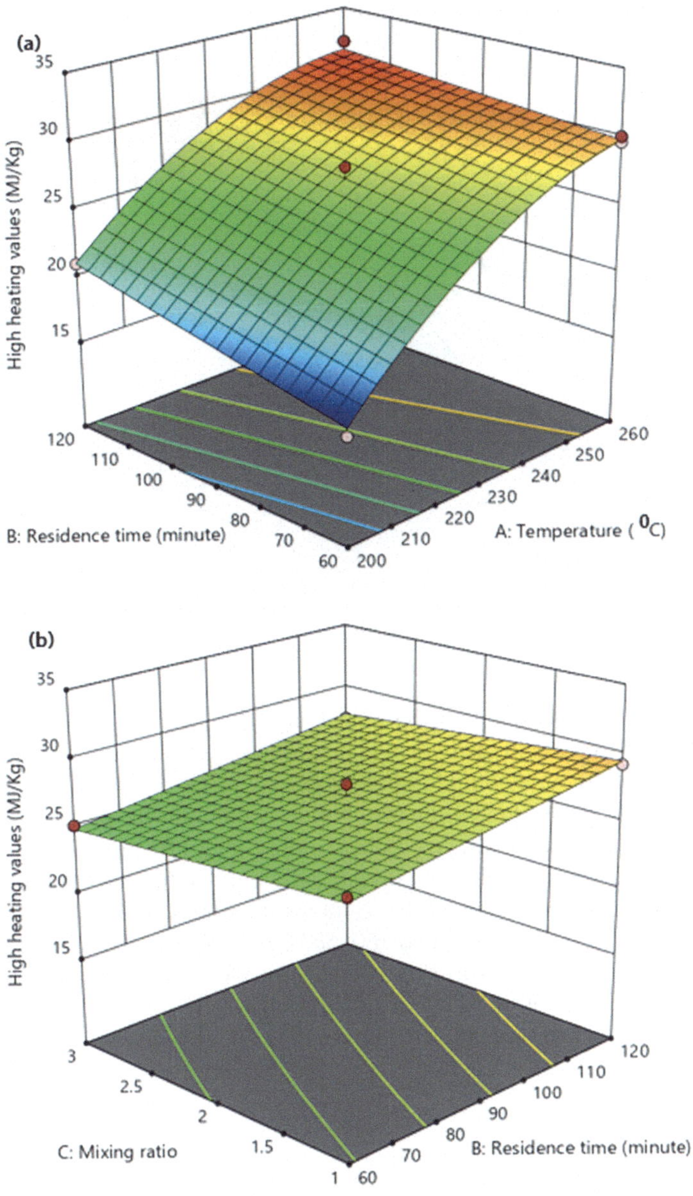

Fig. 6 Response surface plot of High Heating showing, **a** effect of reaction temperature and residence time; **b** effect of mixing ratio and residence time; **c** effect of mixing ratio and temperature

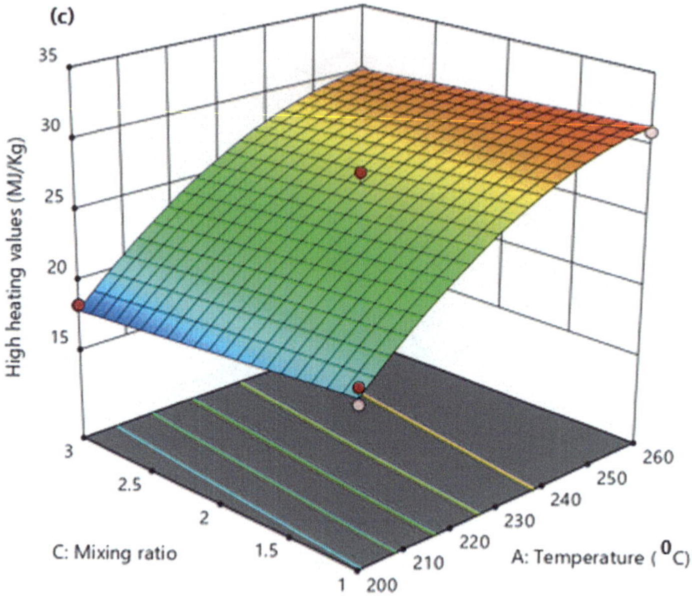

Fig. 6 (continued)

3700–3000 cm^{-1} in addition to a sharp peak at 1050 cm^{-1}, which are allocated to –OH stretching vibration in the hydroxyl or carboxyl group, could be distinguished in the water hyacinth and the hydrochar samples. Correspondingly, the 3500–2850 cm^{-1} band signifying –CH vibration in polysaccharide could be detected in the water hyacinth but it is reduced in the hydrochar sample. The 1700 cm^{-1} band presented a C–O group in carbonyl, quinine, ester, or carboxyl residual on the surface of the dried water hyacinth and the hydrochar samples (Laginhas et al. 2016).

The bands at 1331–1254 cm^{-1} and 760–505 cm^{-1} were attributed to the presence of –C–H and –C–Cl (in the form of –CHCl–) in PVC chains (Yao & Ma 2018). It was clearly visible that these bands in hydrochar were removed in 260 120 0.5, indicating an excellent temperature dechlorinating effect of co-HTC via substitution and elimination reactions.

3.6 *Thermographic Analysis*

Thermal gravimetric analysis (TGA) evaluates weight changes in a substance under temperature control in a controlled environment. The decomposition behavior of the biomass was assessed using a thermogravimetry analyzer (TGA). TGA was conducted using 10 mL/min of air and a heating rate of 10 °C/min. The thermogravimetry (TG) and differential thermogravimetry (DTG) curves in Fig. 8 depict

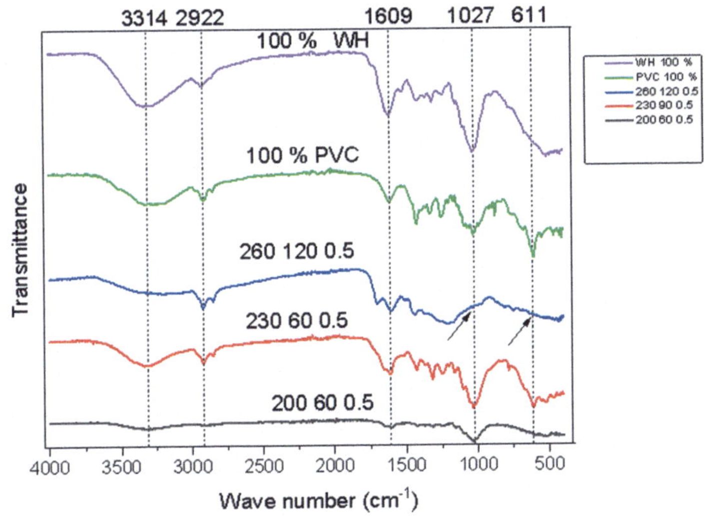

Fig. 7 FT-IR graph for PVC, WH, and Hydrochar produced at 200, 230, and 260 °C

the thermal degradation characteristics of the feedstock and hydrochar. The decomposing profile of the PVC, WH, and hydrochar obtained at 200, 230, and 260 °C are displayed in Fig. 8. The combustion of the PVC occurred in three main stages. The first stage occurred in the temperature range of 280–400 °C. This combustion profile is because of the oxidation of light volatile matter in the PVC (Zhang et al., 2020). These materials include addictive and thickeners that were added to the PVC during synthesis. These materials burn at lower temperatures and are therefore liberated in the first combustion stage. The second stage of combustion occurred at the temperature range of 420–520 °C. This is due to the oxidation of heavy volatile matter in the PVC. The liberation of some fixed carbon occurred within 550–600 °C.

The combustion of WH preceded in three stages, as indicated by the three prominent peaks in Fig. 8. The first stage (25–130 °C) resulted from the removal of moisture and volatile matter from WH. The second combustion stage occurred at the temperature range of 160–400 °C. This stage represents the decomposing of cellulose and hemicellulose. Much of the weight loss in HW occurred in this stage since cellulose and hemicellulose are major components of WH. The degradation of lignin happened in the temperature series of 450–550 °C. Char and inorganic formation from the WH occurred at temperatures beyond 600 °C (Gao et al. 2013).

The TG and TGA curves that were obtained for the hydrochar were similar except for the hydrochar that was produced at the carbonization temperature of 260 °C, which varies slightly. The hydrochar had already been treated at lower temperatures, which resulted in the removal of the most volatile materials. As a result, lower peaks were registered at temperatures below 150 °C. Three prominent peaks were registered for 200 60 0.5. These peaks are at temperatures of 300, 400, and 460 °C. The peak at 300 °C shows incomplete hydrolysis of cellulose and hemicellulose at a lower resident time and a faster decomposition at a lower temperature (Gao et al. 2013).

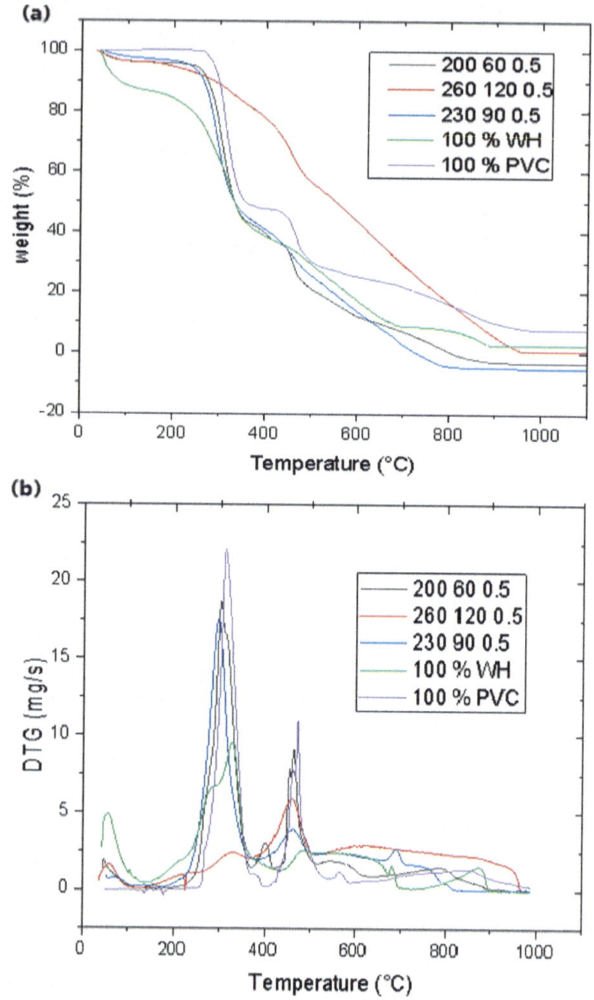

Fig. 8 TGA graph showing the thermal decomposition of raw PVC and WH, and the hydrochars obtained at 200, 230, and 260 °C, **b** DTG graph showing peaks of elimination

Co-dehydration hydrochar peaks were weaker than that of PVC and raw WH. Furthermore, it weakened with HTC temperature, indicating that the co-HTC process could effectively remove inherent moisture from biomass and inhibit moisture re-adsorption (Zhao et al. 2021).

The H/C and O/C ratios are significant criteria to assess the degree of deoxygenation, and the aromatic content during HTC of biomass (Tekin et al. 2014). The raw feedstock and hydrochar H/C and O/C atomic ratios were displayed on the Van Krevelen diagram, as illustrated in Fig. 9. The findings suggest that dehydration is the primary process of coalification of the feedstock, with the O/C and H/C atomic ratios falling with increasing temperature and residence time and the fuel approaching more lignite in nature. The carbon content of hydrochar rose after the hydrothermal process, but the oxygen level fell significantly. Decarboxylation, dehydration, and

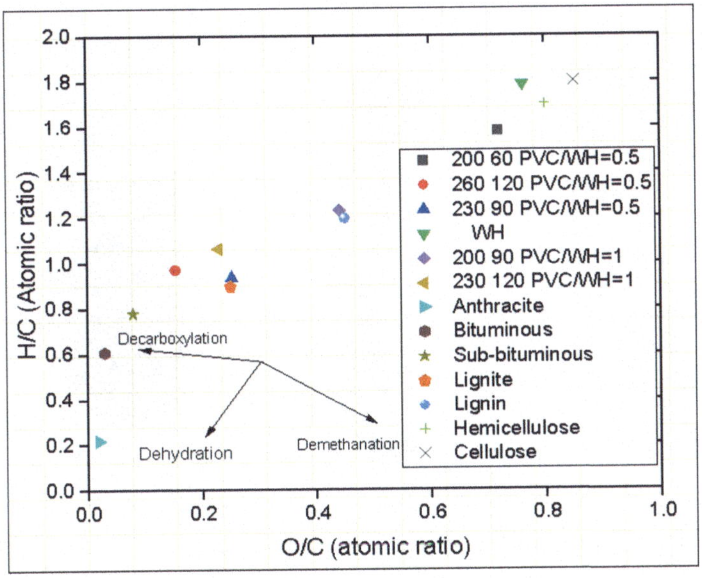

Fig. 9 Van Krevelen diagram comparing the hydrochars to some fuel standards

demethanation reactions were implicated in the reaction pathways (Zhang et al. 2019). The energy densification of hydrochar appears to be caused by changes in the carbon/oxygen (O/C) ratio in the fuel, with the carbon content appearing to grow while the oxygen and ash content appears to decrease. As a result, dehydration was an important response route. At elevated temperatures, the result was more likely to move to the bottom left corner.

3.7 *Physical Characteristics of Hydrochar and Process Water*

The pH and electrical conductivity of (E.C) the feedwater/process water were recorded with a pH meter. The results show a decline in the pH of the process water and an increase in their corresponding E.C. Organic acid formation are important intermediates in the HTC process to catalyze the decomposition of biomacromolecules and hydrochar formation (Wang et al. 2018). The increase in pH is the result of the formation of formic, acetic, lactic, and levulinic acid from biomass.

The color changes of the process water (Fig. 10) also confirmed the changes in the chemical composition of the WH and PVC. Temperature increases result in breaking the bonds in the chemical structure of PVC and the biomass resulting in the release of free radicals into the water, which serves as the solvent. Therefore, the cleavage of the bonds depended heavily on the temperature, making the process water becoming darker as the temperature increased.

Fig. 10 Color changes of process water over temperature changes

4 Conclusion

The main factors that affected the yield of the solid fuel produced by Co-HTC were temperature and residence time. The mixing ratios of the feedstocks had a minimal impact on the yield. Even though a high yield of solid fuel was produced at lower temperatures (180 °C), fuels produced at higher temperatures had high heating values and better combustion characteristics. At high temperatures, most of the volatile and inorganic matter that interferes with burning leading to the production of particulate matter which is undesirable in incinerators was drastically reduced.

High temperatures favored the removal of inorganic chlorine from the PVC. The highest dechlorination efficiency, 94.39% was recorded at a high co-HTC temperature of 260 °C. The chlorine liberated from the PVC structure cleaved with free hydronium ions (+H) that were produced from the ionic dissociation of water molecules, to form hydrochloric acid (HCl). The HCl is soluble in water and remained in the filtrate when the resulting mixtures were filtered and washed.

The thermogravimetric analysis of the solid fuel produced at various temperatures showed that the thermal properties of the fuels were enhanced as the temperatures of the reaction were increased. This was illustrated by the peaks in the TGA and DSC graphs in Fig. 8. As the inorganics and volatile matter were removed from the fuel, the intensities of the peaks were significantly reduced, with some vanishing completely which showed the elimination of specific inorganic material (especially Chlorine).

According to the Fourier Transform Infrared (FT-IR) microscopy, the reactions that ensued in the Co-HTC process were hydrolysis, dehydration and decarboxylation, condensation and polymerization, and aromatization. These results show that temperature played a major role in the coalification of the feedstocks into useful fuels. These results were also confirmed by the reduction in the H/C and O/C ratios when the temperatures of the process were increased, as illustrated in the Van Krevelen diagram in Fig. 9.

Appendix 1

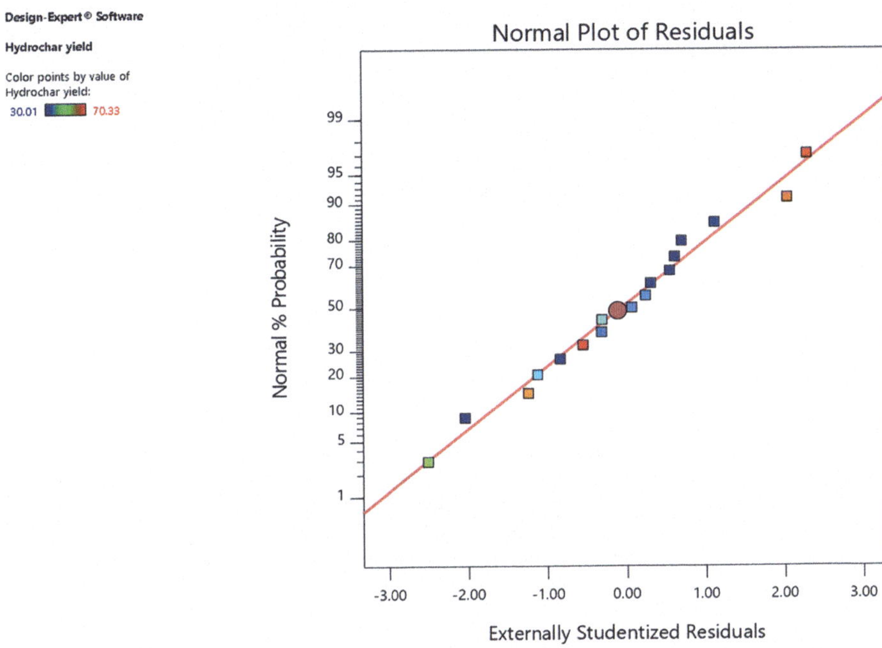

Appendix 2

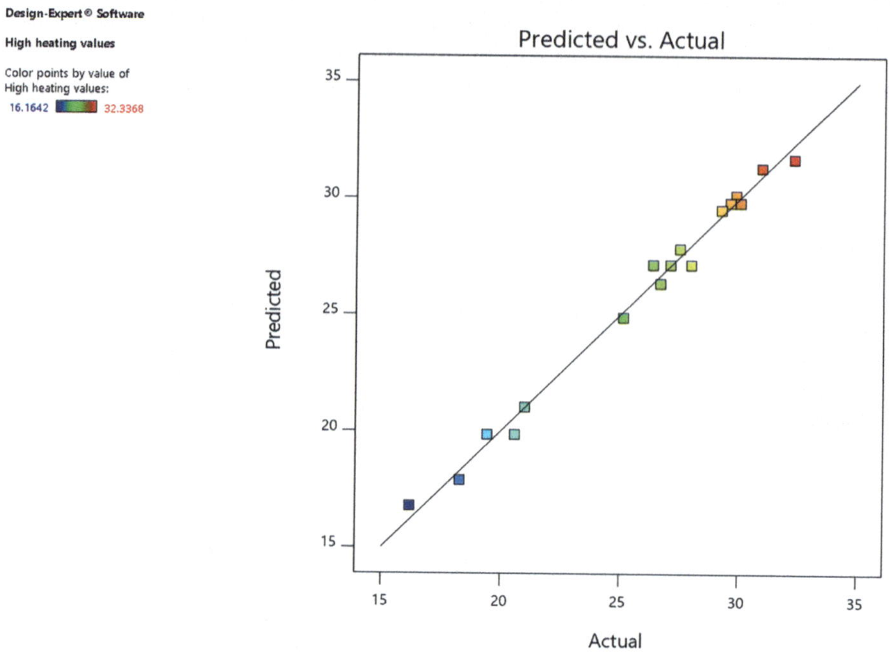

References

Al-Salem, S. M., Antelava, A., Constantinou, A., Manos, G., & Dutta, A. (2017). A review on thermal and catalytic pyrolysis of plastic solid waste (PSW). Journal of Environmental Management, 197(1408), 177–198. https://doi.org/10.1016/j.jenvman.2017.03.084

Anupam, K., Sharma, A. K., Lal, P. S., Dutta, S., & Maity, S. (2016). Preparation, characterization and optimization for upgrading Leucaena leucocephala bark to biochar fuel with high energy yielding. Energy, 106, 743–756. https://doi.org/10.1016/j.energy.2016.03.100

Arellano, O., Flores, M., Guerra, J., Hidalgo, A., Rojas, D., & Strubinger, A. (2016). Hydrothermal carbonization of corncob and characterization of the obtained hydrochar. Chemical Engineering Transactions, 50(ii), 235–240. https://doi.org/10.3303/CET1650040

Azaare, L., Kweku, M., Mark, A., & Kemausuor, F. (2021). Bioresource Technology Reports Co-hydrothermal carbonization of pineapple and watermelon peels : Effects of process parameters on hydrochar yield and energy content. Bioresource Technology Reports, 15(May), 100720. https://doi.org/10.1016/j.biteb.2021.100720

Bote, M. A., Naik, V. R., & Jagadeeshgouda, K. B. (2020). Materials Science for Energy Technologies Review on water hyacinth weed as a potential biofuel crop to meet collective energy needs. Materials Science for Energy Technologies, 3, 397–406. https://doi.org/10.1016/j.mset.2020.02.003

Du, B., Yu, Z., Tian, Y., & Ma, X. (2021). Effects of baking soda on Co-hydrothermal carbonization of sewage sludge and Chlorella vulgaris: Improved the environmental friendliness of hydrochar incineration process. Journal of Environmental Chemical Engineering, 9(6), 106404. https://doi.org/10.1016/j.jece.2021.106404

Fiori, L., Basso, D., Castello, D., & Baratieri, M. (2014). Hydrothermal carbonization of biomass: Design of a batch reactor and preliminary experimental results. Chemical Engineering Transactions, 37, 55–60. https://doi.org/10.3303/CET1437010

Gao, Y., Wang, X., Wang, J., Li, X., Cheng, J., Yang, H., & Chen, H. (2013). Effect of residence time on chemical and structural properties of hydrochar obtained by hydrothermal carbonization of water hyacinth. Energy, 58, 376–383. https://doi.org/10.1016/j.energy.2013.06.023

Huang, N., Zhao, P., Ghosh, S., & Fedyukhin, A. (2019). Co-hydrothermal carbonization of polyvinyl chloride and moist biomass to remove chlorine and inorganics for clean fuel production. Applied Energy, 240(February), 882–892. https://doi.org/10.1016/j.apenergy.2019.02.050

Heidari, M., Dutta, A., Acharya, B., & Mahmud, S. (2019). A review of the current knowledge and challenges of hydrothermal carbonization for biomass conversion. Journal of the Energy Institute, 92(6), 1779–1799. https://doi.org/10.1016/j.joei.2018.12.003

Hosokai, S., Matsuoka, K., Kuramoto, K., & Suzuki, Y. (2016). Modification of Dulong's formula to estimate heating value of gas, liquid and solid fuels. Fuel Processing Technology, 152, 399–405. https://doi.org/10.1016/j.fuproc.2016.06.040

Kang, K., Nanda, S., Sun, G., Qiu, L., Gu, Y., Zhang, T., Zhu, M., & Sun, R. (2019). Microwave-assisted hydrothermal carbonization of corn stalk for solid biofuel production: Optimization of process parameters and characterization of hydrochar. Energy 186, 115795.

Laginhas, C., Nabais, J. M. V., & Titirici, M. M. (2016). Activated carbons with high nitrogen content by a combination of hydrothermal carbonization with activation. Microporous and Mesoporous Materials, 226, 125–132. https://doi.org/10.1016/j.micromeso.2015.12.047

Libra, J. A., Ro, K. S., Kammann, C., Funke, A., Berge, N. D., Neubauer, Y., Ti2rici, M. M., Fühner, C., Bens, O., Kern, J., & Emmerich, K. H. (2011). Hydrothermal carboniza2on of biomass residuals: A compara2ve review of the chemistry, processes and applica2ons of wet and dry pyrolysis. Biofuels, 2(1), 71–106. hAps://doi.org/10.4155/bfs.10.81

Marcus, Y. (1999). On transport properties of hot liquid and supercritical water and their relationship to the hydrogen bonding. Fluid Phase Equilibria, 164(1), 131–142. https://doi.org/10.1016/S0378-3812(99)00244-7

Ning, X., Dang, H., Xu, R., Wang, G., Zhang, J., Zhang, N., & Wang, C. (2022). Co-hydrothermal carbonization of biomass and PVC for clean blast furnace injection fuel production : Experiment and DFT calculation. Renewable Energy, 187, 156–168. https://doi.org/10.1016/j.renene.2022.01.082

Nizamuddin, S., Baloch, H. A., Griffin, G. J., Mubarak, N. M., Bhutto, A. W., Abro, R., Mazari, S. A., & Ali, B. S. (2017). An overview of effect of process parameters on hydrothermal carbonization of biomass. Renewable and Sustainable Energy Reviews, 73(December 2015), 1289–1299. https://doi.org/10.1016/j.rser.2016.12.122

Poerschmann, J., Weiner, B., Woszidlo, S., Koehler, R., & Kopinke, F. (2015). Chemosphere Hydrothermal carbonization of poly (vinyl chloride). Chemosphere, 119, 682–689. https://doi.org/10.1016/j.chemosphere.2014.07.058

Ruiz, H. A., Rodríguez-Jasso, R. M., Fernandes, B. D., Vicente, A. A., & Teixeira, J. A. (2013). Hydrothermal processing, as an alternative for upgrading agriculture residues and marine biomass according to the biorefinery concept: A review. Renewable and Sustainable Energy Reviews, 21, 35–51. https://doi.org/10.1016/j.rser.2012.11.069

Sevilla, M., & Fuertes, A. B. (2009). The production of carbon materials by hydrothermal carbonization of cellulose. Carbon, 47(9), 2281–2289. https://doi.org/10.1016/j.carbon.2009.04.026

Shen, Y. (2020). Biomass and Bioenergy A review on hydrothermal carbonization of biomass and plastic wastes to energy products. Biomass and Bioenergy, 134(August 2019), 105479. https://doi.org/10.1016/j.biombioe.2020.105479

Tekin, K., Karagöz, S., & Bektaş, S. (2014). A review of hydrothermal biomass processing. Renewable and Sustainable Energy Reviews, 40, 673–687. https://doi.org/10.1016/j.rser.2014.07.216

Titirici, M. M., & Antonietti, M. (2010). Chemistry and materials options of sustainable carbon materials made by hydrothermal carbonization. Chemical Society Reviews, 39(1), 103–116. https://doi.org/10.1039/b819318p

Tu, W., Liu, Y., Xie, Z., Chen, M., Ma, L., Du, G., & Zhu, M. (2021). A novel activation-hydrochar via hydrothermal carbonization and KOH activation of sewage sludge and coconut shell for biomass wastes: Preparation, characterization and adsorption properties. Journal of Colloid and Interface Science, 593, 390–407. https://doi.org/10.1016/j.jcis.2021.02.133

Wang, T., Zhai, Y., Zhu, Y., Li, C., & Zeng, G. (2018). A review of the hydrothermal carbonization of biomass waste for hydrochar formation: Process conditions, fundamentals, and physicochemical properties. Renewable and Sustainable Energy Reviews, 90(February), 223–247. https://doi.org/10.1016/j.rser.2018.03.071

Wei, Y., Fakudze, S., Zhang, Y., Ma, R., & Shang, Q. (2022). Co-hydrothermal carbonization of pomelo peel and PVC for production of hydrochar pellets with enhanced fuel properties and dechlorination. 239. https://doi.org/10.1016/j.energy.2021.122350

Yadav, P., & Reddy, S. N. (2020). Bioresource Technology Hydrothermal liquefaction of Fe-impregnated water hyacinth for generation of liquid bio-fuels and nano Fe carbon hybrids. Bioresource Technology, 313(May), 123691. https://doi.org/10.1016/j.biortech.2020.123691

Yao, Z., & Ma, X. (2018). Bioresource Technology Characteristics of co-hydrothermal carbonization on polyvinyl chloride wastes with bamboo. Bioresource Technology, 247(381), 302–309. https://doi.org/10.1016/j.biortech.2017.09.098

Zhang, X., Zhang, L., & Li, A. (2019). Bioresource Technology Co-hydrothermal carbonization of lignocellulosic biomass and waste polyvinyl chloride for high-quality solid fuel production: Hydrochar properties and its combustion and pyrolysis behaviors. Bioresource Technology, 294(June), 122113. https://doi.org/10.1016/j.biortech.2019.122113

Zhang, C., Ma, X., Zheng, C., Huang, T., Lu, X., & Tian, Y. (2020a). Co-hydrothermal Carbonization of Water Hyacinth and Sewage Sludge : Effects of Aqueous Phase Recirculation on the Characteristics of Hydrochar. https://doi.org/10.1021/acs.energyfuels.0c01991

Zhang, C., Ma, X., Huang, T., Zhou, Y., & Tian, Y. (2020b). Co-hydrothermal carbonization of water hyacinth and polyvinyl chloride: Optimization of process parameters and characterization of hydrochar. Bioresource Technology, 314(June), 123676. https://doi.org/10.1016/j.biortech.2020.123676

Zhao, P., Li, Z., Li, T., Yan, W., & Ge, S. (2017). The study of nickel effect on the hydrothermal dechlorination of PVC. Journal of Cleaner Production, 152, 38–46. https://doi.org/10.1016/J.JCLEPRO.2017.03.101

Zhao, P., Huang, N., Li, J., & Cui, X. (2020). Fate of sodium and chlorine during the co-hydrothermal carbonization of high-alkali coal and polyvinyl chloride. Fuel Processing Technology, 199(November 2019), 106277. https://doi.org/10.1016/j.fuproc.2019.106277

Zhao, P., Lin, C., Zhang, J., Huang, N., Cui, X., Tian, H., & Guo, Q. (2021). Moisture re-adsorption characteristics of hydrochar generated from the Co-hydrothermal carbonization of PVC and alkali coal. Fuel Processing Technology, 213(October 2020), 106636. https://doi.org/10.1016/j.fuproc.2020.106636

Green Hydrogen Production and Potentials

Estimation of Green Hydrogen Production Potentials from Exploitable Hydropower Resources for Re-electrification and Climate Change Mitigation in Nigeria

Emmanuel Olorunyomi Aremu, Agnidé Emmanuel Lawin, David Olukanni, and Harrie-Jan Hendricks Franssen

Abstract Renewable energy sources like hydropower energy have the potential to play a major role in advancing the production of green hydrogen which can be used to generate electricity in remote communities where grid electricity is unavailable. Using secondary data sources, Nigeria's outstanding total exploitable hydropower potential stands at 12,240MW from 32 sites. Considering five different hypothetical scenarios, the annual green hydrogen production potential using hydroelectric energy for electrolysis was analyzed. Furthermore, the re-electrification potential using hydrogen to independently power fuel cells was evaluated. The quantity of fossil fuel that could be replaced by using the produced hydrogen for electricity generation, as well as the corresponding greenhouse gas emissions that could be avoided, are also assessed. For the first scenario, using 100% hydroelectric energy, the result reveals that Mambilla has the highest hydrogen potential of about 286,856.3 tons/year, while Kafanchan has the lowest with a potential of about 362.2 tons/year. The total amount of green hydrogen potential from the 32 sites is about 886,646.8 tons/year which could generate about 19,634.35 GWh of electricity per annum for

E. O. Aremu (✉)
Graduate Research Program (GRP) Climate Change and Water Resources, West African Science Service Centre On Climate Change and Adapted Land Use (WASCAL), University of Abomey-Calavi, Cotonou, Benin
e-mail: aremu.e@edu.wascal.org

A. E. Lawin
Laboratory of Applied Hydrology, National Institute of Water, University of Abomey-Calavi, Abomey-Calavi, Cotonou, Benin

D. Olukanni
Department of Civil Engineering, Covenant University, Ota, Ogun State, Nigeria
e-mail: david.olukanni@covenantuniversity.edu.ng

H.-J. H. Franssen
Forschungszentrum Jülich GmbH, Institute of Bio- and Geosciences, IBG-3 (Agrosphere), Jülich, Germany
e-mail: h.hendricks-franssen@fz-juelich.de

M. Narra and S. Narra (eds.), *Innovations in Circular Economy and Renewable Energy in Africa*, World Sustainability Series, https://doi.org/10.1007/978-3-031-68330-5_11

re-electrification. Furthermore, the results also show that a total of 6,019.04 million liters of diesel fuel could be replaced by using the produced hydrogen to power fuel cells. The total estimated carbon dioxide (CO_2) and carbon monoxide (CO) emissions that will be avoided are about 16,251.41 million kg and 46.10585 million kg respectively. The study concludes that the production of green hydrogen from exploitable hydropower sites if developed, will be an opportunity to lead the country toward the utilization of green hydrogen energy as part of its energy mix, and a viable avenue for climate change mitigation.

1 Introduction

Human-induced climate change, especially more frequent and intense extreme events, has had broad negative consequences and linked losses and damages to the environment and humans. The vulnerability has been lessened because of some development and adaptation measures. The most vulnerable individuals and systems have been seen to be disproportionately affected across industries and locations. As natural and human systems are pushed beyond their ability to adapt, the rise in weather and climate extremes has resulted in some irreversible consequences (IPCC 2022).

Hamed et al. (2020) asserted that the world is experiencing challenging times of major difficulties as the global population grows (to over seven billion people), accompanied by worsening economic crises, misuse of natural resources, extreme weather events, uncertainties, and escalating poverty and hunger. Africa has been described as the continent most vulnerable to the impacts of climate change (Serdeczny et al. 2016). Beyond the traditional influence and constraint relationships, the energy-water nexus is currently confronted with new conditions and challenges. Critical global climate change, the continued rise in average global temperature, and frequent extreme weather events have profoundly reshaped global hydrological systems and profoundly reshaped the global quantity and regional allocation of water resources as a result of massive fossil fuel consumption and continuous greenhouse gas (GHG) emissions (Sun et al. 2018). Nigeria is considered to be extremely susceptible to climate change, having a massive geographical area of 923,768 sq km covering many climatic zones. There's no going around the truth that the country is on the brink of endangerment due to a slew of environmental issues, many of which are being exacerbated by climate change (Okedina et al. 2018). In the face of climate change, the country is confronted with an energy crisis that might worsen if no measures are taken. Glassman et al. (2011) pointed out that energy production requires a considerable amount of water, and supplying water also uses a lot of energy.

Nigeria is endowed with natural resources, hydropower potential, and renewable energy resources. Still, the country cannot be said to have ever had an adequate supply of electricity in the history of its electricity generation (Olaoye et al. 2016). According to IEA (2023), their world outlook estimates that only about 60% of the total population in Nigeria have access to electricity in 2019 which indicates just a 20% increase from 40% in 2014. An estimate from the defunct National Electric Power Authority shows that Nigeria's outstanding total exploitable hydropower potential stands at 12,240MW from 32 sites, which, when combined with the existing 1930 MW (Kainji, Jebba, and Shiroro) already developed, brings the country's gross hydropower potential of Large Hydropower (LHP) to 14,750MW (zarma 2006). Olaoye et al. (2016) warned that Nigeria's energy crisis might worsen as the country's population increases and economic development calls for more demand for energy. The effort to save the earth from global warming has seen nations turning to alternative sources to meet their energy demands. Therefore, it is imperative that renewable energy solutions such as green hydrogen energy be harnessed as a strategy for climate change mitigation and be proffered to cater to Nigeria's impending rapid industrialization and growing energy needs. A crucial first step in integrating a hydrogen economy into a country's or region's energy infrastructure is determining the possibility of producing hydrogen. As energy demand is set to increase due to the soaring population and varying climatic conditions, there is a need for a good assessment of green hydrogen potential from exploitable hydropower resources which is considered in this study.

2 Methodology

2.1 Study Area and Data Used

Nigeria, the most populated country in Africa with an estimated 213.4 million people as of 2021, is a developing country with abundant natural and human resources. The nation, which borders the Gulf of Guinea, is situated between latitudes 4° and 14°N and longitudes 3° and 15°E. Its total area is 923,768 km^2, of which 910,768 km^2 are land and 13,000 km^2 are water. The Republic of Niger borders it in the north, the Republic of Chad borders it in the northeast, the Republic of Cameroon borders it in the east, and the Republic of Benin borders it in the west. It has an 853 km long coastline, with the Atlantic Ocean at its southernmost point. The Niger River, the Benue River, and the Chad Basin are just a few of the many landscapes that can be seen there. Some of Nigeria's biggest cities include Lagos, Kano, and Ibadan. The data used for the study as presented in Table 1 is secondary data on the potential capacity of thirty-two (32) exploitable hydropower sites by the defunct National Electric Power Authority (NEPA) obtained from Manohar and Adeyanja (2009). The data of electrical energy (E_e) for five assumed scenarios was estimated for each site using Eq. 1.

Table 1 Exploitable hydropower sites and estimated annual electrical energy available for green hydrogen production

Location	River	Potential capacity (MW)	Ee. S1 (MWh)	Ee. S2 (MWh)	Ee. S3 (MWh)	Ee. S4 (MWh)	Ee. S5 (MWh)
Donko	Niger	225	847,530	847.53	678.024	508.518	339.012
Zungeru I	Kaduna	450	1,695,060	1695.06	1356.048	1017.036	678.024
Zungeru II	Kaduna	500	1,883,400	1883.4	1506.72	1130.04	753.36
Zurubu	Kaduna	20	75,336	75.336	60.2688	45.2016	30.1344
Gwaram	Jamaare	30	113,004	113.004	90.4032	67.8024	45.2016
Izom	Gurara	10	37,668	37.668	30.1344	22.6008	15.0672
Gudi	Mada	40	150,672	150.672	120.5376	90.4032	60.2688
Kafanchan	Kongum	5	18,834	18.834	15.0672	11.3004	7.5336
Kurra ll	Sanga	25	94,170	94.17	75.336	56.502	37.668
Kurra l	Sanga	15	56,502	56.502	45.2016	33.9012	22.6008
Richa II	Daffa	25	94,170	94.17	75.336	56.502	37.668
Richa I	Mosari	35	131,838	131.838	105.4704	79.1028	52.7352
Mistakuku	Kurra	20	75,336	75.336	60.2688	45.2016	30.1344
Korubo	Gongola	35	131,838	131.838	105.4704	79.1028	52.7352
Kiri	Gongola	40	150,672	150.672	120.5376	90.4032	60.2688
Yola	Benue	360	1,356,048	1356.048	1084.838	813.6288	542.4192
Karamti	Kam	115	433,182	433.182	346.5456	259.9092	173.2728
Beli	Taraba	240	904,032	904.032	723.2256	542.4192	361.6128
Garin dali	Taraba	135	508,518	508.518	406.8144	305.1108	203.4072
Sarkin dankon	Suntai	45	169,506	169.506	135.6048	101.7036	67.8024
Gembu	Dongu	140	527,352	527.352	421.8816	316.4112	210.9408
Kashimbila	Katsina-Ala	30	113,004	113.004	90.4032	67.8024	45.2016
Katsina Ala	Katsina Ala	260	979,368	979.368	783.4944	587.6208	391.7472
Makurdi	Benue	1060	3,992,808	3992.808	3194.246	2395.685	1597.123
Lokoja	Niger	1950	7,345,260	7345.26	5876.208	4407.156	2938.104
Onitsha	Niger	1050	3,955,140	3955.14	3164.112	2373.084	1582.056
Ifon	Osse	30	113,004	113.004	90.4032	67.8024	45.2016
Ikom	Cross	730	2,749,764	2749.764	2199.811	1649.858	1099.906
Afokpo	Cross	180	678,024	678.024	542.4192	406.8144	271.2096
Atan	Cross	180	678,024	678.024	542.4192	406.8144	271.2096
Gurara	Gurara	300	1,130,040	1130.04	904.032	678.024	452.016
Mambilla	Danga	3960	14,916,528	14,916.53	11,933.22	8949.917	5966.611
Total		12,240	46,105,632	46,105.63	36,884.51	27,663.38	18,442.25

2.2 Green Hydrogen Potential from Hydropower

The potential for producing green hydrogen may be estimated using one of two basic methods, according to Thapa et al. (2021). The first method makes the assumption that only a certain percentage of the economically feasible hydropower potential is utilized for electrolysis-based H_2 production. This approach was employed in research carried out in Venezuela by Posso and Zambrano (2014). According to the second method, hydrogen production may be assessed based on excess electricity that would have been whittled down due to a decline in demand or enhanced production from higher water inflows during the rainy season at the hydroelectric plants. The first strategy was used in this study. First, it is assumed that the electrical energy linked to the hydropower potential is accessible all year long; Eq. 1 is used to determine the electrical energy as reported by Posso and Zambrano (2014).

$$E_e = \text{HEp} \times \text{CF} \times \text{Operating hours} \quad (\text{MWh}) \tag{1}$$

where E_e is the exploitable electrical energy available throughout the year, HEp is the hydroelectric potential and CF is the capacity factor assumed to be 43%, which is an average value resulting from the operation of the existing Shiroro hydroelectric plant in the Niger Basin in the period 2004–2014 (Gbadamosi and Adedayo, 2015), the operating hours is assumed to be 8760 h (i.e., 365 days × 24 h). The hydropower energy utilized for the assumed five scenarios is obtained using Eq. 2.

$$E_{u.s} = E_e \times \text{Scenario}\,\% \ (\text{MWh}) \tag{2}$$

where $E_{u.s}$ is the electrical energy that is assumed to be utilized for hydrogen production for each of the scenarios. Based on a similar study done in Nepal (Thepa et al., 2021), the following five hypothetical scenarios were assumed for the estimation of hydrogen production potential from the exploitable hydropower sites:

Scenario 1 (S1-H_{2p})—100% utilization of hydropower energy
Scenario 2 (S2- H_{2p})—80% utilization of hydropower energy
Scenario 3 (S3- H_{2p})—60% utilization of hydropower energy
Scenario 4 (S4- H_{2p})—40% utilization of hydropower energy
Scenario 5 (S5- H_{2p})—20% utilization of hydropower energy
Secondly, the hydrogen potential (H_{2p}) is estimated using Eq. 3. For each scenario, it is assumed that a certain percentage of the generated electricity is utilized to produce green hydrogen.

$$H_{2p} = \frac{E_{u.s}}{\text{Electrolyzer Efficiency}} (\text{Kg } H_2/year) \tag{3}$$

According to the research done by Zhou et al. (2020), it was pointed out that the International Energy Agency predicted that "creating all of today's dedicated hydrogen output (69 MtH2) from electricity would equate to an electricity usage of

3600 TWh" (IEA 2019). This equates to a production efficiency of 52 kWh/kgH2 (i.e., 3,600 TWh/69 MtH2 = 52 kWh/kgH2). As a result, for the purposes of this study, the energy consumption of the electrolyzer is considered to be 52 kWh per kg H2, excluding the energy consumption of ancillary equipment needed for the treatment and compression of gases.

2.3 *Green Hydrogen Storage*

Hydrogen gas is either kept in high-pressure (compressed) gas cylinders or as a liquid in cryogenic tanks or as metal hydrides for all practical uses (Bajpaj and Dash 2012; Ayodele et al. 2019). The hydrogen gas in this study is considered to be kept in compressed gas cylinders (tanks). As a result, the volume of compressed hydrogen gas is calculated using the following relationship (Ayodele et al. 2019):

$$V_{H_2\,\text{compressed}} = \frac{A_{H_2} \times \eta_{\text{compression}}}{H_{2CD}} \tag{4.4}$$

where A_{H2} is the amount of H_2 gas produced (kg), η compression is the compression efficiency given as 95% (Lagorse et al. 2008), and H_{2CD} is the density of compressed hydrogen gas given as 36 kg per m^3 (Züttel 2004).

2.4 *Re-electrification Potential Using Green Hydrogen*

Hydrogen energy has the potential to be used to generate electricity in remote communities where grid electricity is unavailable. Furthermore, hydrogen-generated electricity has the potential to replace fossil-based energy from the domestic, industrial and commercial sectors, such as diesel-powered generators (Thapa et al. 2021). Hydrogen produced from hydroelectric energy may be used to generate electricity using fuel cells. There are several different types of fuel cells (FCs), including proton exchange membrane fuel cells (PEMF), molten carbonate fuel cells (MCFC), solid oxide fuel cells (SOFC), and phosphoric acid fuel cells (PAFC). Due to its lower operating temperature, better power density, longer lifespan, quicker start-up, solid membrane, and commercial viability, PEMFC is the most promising category of fuel cells (Bajpai and Dash 2012; Ayodele et al. 2019; Saeed and Warkozek 2015). Because of the benefits described above, PEMFC is used in this study. When hydrogen gas is supplied into a PEMFC, the chemical energy held in the gas is transformed into electricity. As a result, the amount of electrical energy based on the hydrogen gas heating value may be calculated as follows (Ayodele et al. 2019):

$$E_{H_2} = V_{H_2\,\text{Compressed}} \times LHV_{H_2} \times \eta_{FC} \times H_{\text{Density}} \tag{5}$$

where LHV_{H2} is the hydrogen gas's lower heating value, which is reported to be 33.3 kWh per kilogram (Lagorse et al. 2008) or 119.9 MJ/kg (Dincer 2012) and FC represents the fuel cell electricity conversion efficiency, which is assumed at 60% (Alves et al. 2013), while the density of hydrogen is defined as $H_{Density}$.

2.5 *Estimation of Fossil Fuel Replacement*

The majority of urban and rural electricity consumers in Nigeria rely on diesel or gasoline-powered generators to satisfy their energy demands due to the regular power cut. This is due to the fact that the nation's utilities' ability to generate electricity is insufficient to fulfill the country's rising energy needs (Ayodele et al. 2019). Hydrogen generated by hydropower has the potential to replace fossil fuels like diesel and gasoline in the generation of electricity in the country. These fossil fuels emit greenhouse gases and other air pollutants into the atmosphere. It may be possible to reduce greenhouse gas emissions and close the energy supply–demand imbalance in Nigeria by replacing the gasoline and diesel generators with fuel cells powered by hydrogen gas. Using hydrogen to power fuel cells can displace a measurable quantity of diesel or gasoline fuels for the same amount of electricity generation. In this study, the quantity of diesel fuel that would be displaced is examined. So, using its heating values (LHV) in relation to hydrogen gas as indicated in Eq. 6, it is possible to calculate the quantity of fuel (diesel fuel) displaced in liters annually (Ayodele et al. 2019).

$$A_{\text{Fuel}} = \frac{V_{H_2\,\text{Compressed}} \times LHV_{H_2} \times \eta_{FC} \times H_{\text{Density}}}{\left(\frac{LHV_{\text{Fuel}}}{3.6}\right) \times D_{\text{Fuel}} \times \eta_{\text{Fuel}}} \quad (6)$$

where ηFuel is the diesel fuel generator's efficiency, which is considered to be 33%. (Envergent 2010; ICF 2014; Nizami art al. 2017), LHV_{Fuel} is the lower heating value of diesel fuel which is considered to be 42.5 MJ per kg (Dincer 2012; Nizami et al. 2017), D_{Fuel} is the density of diesel fuel whose value is taken as 0.837 kg per liter (MIT 2007) and 3.6 is the conversion factor from MJ to kWh (Ayodele 2019).

2.6 *Estimation of Greenhouse Gases (Carbon Dioxide and Carbon Monoxide) Avoided*

The combustion of fossil fuel in electricity generation, transportation, industries, and other sectors has been a source of concern globally due to the emission of greenhouse gases (CO_2, CH4, CO, and N_2O) which has led to the current global warming. For instance, incomplete combustion of fossil fuel can result in the dangerous emission of hazardous gases such as Carbon monoxide (CO). For this reason, every country is taking a variety of steps to reduce the production of gases and pollutants that contribute to climate change. Although there are several greenhouse gases, nearly 77% of the total global greenhouse gas emissions is made up of CO_2 (Adeoti et al. 2014), hence it is considered in this study. The primary emission from hydrogen-powered fuel cells is water, which poses no environmental risk and has been recognized as a viable solution that could bring remedy to the climate-change crisis. The quantity of GHG emissions (CO_2 and CO) that will be avoided by employing H_2-based fuel cells instead of diesel fuel burning was computed using (Ayodele et al. 2019):

$$E_{GHG} = A_{Fuel} \times SE_F \tag{7}$$

where SE_F is the specific emission factor of the associated greenhouse gas, which might be CO_2 or CO. The CO_2 emission factor of diesel fuel is 2.7 kg per liter, whereas the CO emission factor is 0.00766 kg per liter (Adeoti et al. 2014).

3 Results and Discussion

3.1 *Estimation of Green Hydrogen Production from Hydroelectric Energy*

The results of the annual electrical energy (E_e) that would be available to produce green hydrogen using the five assumed scenarios were estimated for each site using Eq. 1 and presented in Table 1. The results of the estimation of green hydrogen production potential from the electrical energy are displayed in Table 2. The results reveal that the highest hydrogen production potential was recorded in the Mambilla site with a potential of 286,856.3 tons/year, followed by Lokoja and Makurdi sites with potentials of 141,255 tons/year and 76,784.77 tons/year respectively while the lowest three were recorded in Kurra 1, Izom and Kafanchan sites with potentials of 1,086.577 tons/year, 724.3846 tons/year and, 362.1923 tons/year respectively. In summary, for the thirty-two sites, the total green hydrogen production potentials were 886,646.8 tons/year, 709,317.4 tons/year, 531,988.1 tons/year, 354,658.7 tons/year, and 177,329.4 tons/year for the first, second, third, fourth and fifth scenario respectively. The first scenario shows the highest potential while the fifth scenario shows the lowest potential for each of the sites.

3.2 *Volume of Hydrogen Storage and Re-electrification Potential Using Hydrogen*

The results of the volume of compressed hydrogen gas (i.e., hydrogen storage) and re-electrification potential using hydrogen are displayed in Fig. 1a and b respectively. For

Table 2 Estimated Hydrogen potential for the five scenarios from the estimated annual electrical energy

Location	S1-(HP) tons/year	S2-(HP) tons/year	S3-(HP) tons/year	S4-(HP) tons/year	S5-(HP) tons/year
Donko	16,298.65	13,038.92	9779.192	6519.462	3259.731
Zungeru I	32,597.31	26,077.85	19,558.38	13,038.92	6519.462
Zungeru II	36,219.23	28,975.38	21,731.54	14,487.69	7243.846
Zurubu	1448.769	1159.015	869.2615	579.5077	289.7538
Gwaram	2173.154	1738.523	1303.892	869.2615	434.6308
Izom	724.3846	579.5077	434.6308	289.7538	144.8769
Gudi	2897.538	2318.031	1738.523	1159.015	579.5077
Kafanchan	362.1923	289.7538	217.3154	144.8769	72.43846
Kurra ll	1810.962	1448.769	1086.577	724.3846	362.1923
Kurra l	1086.577	869.2615	651.9462	434.6308	217.3154
Richa II	1810.962	1448.769	1086.577	724.3846	362.1923
Richa I	2535.346	2028.277	1521.208	1014.138	507.0692
Mistakuku	1448.769	1159.015	869.2615	579.5077	289.7538
Korubo	2535.346	2028.277	1521.208	1014.138	507.0692
Kiri	2897.538	2318.031	1738.523	1159.015	579.5077
Yola	26,077.85	20,862.28	15,646.71	10,431.14	5215.569
Karamti	8330.423	6664.338	4998.254	3332.169	1666.085
Beli	17,385.23	13,908.18	10,431.14	6954.092	3477.046
Garin dali	9779.192	7823.354	5867.515	3911.677	1955.838
Sarkin dankon	3259.731	2607.785	1955.838	1303.892	651.9462
Gembu	10,141.38	8113.108	6084.831	4056.554	2028.277
Kashimbila	2173.154	1738.523	1303.892	869.2615	434.6308
Katsina Ala	18,834	15,067.2	11,300.4	7533.6	3766.8
Makurdi	76,784.77	61,427.82	46,070.86	30,713.91	15,356.95
Lokoja	141,255	113,004	84,753	56,502	28,251
Onitsha	76,060.38	60,848.31	45,636.23	30,424.15	15,212.08
Ifon	2173.154	1738.523	1303.892	869.2615	434.6308
Ikom	52,880.08	42,304.06	31,728.05	21,152.03	10,576.02
Afokpo	13,038.92	10,431.14	7823.354	5215.569	2607.785
Atan	13,038.92	10,431.14	7823.354	5215.569	2607.785
Gurara	21,731.54	17,385.23	13,038.92	8692.615	4346.308
Mambilla	286,856.3	229,485	172,113.8	114,742.5	57,371.26
Total	886,646.8	709,317.4	531,988.1	354,658.7	177,329.4

the first scenario the results reveal that the highest volume of compressed hydrogen was recorded in the Mambilla site with a value of 7,569.819 tons/year and when used to power fuel cells is able to generate 6,352.59 GWh of electricity, this is followed by the Lokoja and Makurdi sites with a hydrogen storage volume of 3,727.563 tons/year and 2,026.265 tons/year and are able to generate 3,128.021 GWh and 1,700.36 GWh of electricity when used to power fuel cells. The lowest volume of compressed hydrogen gas was recorded in Kurra 1, Izom, and Kafanchan with values of 28.67356 tons/year, 19.11571 tons/year, 9.557853 tons/year, and can generate 24.0617 GWh, 16.0411 GWh, and 8.0205 GWh of electricity respectively using hydrogen-powered fuel cells. In summary, the total volume of compressed hydrogen gas for the all the sites, using the five scenarios was 23,397.62 tons/year, 18,718.1 tons/year, 14,038.57 tons/year, 9,359.049 tons/year, and 4,679.525 tons/year, and can generate total electricity of 19,634.35 GWh, 15,707.48 GWh, 11,780.61 GWh, 7,853.74 GWh, and 3,926.87 GWh for the first to fifth scenario respectively. The highest volume of compressed hydrogen gas and re-electrification potential was recorded for the first scenario while the lowest volume and re-electrification potential was recorded for the fifth scenario.

3.3 *Quantity of Diesel Fuel Replaced and Greenhouse Gases (Carbon Dioxide and Carbon Monoxide) Avoided Using Hydrogen Fuel Cells*

The quantity of diesel fuel replaced using hydrogen to power fuel cells is shown in Fig. 1c while the amount of carbon dioxide (CO_2) and carbon monoxide (CO) that would be avoided is shown in Fig. 2a for CO_2 and Fig. 2b for CO. The results reveal that the total quantity of diesel fuel replaced using hydrogen to power fuel cells for the five assumed scenarios are 6,019.04 million liter/year, 4,815.232 million liter/year, 3,611.424 million liter/year, 2,407.616 million liter/year, 1,203.808 million-liter/year, this diesel replacement with hydrogen would avoid a total carbon dioxide (CO_2) amount of 16,251.41 million kg/year, 13,001.13, million kg/year, 9,750.845 million kg/year, 6,500.563 million kg/year, and 3,250.282 million kg/year, with a total carbon monoxide (CO) of 46.1059 million kg/year, 36.8847 million kg/year, 27.6635 million kg/year, 18.4423 million kg/year, and 9.2212 million kg/year for the first, second, third, fourth and first scenario respectively. For the first scenario, the highest quantity of carbon dioxide and carbon monoxide avoided was recorded in Mambilla site with a value of 5,257.808 million kg/year and 14.9166 million kg/year while the lowest quantity was in Kafanchan site with a value of 6.638647 million kg/year and 0.018834 million kg/year respectively.

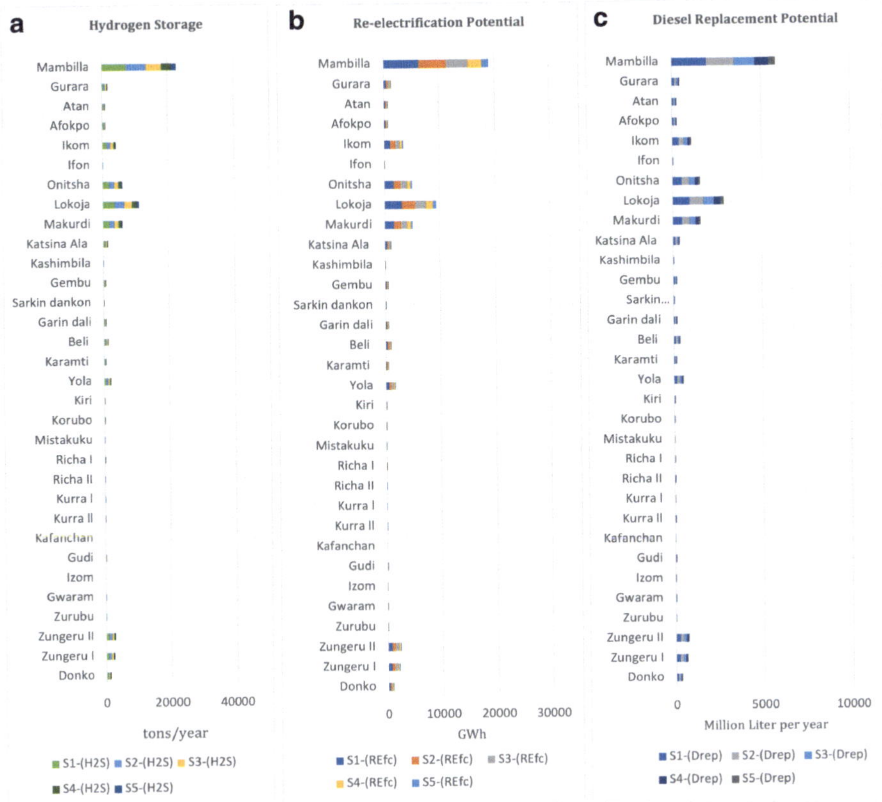

Fig. 1 Estimated, **a** Hydrogen Storage, **b** Re-electrification potential and **c** Diesel replacement amount for the exploitable hydropower sites

4 Conclusion

Green hydrogen production potential from exploitable hydropower sites for re-electrification and climate change mitigation has been presented in this study. From the results, the first scenario with the highest percentage of hydroelectric energy (100%) utilization for green hydrogen production, a total of 886,646.8 tons/year of hydrogen gas could be obtained from the thirty-two sites, and compressed to a total volume of 23,397.62 tons/year and can be used to generate 19,634.35 GWh of electricity using hydrogen to power fuel cells, this will replace 6,019.04 million liter/year of diesel and contribute immensely to the mitigation of climate change by avoiding a total of 16,251.41 million kg/year of CO_2 emissions and 46.10585 million kg/year CO emissions into the atmosphere. The last scenario with the lowest percentage of hydroelectric energy (20%) utilization for green hydrogen production, shows that a total of 177,329.4 tons/year of hydrogen gas could be obtained from the thirty-two sites, and compressed to a volume of 4,679.525 tons/year and will generate 3,926.87

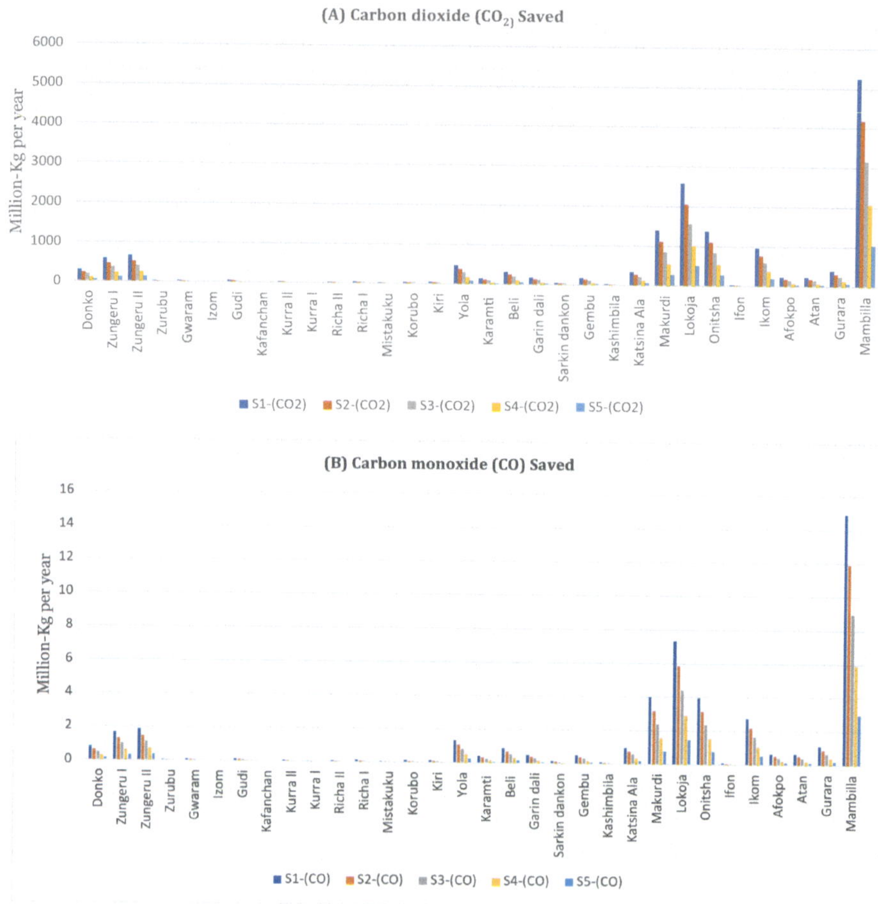

Fig. 2 Estimation of green house gas emission avoided, **a** CO_2 Saved, **b** CO Saved

GWh of electricity, this quantity of hydrogen will replace a total of 1,203.808 million-liter/year of diesel and contribute to the mitigation of climate change by avoiding a total of 3,250.282 million kg/year of CO_2 emissions and 9.221169 million kg/year of CO emissions. This work can serve as an introductory reference for researchers, policy and decision-makers, for the evaluation of green hydrogen production potential from unexploited hydropower sites for rural areas re-electrification and national climate change mitigation efforts in Nigeria. The study concludes that the production of green hydrogen from the thirty-two exploitable hydropower sites if developed, can be a niche opportunity to lead the country toward the adoption of green hydrogen energy as part of its energy mix, especially for rural areas re-electrification and mitigation of climate change.

References

Adeoti, O., Ayelegun, T.A., Osho, S.O. (2014). Nigeria biogas potential from livestock manure and its estimated climate value.

Alves, H.J., Bley Junior, C., Niklevicz, R.R., Frigo, E.P., Frigo, M.S., Coimbra-Araújo, C.H. (2013). Overview of hydrogen production technologies from biogas and the applications in fuel cells. In: International Journal of Hydrogen Energy. pp. 5215–5225.

Ayodele, T.R., Alao, M.A., Ogunjuyigbe, A.S.O., Munda, J.L. (2019). Electricity generation prospective of hydrogen derived from biogas using food waste in south-western Nigeria. Biomass Bioenergy. 127. https://doi.org/10.1016/j.biombioe.2019.105291.

Bajpai, P., Dash, V. (2012). Hybrid renewable energy systems for power generation in stand-alone applications: A review.

Dincer, I. (2012). Green methods for hydrogen production. In: International Journal of Hydrogen Energy. pp. 1954–1971.

Envergent (2010). The Production of Electricity from Wood and Other Solid Biomass.

Gbadamosi, S., Adedayo, O.O. (2015). Evaluation of Operational Efficiency of Shiroro Hydro-Electric Plant in Nigeria. International Journal of Science and Engineering Investigations. 4, 42.

Glassman, D., Wucker, M., Isaacman, T., Champilou, C. (2011). The Water-Energy Nexus Adding Water to the Energy Agenda.

Hamed, A.A., Sayed, M.E., Meshari, L.A.-H. (2020). Assessment of future water, energy, and food nexus by using Weap and Leap models in the state of Kuwait. Journal of Environmental Science. 49, 495–526 (2020). https://doi.org/10.21608/JES.2020.206395

ICF (2014). Diesel Generators: Improving Efficiency and Emission Performance in India.

IEA (2019). The Future of Hydrogen.

IEA (2023), SDG7: Data and Projections, IEA, Paris https://www.iea.org/reports/sdg7-data-and-projections, License: CC BY 4.0

IPCC, (2022). IPCC AR6 WGII Summary For Policymakers 2022.

Lagorse, J., Simões, M.G., Miraoui, A., Costerg, P. (2008). Energy cost analysis of a solar-hydrogen hybrid energy system for stand-alone applications. Int J Hydrogen Energy. 33, 2871–2879. https://doi.org/10.1016/j.ijhydene.2008.03.054

Manohar, K., Adeyanju, A.A. (2009). Hydro power energy resources in Nigeria. Journal of Engineering and Applied Sciences. 4, 68–78

MIT (2007). Units & Conversions Fact Sheet. MIT EL 00–003 Press.

Nizami, A.S., Shahzad, K., Rehan, M., Ouda, O.K.M., Khan, M.Z., Ismail, I.M.I., Almeelbi, T., Basahi, J.M., Demirbas, A. (2017). Developing waste biorefinery in Makkah: A way forward to convert urban waste into renewable energy. Appl Energy. 186, 189–196. https://doi.org/10.1016/j.apenergy.2016.04.116.

Okedina, I.M., Chidinma, E.E., Olatunde, Y.O., Chdinma, E.E. (2018). Energy and climate change: impact of food, water, and energy nexus on economic growth in Nigeria.

Olaoye, T., Ajilore, T., Akinluwade, K., Omole, F., Adetunji, A. (2016). Energy Crisis in Nigeria: Need for Renewable Energy Mix. American Journal of Electrical and Electronic Engineering. 4, 1–8. https://doi.org/10.12691/ajeee-4-1-1.

Posso, F., Zambrano, J. (2014). Estimation of electrolytic hydrogen production potential in Venezuela from renewable energies. Int J Hydrogen Energy. 39, 11846–11853. https://doi.org/10.1016/j.ijhydene.2014.06.033

Saeed, E.W., Warkozek, E.G. (2015). Modeling and Analysis of Renewable PEM Fuel Cell System. In: Energy Procedia. pp. 87–101. Elsevier Ltd.

Serdeczny, O., Adams, S., Baarsch, F., Coumou, D., Robinson, A., Hare, W., Schaeffer, M., Perrette, M., Reinhardt, J. (2016). Climate change impacts in Sub-Saharan Africa: from physical changes to their social repercussions. Regional Environmental Change 2016 17:6. 17, 1585–1600. https://doi.org/10.1007/S10113-015-0910-2

Sun, L., Pan, B., Gu, A., Lu, H., Wang, W. (2018). Energy–water nexus analysis in the Beijing–Tianjin–Hebei region: Case of electricity sector. Renewable and Sustainable Energy Reviews. 93, 27–34. https://doi.org/10.1016/j.rser.2018.04.111

Thapa, B.S., Neupane, B., Yang, H., Seong, Lee, Y.H. (2021). Green hydrogen potentials from surplus hydro energy in Nepal. Int J Hydrogen Energy. 46, 22256–22267. https://doi.org/10.1016/j.ijhydene.2021.04.096

Zarma, I.H. (2006). Hydropower Resources in Nigeria.

Zhou, A., Zhou, W., Manandhar, P.(2020). A Study on the Prospect of Hydropower to Hydrogen in Nepal.

Züttel, A. (2004). Hydrogen storage methods.

Exploring the Potential of Integrating Bio-Hydrogen Generation into Hybrid Waste-To-Energy (W2E) Plant in Ghana

Mutala Mohammed, Francis B. Agyenim, Latifatu Mohammed, Rejart Adomah, Mutala B. Edem, Rafia A. Samii, Juliet Yeboa, Ekua A. Armoo, Eugen Borteley Badu, and Satyanarayana Narra

Abstract Hydrogen (H_2) as an energy carrier to produce electricity has gained interest globally as a consequence of its ability to reduce greenhouse gas (GHG) emissions compared to conventional power generation and improving renewable energy consumption. This study focusses on assessing the potential of a hybrid waste to energy plant for the production of electricity, and other environmental benefits as well as economic impacts in Ghana. The H_2 generation via a biogas steam methane reforming (SMR) process, energy generation capacity, and environmental and economic assessments were determined through a computational analysis. It was found that the total amount of waste generation in the study area is 649.7 kt/y and 253.3 kt/y as the organic fraction of municipal solid waste (OFMSW) with 72% of the waste originating from urban communities. Potential electricity generation of biogas from OFMSW is estimated at 588 GWh/y with Kumasi metropolitan assembly (KMA) contributing 36.8% of the total electricity generation. A total of 40,000 GWh/y of electricity could be available from H_2 yield from biogas. The energy generated has the capacity to offset 1.90 million liters (ML) of diesel and could reduce GHG emission by 14,601 kt CO_2e. This scientific assessment could provide the basis for policymakers to consider the potential of H_2 generation from waste in achieving renewable energy penetration mix and towards a sustainable future.

M. Mohammed (✉) · F. B. Agyenim · L. Mohammed · R. Adomah · R. A. Samii · J. Yeboa · E. A. Armoo
CSIR-Institute of Industrial Research, Box LG 576, Accra, Ghana
e-mail: mutbaby@gmail.com

M. B. Edem
CSIR-Soil Research Institute, Academy Post Office, Private Mail Bag, Kwadaso, Kumasi, Ghana

E. B. Badu
CSIR-Head Office, Box M32, Accra, Ghana

E. A. Armoo · S. Narra
Professorship of Material and Energy Valorisation of Biogenous Residues, Professorship of Waste and Resource Management, University of Rostock, Justus-Von-Liebig-Weg 6, 18059 Rostock, Germany

M. Narra and S. Narra (eds.), *Innovations in Circular Economy and Renewable Energy in Africa*, World Sustainability Series, https://doi.org/10.1007/978-3-031-68330-5_12

1 Introduction

Globally, municipal solid waste (MSW) generation currently stands at 2.01 billion tons annually and this is anticipated to increase to 3.4 billion tons in the next two decades. The drivers of this change are population and urban growth, economic development and changes in standard of life. The organic fraction of municipal solid waste (OFMSW) represents the highest component in MSW generated in developing countries accounting for 53% and 56% in middle- and low-income countries, respectively and 32% in developed countries, contributing to 3.3 billion tons-CO_2eq/y of greenhouse gas (GHG) emissions (Karthikeyan et al. 2018). With Ghana as a developing country, waste generation has accelerated in recent years with an estimated 12,710 tons of MSW generated annually at 0.47 kg/person/d (Miezah et al. 2015). The OFMSW mainly refers to a mixture of food waste, leaves, and yard waste (Cesaro 2021). In an agrarian economy like Ghana, OFMSW accounts for 48% to 69% of the waste generation, depending on the location and are usually disposed of in landfills or incinerated without energy recovery. The disposal of OFMSW in non-engineered landfill has the potential to contribute to GHGs emissions as it represents a large amount of biomass that has the potential to produce methane (CH_4) via anaerobic digestion process (AD) under ambient conditions during storage (Bonk et al. 2015). In Ghana, the waste sector contributes 7.04% of the country's GHGs. Nonetheless, there is a great potential to convert the OFMSW to biogas through AD process in biodigesters to significantly reduce pollution and generates a clean, secure and sustainable energy, with multiple potential benefits for sustainable development. The AD has been used mainly in Ghana to treat MSW, animal manure, abattoir waste, industrial wastewater and sewage sludge and fecal matter. However, the biogas produced is usually flared into the atmosphere with gas unrecovered in majority of these plants due to low production and its associated high costs. The utilization of OFMSW for AD has not seen much development in Ghana. The current AD plant understudy mechanically separates the OFMSW from the inorganic fraction of the MSW. The biogas generated is used to power an internal combustion engine to produce electricity with some consumed within the facility and the remaining power injected into the national grid.

Ghana is one of the countries in the African content with the highest electricity access rate of 87% (Energy Commission, 2023). However, renewable energy has not featured that much in the country's energy generations mix with 1.3% renewables penetration. Likewise, the share of thermal has seen a dramatic increase and presently occupies the highest share (65%) of the country's energy generation mix as against 34.1% from hydro (Energy Commission 2022). Notwithstanding, there is an environmental cost associated with thermal energy generation ranging from air pollution, which is believed to cause millions of premature deaths each year and the release of CO_2 as the prominent driver of global climate change. These and other related adverse impacts associated with the use of fossil fuel have prompted research into more efficient and sustainable technologies and renewable sources of electricity generation such as H_2. According to the revised 2021 National Determined Contributions (NDCs), Ghana is committed to cut its GHG emissions by 64 $MtCO_2e$ by 2030. To achieve this, the government has enacted and introduced policies and legislations including the Renewable Energy Act of 2011 (Act 832) as amended in 2020 (Act 1045), with its accompanying master plan. The master plan provides a framework for the development of the renewable energy sector as well as an investment focused roadmap for the long-term promotion and development of different renewable energy resources in the country. The law aims to achieve renewable energy penetration into the energy mix by 10% by 2030. Ghana's total GHG emissions were estimated to be 59.8 $MtCO_2e$ with the energy sector as the main contributor (46.62%) and IPPU (Industrial Processes and Product Use) as the least contributor (EPA 2022).

Hydrogen (H_2) is a clean and environmentally friendly form of energy. H_2 is considered an ideal energy carrier and fuel due to its outstanding properties such as non-toxicity, long-term storage, and ease of transportation, besides no harmful gas emission on the environment in addition to its high energy yield of 122 kJ/g, which is 2.75-fold higher than hydrocarbon-based energy (Zhao et al. 2020; Dahiya et al. 2021). Researchers have explored and demonstrated the feasibility of generating H_2 from OFMSW by means of biological processes (Li et al. 2008; Kim et al. 2008; Cappai et al. 2014; Gadhe et al. 2014; Guo et al. 2014; Alibardi and Cossu 2016). The biological processes for H_2 production from OFMSW simultaneously offer waste minimization, energy recovery, and valorization (Yeshanew et al. 2018), providing an ecological solution for managing organics (Sharma and Melkania 2018). The high calorific power of OFMSW makes it a suitable substrate for bioenergy generation (Paillet et al. 2021) due to its rich content of carbohydrates and biodegradability (Sekoai and Gueguim Kana 2014) and because it is available all year and apparently free of cost (Escamilla-Alvarado et al. 2015). The OFMSW could be suitable for H_2 production because it is rich in carbohydrates and easily hydrolysable waste. The characteristics of MSW such as composition, moisture, and traces of inorganic residues, might vary depending on the regional economy and culture of each country and are usually considered in the design and operation of H_2 production from OFMSW. In a study by Francis et al. (2015), they concluded that the introduction of bioenergy into Ghana's energy mix could reduce GHG emissions by 6 $MtCO_2eq$ by 2030, accounting for a 14% reduction in a business as usual (BAU) scenario.

Despite the myriad of studies in estimating the quantity of waste generation in major cities in Ghana including Kumasi as well as its biogas potential, there is a dearth of studies regarding the assessment of H_2 energy generation potential from OFMSW. Fabian et al. (2019), Francis et al. (2014), (2015) and Arthur et al. (2011) and Mohammed et al. (2017) have all conducted studies in assessing the potential of biogas from biomass resources in Ghana for energy generation. Nevertheless, none of these studies has considered evaluating the potential of H_2 energy generation from these biomass resources. The present study determines the biogas potential of OFMSW, evaluate the GHG emissions reductions by replacing diesel fuel with green H_2 produced from the digestion of OFMSW and analyze the environmental benefits of H_2 fuel as renewable energy source in a hybrid waste to energy (W2E) plant located in *Gyankobaa*, a suburb of Atwima Nwabiagya Municipal Assembly (ANMA) in the

Ashanti Region of Ghana. The assessment considered resource generation within a 30 km radius of the plant location.

2 Materials and Methods

2.1 Study Area—Ashanti Region

Ashanti region lies in the southern part of Ghana occupying 24,389 km^2 (9,417 mi^2) with Kumasi as its capital. It lies between longitudes 0.15W and 2.25W, and latitudes 5.50N and 7.46N. The region is the third largest administrative region in Ghana forming 10.2% of the total land area of Ghana. With regards to population, it is the second most populated region with a population of 5,432,485 accounting for 17.6% of Ghana's total population with a population density of 1,235.8 persons per sq. km (GSS 2021). Administratively, the region is divided into Forty-three (43) metropolitan, municipal and district assemblies (MMDAs). These are one (1) Metropolis, Eighteen (18) Municipalities and Twenty-Four (24) Districts. Figure 1a and b show the map of Ashanti region and the MMDAs in the 30 km radius zone from the site.

The hybrid W2E demonstration plant is sited in *Gyankobaa* in the ANMA of the Ashanti Region of Ghana. The municipality has about 64 settlements with Nkawie as its administrative capital. It is mainly rural with 35.3% urban and 64.7% rural. The municipality lies approximately between latitude 6° 32'N and 6° 75'N, and between longitude 1° 36' and 2° 00' West. It is situated in the western part of the Ashanti Region and shares common boundaries with Ahafo Ano Southwest and East, and Atwima Nwabiagya North District to the north, Atwima Mponua District to the west, Amansie West and Atwima Kwanwoma District to the South and Kwadaso Municipality to the east. Agriculture is the dominant economic activity followed by trading in wares and other products. It covers an estimated area of 71 km^2. The

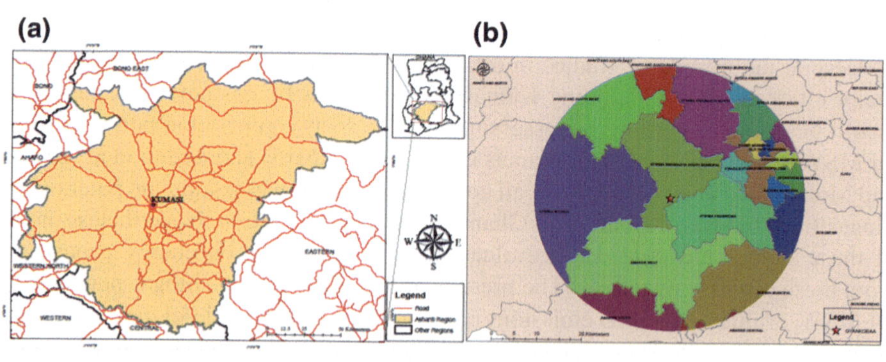

Fig. 1 Map of **a** Ashanti Region, **b** MMDAs in the 30 km buffer around *Gyankobaa*

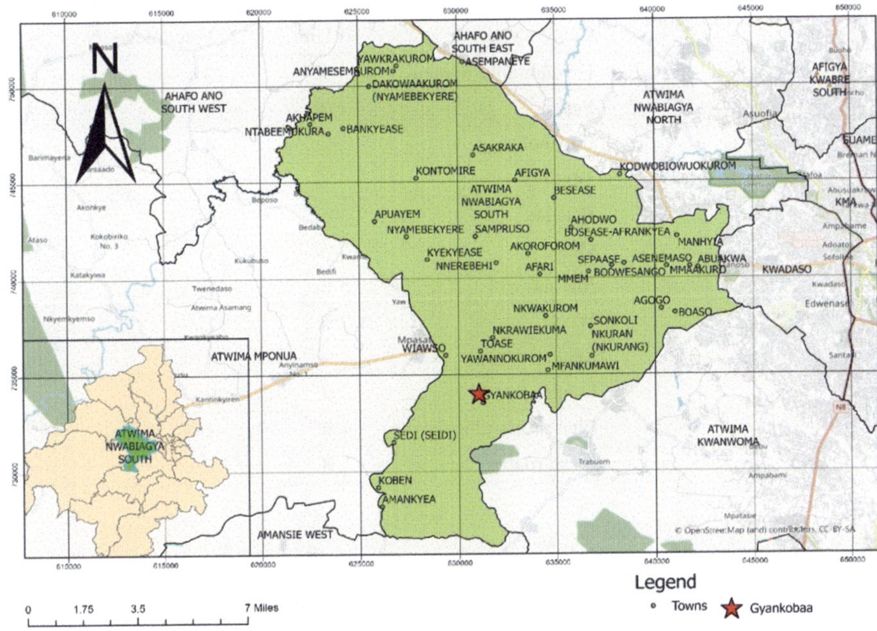

Fig. 2 Map of Atwima Nwabiagya municipal assembly

project site (Gyankobaa) is situated in the southern part of the municipality, close to Amansie West and Atwima Kwanwoma (Fig. 2).

2.2 *Methods*

2.2.1 Quantification of Organic Waste as Feedstock for AD Process

Planning for waste management is always a daunting task for authorities at the MMDAs due to the unreliability and scarcity of data on waste generation in Ghana. With regards to waste generation, a theoretical waste generation was quantified based on population and per capita waste generation daily. Per capita per day waste generation rate in Ghana varies between 0.47 and 0.70 kg for rural and urban centers, respectively. Waste characterization was also carried out to ascertain the composition of waste generated. A profile of waste composition is shown in Table 1. Additionally, the ultimate analysis of the OFMSW is presented in Table 2.

Table 1 Waste composition in the study area

Category	Unit	Percentage (%)
Fermentable wastes	%	37
Papers	%	1
Cardboards	%	4
Textiles	%	5
Plastics	%	18
Glass	%	1
Metals	%	1
Wood	%	1
Special wastes	%	3
Others	%	9
Fine ores (<25 mm)	%	18
Fecal matter	%	2

Table 2 Elemental analysis of OFMSW and molar mass of the respective elements

Element	Unit	OFMW	Molar mass
Carbon	%	49.53925	12.01
Hydrogen	%	4.445	1.01
Nitrogen	%	1.3835	14.01
Oxygen	%	43.6595	16.00
Sulfur	%	0.5765	32.06

2.3 Evaluation of Biogas Recovery from OFMSW

Organic waste or biodegradable materials are considered a good source of H_2 that can be produced simply and efficiently through AD. Depending on the type of organic material used as a feedstock for AD process, biogas as a by-product usually comprises of 40–75% CH_4 and 25–55% carbon dioxide (CO_2) as well as other traces of gases such as carbon monoxide (CO) and hydrogen sulfide (H_2S). However, H_2 as an intermediate product of AD process has gained an interest in recent years in optimizing further the AD process in achieving H_2 production from organic waste. Processes such as adsorption on activated carbon, silica gel permeation, physical adsorption with caustic soda or cryogenic methods can be used to clean biogas. An estimation of the capacity of the biogas digester is crucial for proper biogas steam methane reforming (SMR) process design. In this study, the theoretical biogas production from the OFMSW as a feedstock was evaluated using the Buswell's equation as shown in Table 3.

It has been established, that the volume of theoretical biogas produced in Eq. (10) is lower than the actual biogas yield in practice as a result of the incomplete decomposition of the organic material in the digester with 10% remaining as undecomposed

Table 3 Summary of the equations in estimating biogas and electricity potential from OFMSW

No	Parameter	Equation	References
(8)	Theoretical CH_4 yield	$C_aH_bO_cN_d + \left(a - \frac{b}{4} - \frac{c}{2} + \frac{3d}{4}\right)H_2O \Rightarrow \left(\frac{a}{2} - \frac{b}{8} + \frac{c}{4} + \frac{3d}{8}\right)CO_2 + \left(\frac{a}{2} + \frac{b}{8} - \frac{c}{4} - \frac{3d}{8}\right)CH_4 + dNH_3$ $C_aH_bO_cN_d + \alpha_1 H_2O \Rightarrow \alpha_2 CH_4 + \alpha_3 CO_2 + \alpha_4 NH_3$ *a, b, c,* and *d* are the number of Carbon, Hydrogen, Oxygen and Nitrogen respectively	Buswell and Hatfield (1936)
(9)	Number of C,H,O,N	$Mole\ ratio = \frac{E_{comp}[C,H,O,N]}{M[C,H,O,N]}$ E_{comp} and M are the element composition (carbon-C, hydrogen-H, oxygen-O, and nitrogen-N) obtained from the ultimate analysis of organic waste and M is the molar mass of the elements	Salami and Susu (2011)
(10)	Volume of theoretical biogas (T_{bio}), (m^3/ton)	$T_{bio} = CH_{4(t)} + CO_{2(t)}$ $CH_{4(t)}$ and $CO_{2(t)}$ are the theoretical CH_4 and CO_2 yield (m^3/ton) at a standard temperature and pressure (0°C at 1 atm)	Ayodele et al. (2019) and Nielfa et al. (2015)
(11)	Theoretical CH_4 (m^3/ton)	$T_{CH_4(vol)} = \frac{\left[\frac{a}{2} + \frac{b}{8} - \frac{c}{4} - \frac{3d}{8}\right]x22400}{12a+b+16c+14d}$ *a, b, c,* and *d* are the number of Carbon, Hydrogen, Oxygen and Nitrogen respectively	
(12)	Theoretical CO_2 (m^3/ton)	$T_{CO_{2(t)}} = \frac{\left[\frac{a}{2} - \frac{b}{8} + \frac{c}{4} + \frac{3d}{8}\right]x22400}{12a+b+16c+14d}$ a, b, c, and d are the number of Carbon, Hydrogen, Oxygen and Nitrogen respectively	
(13)	Actual biogas generation from OFMSW (B_{act})	$B_{act} = [Q_{msw} + OFMSW_{comp}]x[T_{bio} + \beta]$ Q_{msw} = total quantity of MSW generated in the study area $OFMSW_{(comp)}$ = average percentage composition of organic waste in the study area (Table 1), and β = fraction of organic matter utilized for cell tissue synthesis, and it is assumed to be 85% (Salami et al. 2011)	Cudjoe et al. (2022)
(14)	Amount of CH_4 from purified biogas	$CH_{4(\mathcal{P})} = B_{act} + \%\chi$ $\%\chi$ is the percentage to which the biogas is upgraded and is assumed to be 75.7% (Braga et al. 2013)	

material. According to Ayodele et al. (2019), about 5–10% of the organic matter synthesizes the organisms' cell tissue affecting microbial degradation.

For the biogas to be used for SMR process, the obtained biogas must be cleaned and upgraded. According to literature, the chemical composition of biogas consists of 93–96% of CH_4, 4–7% of CO_2, and < 20 ppm of H_2S (Alves et al. 2013). With CO_2 as the only impurity in the products of AD, the amount of CH_4 from purified biogas is determined in Eq. (14).

2.4 H_2 Generation by Biogas SMR

Steam reforming, partial oxidation of oil, and coal gasification are all methods for producing H_2 from fossil fuels and biomass. Other non-biological ways for producing H_2 from H_2O include thermal and thermochemical reactions, electrolysis, and photolysis. However, SMR of natural gas or CH_4, with a conversion efficiency of 74–85% is currently the most widely used method and cost-effective way to produce H_2 by researchers and industries, respectively (Boyano et al. 2012; Gross 2020). The SMR process occurs in two reactors i.e., reform reactor and shift reactor. Catalytic reforming reactions of hydrocarbons, like CH_4, naphtha or ethanol with water (pre-vaporized by a steam generator) occurs in the reform reactor with H_2 (main objective of the reform) and CO as the main products of the reaction (Eq. (15)). In the shift reactor, all CO produced in the reform reactor, react with water steam to produce H_2 and CO_2 (Eq. (16)). Consequently, H_2 and CO_2 become the main products at the end of the reform process (Silveira et al. 2009; Martelli et al. 2012). Table 4 below provides a summary of reactions of the SMR and H_2 production from purified biogas.

Table 4 Summary of equations employed in estimating H_2 potential

No	Parameter	Equation	ReferenceS
(15)	CH_4 steam reactor	$CH_4 + H_2O \leftrightarrow CO + 3H_2 \quad \Delta H_{298} = +206.2\ kJ/mol$	Saebea et al. (2014)
(16)	Shift reaction/ Water to gas reaction	$CO + H_2O \leftrightarrow CO_2 + H_2 \quad \Delta H_{298} = -41\ kJ/mol$	
(17)	Reform process	Equation (8) + (9) $CH_4 + 2H_2O \leftrightarrow CO_2 + 4H_2 \quad \Delta H_{298} = +165.2\ kJ/mol$	
(18)	H_2 from purified biogas (kg)	$H_{2(Q)} = CH_{4(P)}\ X\ \rho_{CH_4}\ X\ \Delta_{(BOILER)} X\ \Delta_{(REFORMER)} X\ 0.5$ ρCH4 is the density of CH_4, which is considered as 0.717 kg/m^3 (Dan et al., 2020), $\otimes_{(boiler)}$ and $\otimes_{(reformer)}$ are the boiler efficiency and reformer efficiency specified as 80%	Braga et al. (2013)

Table 5 Summary of equations employed in quantifying energy generation from H_2

No	Parameter	Equation	Reference
(19)	Power generation from H_2 (kWh)	$H_{2(power)} = \delta X\ LHV_{H_2}\ \rho_{H_2}\ X\ H_{2(Q)}$ LHV_{H2} is the lower heating value of H_2 which is considered as 33.3 kWh/kg (Dufo-L´opez et al. 2008), ρ_{H2} is the density of H_2, which is given as 0.09 kg/m^3 (Ayodele et al. 2019), and δ is the electricity conversion efficiency of fuel cells and is assumed to be 60% (Alves et al. 2013)	Ayodele et al. (2019)

In Eq. (16), 1 kg of steam reformed CH_4 produces 0.5 kg of H_2. Consequently, it is assumed that all the energy utilized by the reformer is generated from a boiler fed by biogas. The amount of H_2 produced is determined by the combined efficiencies of the boiler and reformer. The H_2 produced ($H_{2(Q)}$) from the purified biogas by SMR in kg is determined as shown in Eq. (18).

2.5 *Electricity Generation Potential of H_2 Gas*

H_2 is increasingly gaining great interest as an energetic carrier due to its high heating value and the low environmental impact of its use as fuel. It can be used to generate electricity by a fuel cell and contributes to increase the distributed generation of energy and reduce emissions of pollutants into the environment (Martinez-Burgos et al. 2021; Yukesh Kannah et al. 2021; IEA 2019a, b). Consequently, there is a significant potential of generating energy from OFMSW coupled with its high ecological efficiency. The quantity of electricity generation from H_2 in kWh is estimated according to Eq. (19) as shown in Table 5.

2.6 *Environmental Assessment of H_2 Power Generation*

According to the Energy Commission, Ghana is still far behind in achieving its renewable energy generation mix target of 10% by 2030 (Energy Commission 2022). The abundance of renewable energy sources such as solar, wind, biomass etc. provides a huge potential for the country to increase its share of renewable energy in the energy generation mix. Hydrogen generation from these available renewable resources could replace fossil fuels in electricity generation and transportation while emitting no hazardous pollutants. This study therefore assessed the environmental benefits of valorizing OFMSW into H_2 for energy generation by evaluating the quantity of fossil fuel that could be replaced by H_2 energy generation from OFMSW and the possible GHG emissions that can be avoided using Eq. (20) as provided in Table 6.

Table 6 Fossil fuel (diesel) replacement by H_2

No	Parameter	Equation	Reference
(20)	Fossil fuel replacement (liters)	$F_{(Diesel)} = \frac{H_{2(power)}}{\rho_{diesel} X \left(\frac{LHV_{(Diesel)}}{\varphi}\right) X\ GEN_{(ef)}}$ $F_{(\text{Diesel})}$ is the quantity of diesel fuel consumption replaced by electricity from the project, $\rho_{(\text{diesel})}$ is the density of diesel fuel and is taken as 0.84 kg/liter (Cudjoe et al. 2022), $LHV_{(\text{Diesel})}$ is the lower heating value of diesel fuel, which is given as 42.5 MJ/kg (Demirbas et al. 2014), $GEN_{(\text{ef})}$ is the efficiency of the diesel powered generator and is taken as 33% (Nizami et al. 2017), and φ is the conversion factor from MJ to kWh	Cudjoe et al. (2022)

2.7 Global Warming Potential (GWP) of H_2 Power Generation

It is well known that to combat climate change and limit global warming potential the world needs to make efforts to ensure the transition to a low-carbon energy system as the combustion of fossil fuel leads to the emission of GHGs that causes atmospheric pollution and is related to the phenomenon of global warming and climate change (Abokyi et al. 2019). The global warming mitigation potential of H_2 for energy generation was used for the environmental assessment. Based on the amount of diesel fuel that could be replaced by energy from H_2, the global warming reduction potential is estimated using Eq. (21), whereas Eqs. (22), (23) and (24) are used to estimate the CO_2 equivalent of CO_2, CH_4 and N_2O (Table 7).

3 Results and Discussion

3.1 MSW and Electricity Generation Potential

Per capita per day waste generation rate varies between 0.47 and 0.70 kg for the 20 MMDAs that fall within the 30 km radius. However, 0.70 kg/capita/day was applied as the per capita generation rate per day for urban communities with 0.47 kg/capita/day for rural centers. It is clear that the total waste generation per day varies widely from one municipality to another, depending on the population against the per capita waste generation (Fig. 3). The total waste generation potential is estimated at 1,780

Table 7 Summary of equations for GWP of H_2

No	Parameter	Equation	Reference
(21)	Total global warming reduction potential (kg CO_2eq)	$GWP_{(TOTAL)} = CO_{2e}(C)+CO_{2e}(CH)+CO_{2e}(N)$	Cudjoe et al. (2022)
(22)	Equivalent of CO_2 (kg CO_2eq)	$CO_{2e}(C) = F_{(Diesel)}\ X\ \in_{CO_2}\ X\ CO_{2(GWP)}$	
(23)	Equivalent of CH_4 (kg CO_2eq)	$CO_{2e}(CH) = F_{(Diesel)}\ X\ \in_{CH_4}\ X\ CH_{4(GWP)}$	
(24)	Equivalent of N_2O (kg CO_2eq)	$CO_{2e}(N) = F_{(Diesel)}\ X\ \in_{N_2O}\ X\ N_2O_{(GWP)}$ $GWP_{(TOTAL)}$ is the total global warming reduction potential, CO_{2e}(C), CO_{2e}(CH), and CO_{2e}(N) are the CO_2 equivalent of CO_2, CH_4, and N_2O, ϵ_{CO2}, ϵ_{CH4}, and ϵ_{N2O} are the specific emission factors of diesel fuel for the respective GHGs considered and are given as 2.7 kg/liter for CO_2, 3.612×10^{-4} kg/liter for CH_4, and 2.167×10^{-5} kg/liter for N_2O (IPCC 2006), $CO_{2(GWP)}$, $CH_{4(GWP)}$, and $N_2O_{(GWP)}$ are global warming potential of the GHGs and are taken as 1 kg CO_{2e} for CO_2 (Cudjoe et al. 2022), 298 kg CO_{2e} for N_2O (Cudjoe et al. 2022), and 32 kg CO_{2e} for CH_4 (Ayodele et al. 2020)	

tons/day with the urban centers consisting of 7 MMDAs representing an approximately 72% of the total waste generation in the 30 km radius zone. Even though, organic waste is the predominant waste in the study area with a potential of 470 tons/day of OFMSW, this contradicts the findings of Miezah and colleagues who estimated 61% as the organic fraction of MSW generated in Ghana (Miezah et al. 2015) and 55% reported by Ketibuah et al. (2004) on household waste characterization in Kumasi, respectively. This is attributed to the sampling techniques employed as waste samples were collected in communal containers for analysis in this study as compared to waste collected directly from households, implying that the waste could have undergone some form of decomposition prior to sampling for analysis. Additionally, the results show that KMA has the highest potential for MSW generation of 186.17 kt/y contributing 28.7% of the total waste to be generated due to its high per capita waste generation and high population as well, whereas Afigya Kwabre North District Assembly was the least contributor to the total waste generation in the study area. The results revealed in this study are consistent with the findings

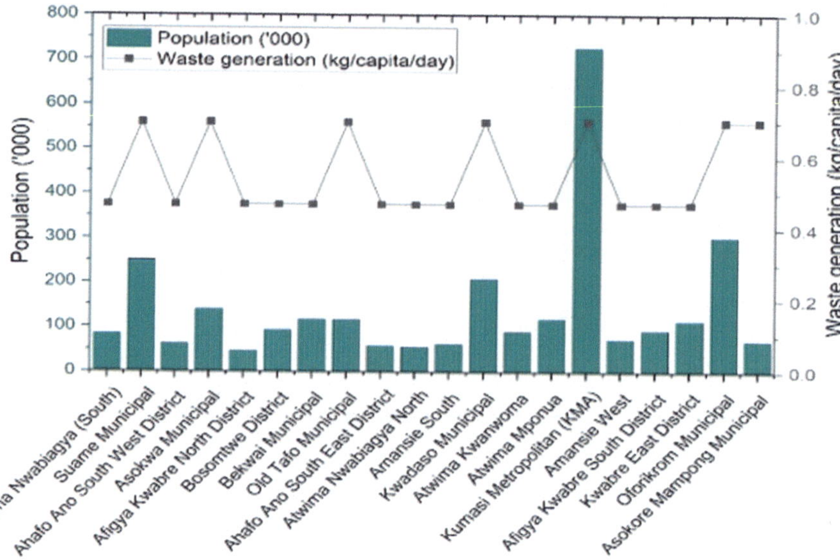

Fig. 3 Population against per capita waste generation in MMDAs

of Ogunjuyigbe et al. (2017), who found a direct correlation established between population and high MSW generation in some regions in Nigeria. The total amount of OFMSW collectable in the study area is 253.36 kt/y with a daily output of 694 tons.

Biogas for electricity generation is a well-established process that has been extensively used globally especially in developed countries with majority of such plants located in Europe. Electricity generation from biogas depends on potential biogas generation, and the quality of the feedstock coupled with other related parameters. Total biogas generation potential in the study area is estimated at 17,515 million m^3/y with the urban MMDAs contributing 95% of the actual biogas that could be produced from the OFMSW. It is further revealed that the total volume of purified biogas that could be generated is 13,259 million m^3/y. Kumasi metropolis had the highest biogas generation potential of 31.41 million m^3/d and a purified biogas potential of 23.78 million m^3/d. Electricity generation potential of biogas from OFMSW in the study is 40,000 GWh/y with a daily generation of 109.6 GWh. From the analysis, it is revealed that electricity generation is positively correlated with the potential waste generation and OFMSW. As shown in Fig. 4, KMA had the highest electricity generation potential of 26,181 GWh/y followed by Oforikrom municipal assembly (4,527 GWh/y) with Afigya Kwabre North district generating the least electricity of 47.61 GWh/y. Atwima Nwabiagye municipal assembly has electricity generation potential of 157.68 GWh/y representing 0.39% of the total electricity generation potential in the study area. This is primarily due to KMA higher population and per capita waste generation which resulted in higher waste generation and OFMSW as compared to the other MMDAs. This is corroborated by the findings of Cudjoe et al (2020), which

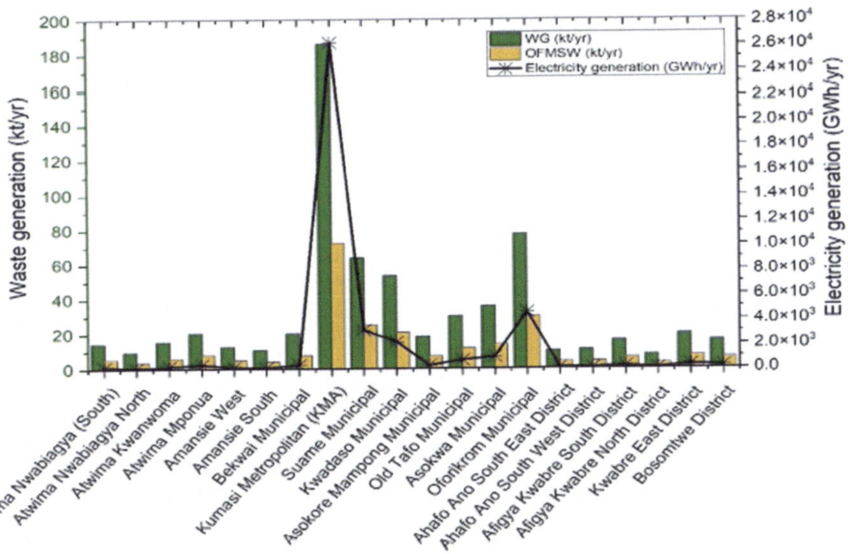

Fig. 4 Waste generation (t/d) and electricity generation (GWh/y)

concluded that high MSW generation is directly linked to high electricity generation potential in some provinces in China with the application of AD technology.

3.2 H_2 Energy Generation Potential Through SMR

It is widely acknowledged that biogas has great potential to produce H_2 through AD process. This study examined the potential H_2 gas generation from biogas using SMR. The quantity of H_2 produced in the MMDAs varied between 3.62 and 1991.25 kt annually with daily generation of 0.10–5.45 kt. The MMDAs in the urban centers constituted 95% of the total generation with the remaining 13 MMDAs primarily in the rural settings contributing the remaining generation potential. The above findings demonstrate a positive correlation between potential H_2 generation and population and per capital waste generation as well. The volume of purified CH_4 generation and H_2 yield are presented in Fig. 5. From the analysis, a total of 5,470 GWh/y of electricity could be available from H_2 yield from biogas derived from OFMSW in the study area. Similarly, the urban centers alone constitute 95% of the total H_2 yield from the biogas produced with an available electricity of 5,177.22 GWh/y (14.18 GWh/d). Similarly, KMA had the highest contribution of 37% followed by Oforikrom municipality of 12% with Afigya Kwabre North as the least contributing district of 0.98%.

Despite Ghana been an oil producing country, majority of the oil products consumed in the country is imported. According to the Energy Commission, Ghana

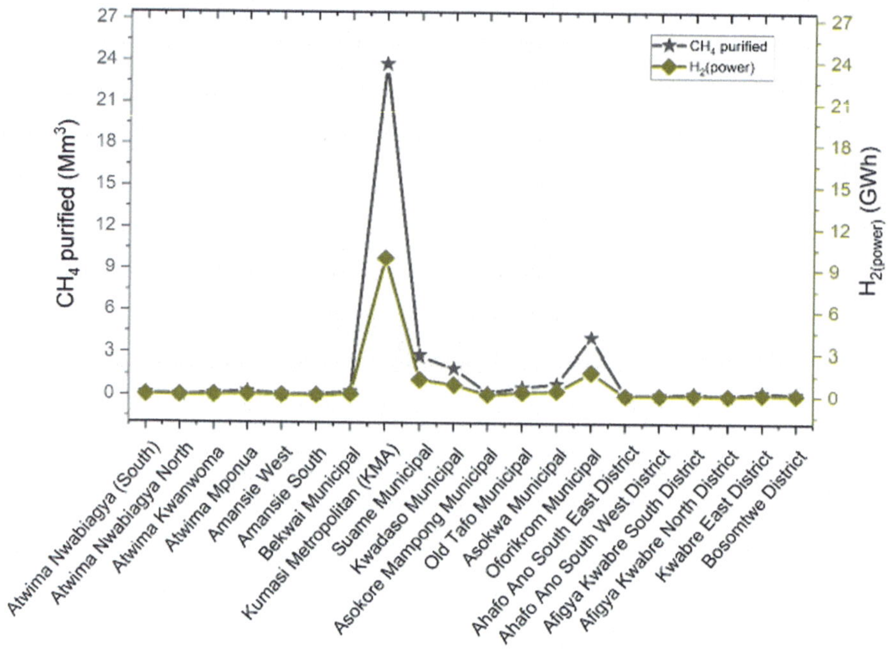

Fig. 5 Quantity of purified CH_4 (m^3) and H_2 (kg) from OFMSW

is estimated to have produced only 2.37% of its total gasoline consumed with the rest being imported in 2021. In total, 1,812 kt of gasoline products were consumed in 2021 (Energy Commission 2022). The environmental implications of the use of gasoline including the emission of GHGs with its cascading effect on climate change cannot be overemphasized. The energy sector in Ghana was the major source of GHG emissions in 2022 accounting for 46.66% of the country's total GHG emissions of 59.8 $MtCO_2e$, with fuel combustion representing 32.3% of the emission in the energy sector. From the analysis presented in this study, the development of H_2 in the study area could replace 0.45% of the total gasoline imported to Ghana based on the assumption that, 1 kg of liquid H_2 is equivalent to 2.78 kg of gasoline (Posso et al. 2016). It must be emphasized that the study area, i.e., the 30 km radius zone represents 53.23% of the total population of the entire Ashanti region. This suggests that a comprehensive assessment of the H_2 generation potential of the region could results in higher production taking into cognizance the characteristics of the waste and the region's population. Through H_2 generation initiative, the high importation of petroleum products into the country could be decreased while increasing production. The findings from the study indicate that OFMSW in Ghana has a great potential for bio-methane and H_2 generation due to the considerable high proportion of OFMSW in the waste component which are easily biodegradable. Though the findings are consistent with other research reported in the study area, the proportion of OFMSW differs from the work of Miezah et al. (2015), Kuffour (2020) and Osei-Mensah et al.

(2014). Nonetheless, it corroborates the work of Gioele Di Marcoberardino et al. (2018) and Alves et al. (2013), who concluded that H_2 is a potential energy carrier capable of replacing fossil fuels in power generation. This assessment could be used to demonstrate the potential of generating H_2 from organic waste for energy generation and could be seen as one of the renewable energy sources that could contribute significantly to achieve the target set in the Renewable Energy Act, 2011 (Act 832) as amended by the Renewable Energy Amendment Act, 2020 (Act 1045) to achieve 10% renewable energy penetration into the energy mix by 2030. Coupled with the energy generation, H_2 from OFMSW could be an integrated approach to organic waste management in the country.

3.3 *Environmental Impact Analysis*

From environmental perspectives, the generation of power from H_2 has the capacity to displace fossil fuel which could be seen as an innovative solution to decrease fossil fuel consumption with cascading impact on global warming reduction. The global warming reduction potential at the various MMDAs level ranged from 0.43 kt CO_2e to 237.10 kt CO_2e. Kumasi metropolis, Oforikrom, Suame and Kwadaso municipalities had the highest global warming reduction potential of 237.10 kt CO_2e, 41 kt CO_2e, 27.99 kt CO_2e and 19.51 kt CO_2e, whereas the lowest global warming reduction potential of 0.43 kt CO_2e, 0.63 kt CO_2e, 0.70 kt CO_2e and 0.79 kt CO_2e was recorded in Afigya Kwabre North, Atwima Nwabiagya North, Ahafo Ano South East and Ahafo Ano South West districts, respectively. With this analysis, Ghana could reduce its global warming by 362 kt CO_2e corresponding to 1.87% of the fuel combustion in the country. In a study by Seglah et al. (2023), it was found that H_2 generation from food waste in Kumasi could offset 3.83% of Ghana's GHGs. The differences in the results obtained in this study and Seglah et al. (2023) could be attributed to the percentage of food waste composition and the method employed. Nonetheless, the study established that electricity generation from H_2 derived from OFMSW could reduce global warming potential in Ghana caused by GHGs including CO_2, CH_4, and N_2O.

4 Conclusion

Hydrogen gas production from OFMSW can be harnessed in Ghana to aid Ghana in its renewable energy generation mix by 2030. Potential H_2 generation and environmental impacts of valorization of OFMSW in a hybrid waste to energy plant have been analyzed. The study demonstrates that waste generated 253.36 kt annually and 694 tons daily capable of producing 17,515 million m^3/y. The amount of purified CH_4 generated has a total of 5,470 GWh/y of electricity available from H_2 yield from biogas derived from OFMSW, with the potential to reduce GHG emission by displacing diesel. With Ghana's large biodegradable potential compared to its fossil fuel consumption, substantial quantities of H_2 can be produced from OFMSW to augment its energy needs and displace high proportion of its fossil fuel consumption. H_2 could meet the country's fossil fuel demand in a future scenario and eventually replace petrol as the nation's primary fuel. The results further support the concept of integrating H_2 production into W2E plant and highlight the value of such innovative technologies in facilitating the transition towards a green H_2 and sustainable economy.

References

Abokyi E, Appiah-Kinadu P, Abokyi F, Oteng-Abayie FE (2019). Industrial growth and emissions of CO2 in Ghana: The role of financial development and fossil fuel consumption. Energy Rep 5:1339–53. https://doi.org/10.1016/j.egyr.2019.09.002.

Alibardi L, Cossu R (2016) Effects of carbohydrate, protein and lipid content of organic waste on hydrogen production and fermentation products. Waste Manag 47:69–77. https://doi.org/10.1016/j.wasman.2015.07.049

Alves, H.J., Bley Junior, C., Niklevicz, R.R., Frigo, E.P., Frigo, M.S., Coimbra-Araújo, C.H. (2013). Overview of hydrogen production technologies from biogas and the applications in fuel cells. In: International Journal of Hydrogen Energy. pp. 5215–5225.

Arthur R, Baidoo MF, Antwi E (2011) Biogas as a potential renewable energy source: a Ghanaian case study. Renew Energy 36:1510e6. https://doi.org/10.1016/j.renene.2010.11.012

Ayodele, T.R., Alao, M.A., Ogunjuyigbe, A.S.O., Munda, J.L. (2019). Electricity generation prospective of hydrogen derived from biogas using food waste in south-western Nigeria. Biomass Bioenergy. 127. https://doi.org/10.1016/j.biombioe.2019.105291.

Bonk F, Bastidas-Oyanedel J-R, Schmidt JE (2015) Converting the organic fraction of solid waste from the city of Abu Dhabi to valuable products via dark fermentation – economic and energy assessment. Waste Manag 40:82–91. https://doi.org/10.1016/j.wasman.2015.03.008

Boyano A, Morosuk T, Blanco-Marigorta AM, Tsatsaronis G (2012) Conventional and Advanced exergoenvironmental analysis of a steam methane reforming reactor for hydrogen production. Journal of Cleaner Production 20:152–60. https://doi.org/10.1016/j.jclepro.2011.07.027

Braga LB, Silveira JL, Da Silva ME, Tuna CE, Machin EB, Pedroso DT (2013) Hydrogen production by biogas steam reforming: a technical, economic and ecological analysis, Renew. Sustain. Energy Rev. 28;166–173, https://doi.org/10.1016/j.rser.2013.07.060

Buswell AM, Hatfield WD (1936) Anaerobic fermentations, State Illinois, Dept. of Registration and Education, Bull. 32.

Cappai G, De Gioannis G, Friargiu M, Massi E, Muntoni A, Polettini A, Pomi R, Spiga D (2014) An experimental study on fermentative H-2 production from food waste as affected by pH. Waste Manag 34:1510–1519. https://doi.org/10.1016/j.wasman.2014.04.014

Cesaro A (2021) The valorization of the anaerobic digestate from the organic fractions of municipal solid waste: Challenges and perspectives. J Environ Manag 280:111742. https://doi.org/10.1016/j.jenvman.2020.111742

Energy Commission (2022), 2022 National Energy Statistics, (https://energycom.gov.gh/files/2022%20Energy%20Statistics.pdf). Accessed 5 July 2023

Cudjoe D, Han MS, Nandiwardhana AP (2020) Electricity generation using biogas from organic fraction of municipal solid waste generated in provinces of China: Techno-economic and environmental impact analysis. Fuel Process Technol 203:106381. https://doi.org/10.1016/j.fuproc.2020.106381.

Cudjoe D, Chen W, Zhu B (2022) Valorization of food waste into hydrogen: energy potential, economic feasibility and environmental impact analysis, Fuel 324;124476, https://doi.org/10.1016/j.fuel.2022.124476

Dahiya S, Chatterjee S, Sarkar O, Mohan SV (2021) Renewable hydrogen production by dark fermentation: current status, challenges and perspectives. Bioresour Technol 321:124354. https://doi.org/10.1016/j.biortech.2020.124354

Demirbas A, Baluabaid MA, Kabli M, Ahmad W (2014) Diesel fuel from waste lubricating oil by pyrolytic distillation. Pet Sci Technol 33(2):129–38. https://doi.org/10.1080/10916466.2014.955921.

Dufo-L´opez R, Bernal-Agustín LJ (2008) Multi-objective design of PV–wind–diesel–hydrogen–battery systems. Renewable Energy 33(12):2559–72. https://doi.org/10.1016/j.renene.2008.02.027.

Environmental Protection Agency (EPA) (2023) Ghana's Fifth National Greenhouse Gas Inventory, 2021 National Greenhouse Gas Emissions Report. https://unfccc.int/sites/default/files/resource/gh_nir5_15052022_final.pdf. Accessed 12 March 2023

Escamilla-Alvarado C, Poggi-Varaldo HM, Ponce-Noyola T, Rios-Leal E, Robles-Gonzalez I, Rinderknecht-Seijas N (2015) Saccharification of fermented residues as integral part in a conceptual hydrogen-producing biorefinery. Int J Hydrog Energy 40:17200–17211. https://doi.org/10.1016/j.ijhydene.2015.06.164

Fabian P, Sebastian P, Gregor S, Agyemang NSD, Stefan P (2019) Biomass sources for a sustainable energy supply in Ghana e a case study for Sunyani. Renew Sustain Energy Rev 107:413e24

Francis K, Andreas K, Sune TT, Cudjoe BE, HanneØstergård (2014) Assessment of biomass residue availability and bioenergy yields in Ghana. Resour Conserv Recycl 86:28e37

Francis K, Ivan N, Gordon M (2015) Prospects for bioenergy use in Ghana using long-range energy alternatives planning model. Energy 93:672e82

Gadhe A, Sonawane SS, Varma MN (2014) Ultrasonic pretreatment for an enhancement of biohydrogen production from complex food waste. Int J Hydrog Energy 39:7721–7729. https://doi.org/10.1016/j.ijhydene.2014.03.105

Ghana Statistical Service (GSS): Population and Housing Census (2021) Population and Housing Census Report (2021)

Gioele Di Marcoberardino DV, Spinelli F, Marco Binotti ID, Manzolini G (2018) Green hydrogen production from raw biogas: a techno-economic investigation of conventional processes using pressure swing adsorption unit. Processes 6:19, https://doi.org/10.3390/pr6030019

Gross S (2020) Why are fossil fuels so hard to quit? In: Brookings Inst. https://www.brookings.edu/essay/why-are-fossil-fuels-so-hard-to-quit/#:~:text=We understand today that humanity's, climate of our entire planet. Accessed 2 April 2023

Guo XM, Trably E, Latrille E, Carrere H, Steyer J-P (2014) Predictive and explicative models of fermentative hydrogen production from solid organic waste: role of butyrate and lactate pathways. Int J Hydrog Energy 39:7476–7485. https://doi.org/10.1016/j.ijhydene.2013.08.079

IEA (2019a) The future of hydrogen. Seizing today's opportunities. https://iea.blob.core.windows.net/assets/9e3a3493-b9a6-4b7d b4997ca48e357561/The_Future_of_Hydrogen.pdf. Accessed 12 June 2023

IEA (2019b). The Future of Hydrogen.

IPCC (2006) Guidelines for National Greenhouse Gas Inventories, http://www.ipccnggip.iges.or.jp/public/2006gl/vol2.html. Accessed 10 February 2023.

Karthikeyan OP, Trably E, Mehariya S, Bernet N, Wong JWC, Carrere H (2018) Pretreatment of food waste for methane and hydrogen recovery: a review. Bioresour Technol 249:1025–1039. https://doi.org/10.1016/j.biortech.2017.09.105

Ketibuah E, Asase M, Yussif S, Mensah MY, Fischer K (2004) Comparative Analysis of Household Waste in the Cities of Stuttgart and Kumasi – Options of Waste Recycling and Treatment in Kumasi. Proceedings of the 19th International CODATA Conference Berlin, Germany, 7–10 November

Kim JK, Nhat L, Chun YN, Kim SW (2008) Hydrogen production conditions from food waste by dark fermentation with Clostridium beijerinckii KCTC 1785. Biotechnol Bioprocess Eng 13:499–504. https://doi.org/10.1007/s12257-008-0142-0

Kuffour RA (2020) Possibility of improving solid waste management in senior high schools in the Ashanti region of Ghana. Afr. J. Environ. Sci. Technol., 14(8):231-240, https://doi.org/10.5897/AJEST2020.2859

Li M, Zhao Y, Guo Q, Qian X, Niu D (2008) Bio-hydrogen production from food waste and sewage sludge in the presence of aged refuse excavated from refuse landfill. Renew Energy 33:2573–2579. https://doi.org/10.1016/j.renene.2008.02.018

Martelli E, Nordb LO, Bollandb O (2012) Design criteria and optimization of heat Recovery steam cycles for integrated reforming combined cycles with CO_2 capture. Applied Energy 92:255–68. https://doi.org/10.1016/j.apenergy.2011.10.043

Martinez-Burgos WJ, de Souza Candeo E, Pedroni Medeiros AB et al (2021) Hydrogen: current advances and patented technologies of its renewable production. J Clean Prod 286:124970. https://doi.org/10.1016/j.jclepro.2020.124970

Miezah K, Obiri-Danso K, Kadar Z, Fei-Baffoe B, Mensah MY (2015) Municipal solid waste characterization and quantification as a measure towards effective waste management in Ghana. Waste Manag 46, 15–27. https://doi.org/10.1016/j.wasman.2015.09.009

Mohammed M, S Egyir I, Donkor AK, Amoah P, Nyarko S, Boateng KK, Ziwu C (2017) Feasibility study for biogas integration into waste treatment plants in Ghana. Egyptian Journal of Petroleum 26:695e703. https://doi.org/10.1016/j.ejpe.2016.10.004

Nielfa A, Cano Fdz-Polanco RM (2015) Theoretical methane production generated by the co digestion of organic fraction municipal solid waste and biological sludge, Biotechnol. Rep. 14 21, https://doi.org/10.1016/j.btre.2014.10.005.

Nizami, A.S., Shahzad, K., Rehan, M., Ouda, O.K.M., Khan, M.Z., Ismail, I.M.I., Almeelbi, T., Basahi, J.M., Demirbas, A. (2017). Developing waste biorefinery in Makkah: A way forward to convert urban waste into renewable energy. Appl Energy. 186, 189–196. https://doi.org/10.1016/j.apenergy.2016.04.116.

Ogunjuyigbe ASO, Ayodele TR, Alao MA (2017) Electricity generation from municipal solid waste in some selected cities of Nigeria: an assessment of feasibility, potential and technologies, Renew. Sustain. Energy Rev. 80:149–162, https://doi.org/10.1016/j.rser.2017.05.177

Osei-Mensah P, Adjaottor AA, Owusu-Boateng G (2014) Characterization of solid waste in the Atwima Nwabiagya District of the Ashanti Region, Kumasi-Ghana. Int. J. Waste Manage. Technol. 2(1):1-14

Paillet F, Barrau C, Escudié R, Bernet N, Trably E (2021) Robust operation through effluent recycling for hydrogen production from the organic fraction of municipal solid waste. Bioresour Technol 319:124196. https://doi.org/10.1016/j.biortech.2020.124196

Posso F, Sanchez J, Espinoza JL, Siguencia J (2016) Preliminary estimation of electrolytic hydrogen production potential from renewable energies in Ecuador, Int. J. Hydrogen Energy 41(4):2326–2344, https://doi.org/10.1016/j.ijhydene.2015.11.155

Saebea D, Authayanun S, Patcharavorachot Y, Arpornwichanop A (2014) Thermodynamic analysis of hydrogen production from the adsorption-enhanced steam reforming of biogas, Energy Proc. 61:2254–2257, https://doi.org/10.1016/j.egypro.2014.12.120

Salami L, Susu AA (2011) Characterization study of solid wastes: a case of Lagos State. Int J Appl Sci Technol 3:47e52.

Seglah PA, Wang Y, Wang H, Neglo KAW, Zhou K, Sun N, Shao J, Xie J, Bi Y, Gao C (2023) Utilization of food waste for hydrogen-based power generation: Evidence from four cities in Ghana. Heliyon 9:e14373. https://doi.org/10.1016/j.heliyon.2023.e14373

Sekoai PT, Gueguim Kana EB (2014) Semi-pilot scale production of hydrogen from organic fraction of solid municipal waste and electricity generation from process effluents. Biomass Bioenergy 60:156–163. https://doi.org/10.1016/j.biombioe.2013.11.008

Sharma P, Melkania U (2018) Effect of bioaugmentation on hydrogen production from organic fraction of municipal solid waste. Int J Hydrog Energy 43:7290–7298. https://doi.org/10.1016/j.ijhydene.2018.03.031

Silveira JL, Braga LB, Souza ACC, Antunes JS, Zanzi R (2009) The benefits of ethanol use for hydrogen production in urban transportation. Renewable & Sustainable Energy Reviews 13:2525–34.

Yeshanew MM, Paillet F, Barrau C, Frunzo L, Lens PNL, Esposito G, Escudie R, Trably E (2018) Co-production of hydrogen and methane from the organic fraction of municipal solid waste in a pilot scale dark fermenter and methanogenic biofilm reactor. Front Environ Sci 6:41. https://doi.org/10.3389/fenvs.2018.00041

Yukesh Kannah R, Kavitha S, Preethi et al (2021) Techno-economic assessment of various hydrogen production methods—A review. Bioresour Technol 319:124175. https://doi.org/10.1016/j.biortech.2020.124175

Zhao M, Liu Z, Xu J, Liu H, Dai X, Gu S, Ruan W (2020) Dosing effect of nano zero valent iron (NZVI) on the dark hydrogen fermentation performance via lake algae and food waste co-digestion. Energy Rep 6:3192–3199. https://doi.org/10.1016/j.egyr.2020.11.200

Uncited References

Adeoti, O., Ayelegun, T.A., Osho, S.O. (2014). Nigeria biogas potential from livestock manure and its estimated climate value.

Ayodele TR, Ogunjuyigbe ASO, Alao MA (2018) Economic and environmental assessment of electricity generation using biogas from organic fraction of municipal solid waste for the city of Ibadan, Nigeria. J Clean Prod 203:718e35. https://doi.org/10.1016/j.jclepro.2018.08.282

Envergent (2010). The Production of Electricity from Wood and Other Solid Biomass.

ICF (2014). Diesel Generators: Improving Efficiency and Emission Performance in India.

MIT (2007). Units & Conversions Fact Sheet. MIT EL 00–003 Press.

Advances in Municipal Solid Waste Management and Industrial Use of Residue Materials

Addressing Municipal Solid Waste Mismanagement Through Data-Driven Action Planning: A Case Study in Mukono, Uganda

Haniyeh Jalalipour and Pauline Namutebi

Abstract Efficient and effective management of municipal solid waste (MSW) poses a significant challenge in Uganda, especially in impoverished urban and peri-urban areas where the growing population strains the already struggling waste infrastructure. In the greater Kampala region, only 40% of the daily 250 tonnes of waste generated is properly collected, while the remaining 60% is abandoned, burned, or improperly disposed of. This mismanagement exposes communities to severe health risks and contributes to environmental pollution, thereby increasing the prevalence of diseases. Municipalities in the region face numerous obstacles, including the absence of effective frameworks, regulations, monitoring, and insufficient resources. However, a critical hurdle in addressing these issues is the lack of reliable information, hindering the development of sustainable solid waste management plans.

This study emphasizes the significance of collecting accurate waste data and conducting a comprehensive baseline study using available waste characterization methodologies and tracking tools. It advocates for enhancing data-driven MSW planning through capacity building among local stakeholders. By conducting a waste management baseline study using tools such as the Waste Wise Cities Tool (WaCT), Waste Flow Diagram (WFD), and Wasteaware Benchmark Indicators (WABIs), the municipality can identify key intervention areas and address policy and infrastructure investment gaps that contribute to open waste burning in the city. Strengthening the technical capacity of local partners through these initiatives facilitates data collection and the implementation of evidence-based strategies to mitigate the negative effects of openly burning municipal solid waste on public health and safety. The skills developed through this process can be leveraged for future data collection and replicated by other practitioners in the region. Additionally, evidence-based strategy and action planning can attract potential donors to support impactful projects on the ground.

H. Jalalipour (✉)
Professorship of Waste and Resource Management, University of Rostock, Justus-von Liebig-Weg 6, 18059 Rostock, Germany
e-mail: haniyeh.jalalipour@uni-rostock.de

P. Namutebi
Fireball Energy Uganda Limited, Kabembe, Mukono District, Uganda

M. Narra and S. Narra (eds.), *Innovations in Circular Economy and Renewable Energy in Africa*, World Sustainability Series, https://doi.org/10.1007/978-3-031-68330-5_13

Keywords Waste management · Open burning · Capacity building · Waste characterization · Action plan

1 Introduction

The collection, transportation, and proper disposal of municipal solid waste remain a big challenge in Uganda, especially in the urban poor and peri-urban areas (Castellani et al. 2022). The increasing population in these areas not only increases the waste generated per capita but also increases the pressure on an already failing solid waste management infrastructure. A partial amount of waste is collected by the managing authority and private waste collectors, while the remaining is often left lying on the streets, open burned, or drained (KCCA 2020). Mismanagement of municipal solid waste increases public health risks and environmental pollution. Communities with direct exposure to polluted air and water often suffer from diseases such as asthma and cholera.

The majority of households lack sufficient understanding regarding appropriate methods for managing solid waste (Majak 2020). Economic constraints in these households contribute to their inability to cover service expenses (Mukama et al. 2016). Local authorities periodically collect solid waste from households and public bins, resulting in a situation in which many households opt to improperly dispose of waste in open areas. The open burning of waste both in dump sites and residential areas is frequent and poses a serious threat to surrounding populations. The burning of large quantities of unselective waste occurs often close to the ground and is characterized by large black smoke clouds from slow burning that lasts for substantial amounts of time.

The existing conditions merit immediate action on a national scale. Recently, the Greater Kampala Metropolitan area launched a new solid waste management strategy that aims to collect and safely manage up to 65% of generated waste, divert at least 35% of the collected waste from reaching landfill, reduce per capita plastic emissions by 30%, reduce GHG by 15% and create up to 50,000 jobs by 2030 (GGGI 2023). However, the municipality doesn't have effective frameworks in place to address the faced issues. The absence of reliable data is one of the main challenges that hinder solid waste management development. City officials, as the leading local stakeholders, do not have the capability to collect solid waste-related data and promote the system.

Therefore, prior to developing a waste management plan, it is essential to raise local stakeholders´ awareness of the need to mitigate the mismanagement of waste, strengthen the knowledge and technical capacity of local stakeholders for sustainable resource and waste management and identify key intervention areas of municipal solid waste management. This study aims to highlight the significance of collecting accurate waste data and conducting a comprehensive baseline study using available waste characterization methodologies and tracking tools. It advocates for enhancing data-driven MSW planning through capacity building among local stakeholders.

2 Methodology

2.1 *Study Area*

According to the Uganda Bureau of Statistics (2019) report, Mukono district has a total population of approximately 701,400 people. Mukono is part of the Greater Kampala Metropolitan area. This area includes Kampala Central Business District, Wakiso District, and Mukono District. The urbanization rate is estimated at 27.3% in Mukono district and subsistence farming served as the primary source of income for the vast majority of families.

2.2 *Research Design*

In this research study, the Mukono district in Uganda was investigated, with a specific focus on waste management practices. It was found that there is limited reliable data and literature concerning this topic, highlighting the necessity for comprehensive research in this critical domain. The initial step involved generating a waste flow diagram based on specific assumptions in the greater Kampala region. This diagram was utilized as a foundational tool to gain a better understanding of the existing waste management processes and their implications. Subsequently, an in-depth analysis of stakeholders interested in exploring alternative solid waste management (SWM) systems was conducted. This led to the development of a comprehensive stakeholder matrix, allowing for the identification of key actors and their potential roles in promoting sustainable waste management practices. Moreover, the study sought to identify and address the challenges that pose serious risks to both the environment and human health in these regions. By identifying and comprehending these challenges, activities were suggested to collect reliable waste data.

3 Results and Discussion

3.1 *Current Waste Management in the Greater Kampala Region*

The Kampala Waste Management Plan (KCCA 2020) estimates the generation rate per capita at 0.37 kg per day in the Mukono district which equals 250 tons per day of municipal solid waste generation. Yusuf et al. (2019) calculated the annual production of solid waste in Mukono municipality to be 25,023.6 tons, assuming a generation rate of 0.6 kg/person/day (Fig. 13.1). Organic waste accounts for a significant portion of MSW production (Majak 2020).

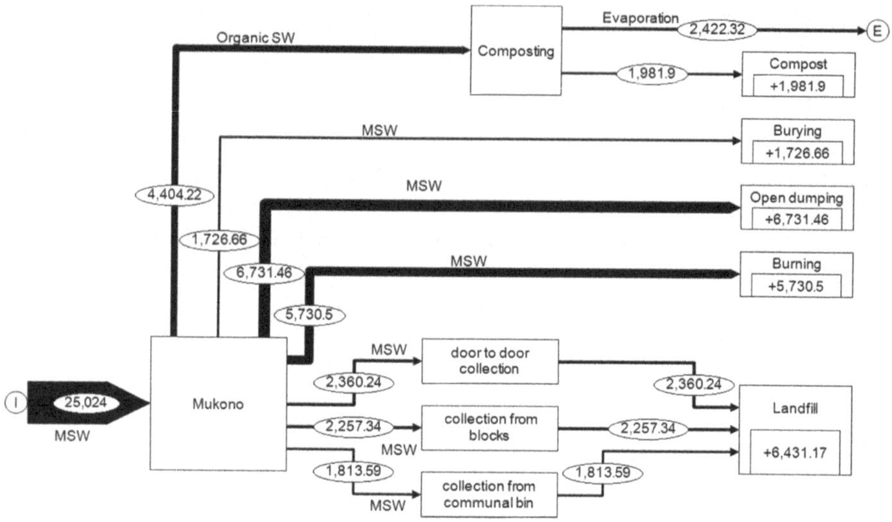

Fig. 13.1 Quantitative waste flow analysis of Mukono district (data from Majak 2020) visualized by the authors through the Software Stan® (TU Wien, Austria)

According to Majak (2020), 26.9% of households in Mukono municipality open dump their wastes, while 25.7% use a collection system and a landfill, 22.9% dispose of their solid waste by burning, 6.9% dispose of their solid waste by burying in the soil, and 17.6% use composting method for waste disposal. Collection methods include private waste collection from door to door (36.7%), Collection from blocks of houses (35.1%), and collection from communal bins (28.2%). The waste collected by the municipality and licensed private waste collectors is transported to an open dumpsite with about 15 hectares of area located at Katikolo village (KCCA 2020).

3.2 *Stakeholder Matrix*

The stakeholder map identified the key individuals, institutions, and groups involved in waste management, environmental conservation, and local governance in Mukono district as shown in Fig. 13.2. The map showcases the power, influence, and interests of the different stakeholders involved and impacted by waste management. This served as a guide to understand their roles, conflicting interests, and potential contributions to the data collection and baseline study initiative. The key stakeholders in the Mukono district to ensure comprehensive data gathering and better waste management practices include the municipal council collector, private waste collators in the landfill, waste pickers, and plastic manufacturers involved in plastic recycling.

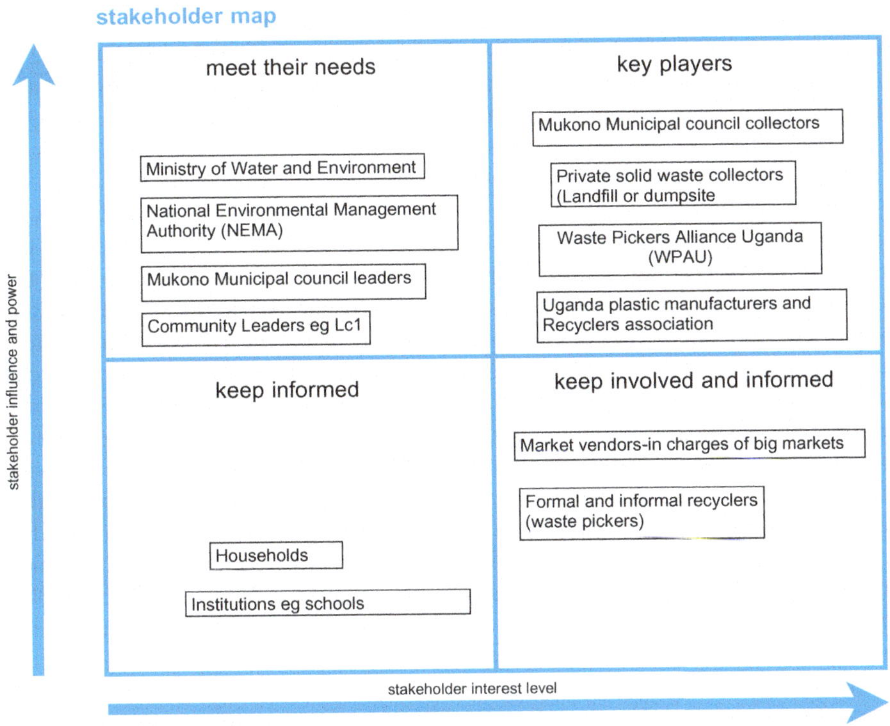

Fig. 13.2 Stakeholders interested in exploring alternative SWM systems in the study area

3.3 *Addressing Health and Safety Concerns*

Access to basic urban services and working under safe conditions are critical to sustainable urban development. In Uganda, under existing conditions, the lack of an effective solid waste management framework results in immense mismanagement of municipal waste including open burning, dumping, and draining (Castellani et al. 2022). Vulnerable populations are affected the most by facing social, economic, political, and environmental challenges. They usually reside near illegal dumpsites and many people earn a living from waste. Further, the absence of a monitoring system deteriorates the environmental control at the waste management facilities and the health and safety conditions of the workers including informal waste pickers. In order to identify the key intervention areas and plan an effective municipal solid waste management system, accurate data is vital. Therefore, priority should be given to generating waste data that enables the local stakeholders to make data-driven decisions. This results in increasing knowledge and raising awareness on mitigating the open burning of waste as well as generating opportunities to better manage resources, and create business and livelihood opportunities.

In this regard, the Waste Wise Cities Tool (WaCT) may be used to conduct a rapid assessment of a city's municipal solid waste management system (UN-Habitat 2021). WaCT was developed by UN-Habitat, in 2018 to enable city officials to measure SDG indicator 11.6.1 "Proportion of municipal solid waste collected and managed in controlled facilities out of total municipal solid waste generated, by the city," food waste generation (SDG 12.3.1) and resource recovery systems (SDG 12.5.1).

The Waste Flow Diagram (WFD), a tool compatible with WaCT (GIZ, University of Leeds, Eawag-Sandec, Wasteaware, 2020), can be used to determine the output of plastic leakage. It uses a rapid, observation-based evaluation to trace the flow of plastic waste and quantify plastic leakage from MSW management systems. In addition, to evaluate municipal solid waste management and recycling system performance in a city, in an organized way, Wasteaware benchmark indicators can be used along with the WaCT (Wilson et al. 2015). It enables a city to assess its own performance in terms of providing solid waste management services, to provide information for decision-making on priorities for the limited funds available for service improvements, to identify regional strengths that can be built on and weak points that need to be addressed, and to track changes over time. Further, it is recommended to conduct workshops and offer innovative training sessions, such as the UN-Habitat online course "From Data to Tangible Impact—Achieving Waste SDGs by 2030." This approach will effectively raise awareness among local stakeholders regarding the importance of addressing the issue of open burning of waste.

These methods will streamline data collection and facilitate evidence-based strategies, effectively mitigating the adverse effects of open burning of municipal solid waste on public health and safety. The skills developed by city officials and the implementation team can be leveraged for future data collection efforts, making it replicable by other practitioners in the region. By adapting an evidence-based strategy, ensuring a comprehensive approach to address waste management challenges. This, in turn, will create a compelling project proposal, inviting potential donors to contribute significantly to the cause, and ultimately making a positive impact on the ground.

4 Conclusion

This article highlights the crucial importance of two key factors for sustainable waste management. Firstly, the collection of accurate waste data and the execution of a thorough baseline study through established waste characterization methods and tracking tools are fundamental. These steps serve as the foundation upon which effective waste management strategies can be built, ensuring that decision-makers have a clear understanding of the composition and trends in MSW.

Secondly, the enhancement of data-driven MSW planning is emphasized, achieved through empowering local stakeholders with capacity-building initiatives. By equipping these stakeholders with the skills and knowledge to analyze and interpret waste data, a more collaborative and informed approach to waste management can be fostered. This enables communities to tailor their strategies to the unique challenges and opportunities presented by their waste profiles, thereby promoting more sustainable and efficient waste management practices. Overall, it underscores that accurate data and informed stakeholders are cornerstones in the pursuit of effective and sustainable waste management solutions.

References

Castellani P, Ferronato N and Torretta V (2022) Setting priorities to achieve Sustainable Development Goals through appropriate waste management systems in Uganda. Environmental Development 44: 100764.

GIZ, University of Leeds, Eawag-Sandec, Wasteaware (2020) Waste Flow Diagram (WFD): A Rapid Assessment Tool for Mapping Waste Flows and Quantifying Plastic Leakage. Version 1.0. February 2020. Authors: Velis CA, Cottom J., Zabaleta, I; Zurbruegg, C; Stretz, J; Blume, S. University of Leeds.[Online] Leeds, UK.

Global Green Growth Institute (GGGI). 2023. Greater Kampala Metropolitan Area´s Solid Waste Management Strategy. Accessed from: https://gggi.org/validation-workshop-of-the-greater-kampala-metropolitan-area-solid-waste-management-strategy/.

Kampala Capital City Authority (KCCA). 2020. Kampala Waste Management Plan, Accessed from: https://www.kcca.go.ug/media/docs/IFC-KWM-PIM-12-MARCH.pdf

Majak, J. M. 2020. Influence of household characteristics on solid waste management practices at the household level in Mukono town, Uganda. Master of Science Degree in Environment and Natural Resources, Africa Nazarene University. Nairobi, Kenya.

Mukama T, Ndejjo R, Musoke D, et al. (2016) Practices, Concerns, and Willingness to Participate in Solid Waste Management in Two Urban Slums in Central Uganda. Journal of environmental and public health 2016: 6830163.

Uganda Bureau Of Statistics (UBOS). 2019. UBOS Statistical Abstract, Accessed from: https://www.ubos.org/wp-content/uploads/publications/01_20202019_Statistical_Abstract_-Final.pdf

UN-Habitat (2021) Waste Wise Cities Tool: Step-by-Step Guide to Assess a City's Municipal Solid Waste Management Performance through SDG indicator 11.6.1 Monitoring. Available at: https://unhabitat.org/sites/default/files/2021/02/Waste%20wise%20cities%20tool%20-%20EN%203.pdf.

Wilson DC, Rodic L, Cowing MJ, et al. (2015) 'Wasteaware'benchmark indicators for integrated sustainable waste management in cities. Waste management 35: 329–342.

Yusuf AA, Peter O, Hassan AS, et al. (2019) Municipality solid waste management system for Mukono District, Uganda. Procedia Manufacturing 35: 613–622.

Cashew Nutshell and Raw Coal Co-grinding for Sustainability Goals in Clinkerization Pyro-Processes

E. Beguedou, S. Narra, K. Agboka, E. A. Armoo, and M. K. Damgou

1 Introduction

Many people think that the cashew tree, *Anacardium occidentale*, originated in South America and was spread to Africa by Portuguese traders. The tree is mostly planted for its nuts, which are a delicacy all over the world, and flourishes in tropical areas of the world. The cashew peduncle is initially green, becoming red, yellow, or orange as it reaches full maturity. The fruit consists of an apple with a fleshy peduncle that is edible, along with a nut or kernel that is protected by a shell at the fruit's base. According to several sources the peduncle weighs between 50 and 140 g. According to Sivagurunathan et al. (2010), 100 g of fresh peduncle will yield between 10.8 and 29.9 ml of syrup leaving behind 72.5–84.5 g of residue. This means that even when the peduncle is used for syrup production, a significant percentage of the peduncle is returned as waste material. This indicates that a sizeable portion of the peduncle is wasted even when it is utilized to make syrup.

Cashew is the most promising cash crop in the sub-humid regions of Togo. For a total area of 33,196 ha recorded in Togo as part of the national census of agriculture (MAEP 2013), about 14,969 hectares are found in the Plateau region (45.1%), 12,208 hectares in the Central region (36.8%) and 3,543 ha in the Kara region (about 10.7%).

E. Beguedou (✉) · S. Narra · E. A. Armoo
Professorship of Material and Energy Valorisation of Biogenous Residues, Professorship of Waste and Resource Management, University of Rostock, 18059 Rostock, Germany
e-mail: essossinam.beguedou85@gmail.com

S. Narra · K. Agboka
German Biomass Research Centre gGmbH, Torgauer Straße 116, 04347 Leipzig, Germany

M. K. Damgou
University of Lomé, West African Service Center for Climate Change and Adaptive Land Use (WASCAL), P.O. Box 1515, Lomé, Togo

Department of Physique, Centre d'Excellence Regional Pour la Maîtrise de l'Electricité (CERME), University of Lomé, P.O. Box 1515, Lomé, Togo

M. Narra and S. Narra (eds.), *Innovations in Circular Economy and Renewable Energy in Africa*, World Sustainability Series, https://doi.org/10.1007/978-3-031-68330-5_14

Most plantations are found in the prefectures of Tchamba, Est-Mono, Ogou and Blitta (Table 1). New plantations have been set up in Blitta Prefecture over the last 5 years. Few of them are already in production. Their actual extent would require an update of the Togolese agricultural census (MAEP 2013).

Only one type of combustible residue is produced from cashew processing. These are the hulls that come from the shelling of nuts in the processing units (Sawadogo et al. 2018). To date, only two structures known as "CAJOU ESPOIR" and "CAJOU CENTRAL" are involved in this transformation.

The transformation plants are in Tchamba, Sokode and Blitta. With a processing capacity of less than 16,000 tons of cashew per year, the tree plants are not able to handle all the domestic production, which was 27,360 tons in 2021. The surplus is exported indirectly via Benin or directly via the port of Lomé to Asian countries for processing (MAEP 2013).

The hulls produced in both plants are estimated at 8,290 tons, of which approximately 60% would be available for sale. This equates to 1,974 tons of currently available cashew hulls (Table 2) shows out a potential of approximately 11,491 tons of movable cashew hulls if all domestic production were processed on site.

Cashew Shells is available through cashew plants in the Central area of Togo about 390 km from Lomé, pyro industries or cement industries, collecting from different cashew plants (Cajou du centre, Cajou Espoir), currently this is the biggest availability in the entral region of Togo. Material is received ready to be used.

Togo cashew nuts industry has immensely grown over the years, recording a 54% increase of export value to the oversea market in years before.

No large cashew nutshell was recovered in Togo before this industrial sustainable project. The aim of this this industrial sustainable project is to highlight the co-grinding processing, the clinker quality impact in the firing of the mix fuels and the social economy output of the using at high scale of cashew nutshell.

Its exploitation in clinkarization pyro-processing is not only enhance the economic returns to the cashew industry but also contribute significantly to conservation of

Table 1 Area of productive cashew plantations in the study area (MAEP 2013)

Zone	Blitta	Sotouboua	Tchamba	Tchaoudjo	Est-Mono	Ogou	Anié
Area (ha)	366	1 173	7 859	2 682	8 688	2 294	3 670
	26 732						

Table 2 Cashew shell production by cashew hope processing plants

Plants	Cashew transformed (T)	Cashew nutshell produced (T)	Avalable cashew nutshell (T)
Cajou central	1 2000	8400	5040
Cajou espoir	4700	3290	1974
Total	16,700	11,690	7014

the environment. It is recommended that further work be done to scale-up production of CNS based products and demonstrate feasibility of the same. This statement can complete only if the CNS in CNS-charcoal is carbonized.

Production and local processing of the cashew nuts accompanied by recovery of cashew nutshell should be enhanced. Cashew nutshell is becoming the main by-product from cashew nuts processing because of this possible co-grinding in vertical coal mill.

The plant Nabors who used to get cashew nutshell as waste for free to make charcoal for home use, this social help become more difficult because of this good performance, the cashew processing plant are no longer interest to social assistant because of plant cost optimization by selling cashew nutshell. If the aim of sustainable project is to deal with the problem of family scarcity energy, now we assisting to the opposite effect with hutch consequences on the center Togolese population.

2 Materials and Methods

2.1 Cashew Nutshell and Raw Coal Analysis

As the data shows in Table 3, the raw coal LHV is about 23,500 kJ/kg., the sued cashew nutshell LHV is about 26537 kJ/KG the moisture of the coal is high due to open storage, for CNS is still in the target. The mixing percentage is 5.7% of total raw coal.

2.2 Vertical Coal Mine

Vertical coal mine (VCM) is a vertical mill with three grinding rollers and one grinding track. A grinding table rotates round a vertical axis in driven by a planetary gearbox. The three grinding rollers, mounted in fixed position, are passed down on the table under hydraulic loading. Raw coal enters through the center of the mill and is deposited on the grinding table. Centrifugal force feeds the coal radially outward where it is ground between the roller and grinding track.

The grinding force, is statically determined by three grinding rollers, which is generated by hydraulic loading. It starts from loading, to grinding rollers, grinding track, towards the foundation.

The pulverized coal is flowed by centrifuged force into the zone over the nozzle ring surrounding the table, where the coal particles are swept upwards by the stream of hot primary air into the classifier.

The intimate contact with the hot primary air causes the moisture contaminated in the coal to evaporate spontaneously. The oversize particles rejected by the classifier fall back onto the grinding table for further grinding. The fine particles, i.e. finished

Table 3 Fuels analysis

Biomass	Moisture	Q(net) [KJ/kg]	SiO_2	Al_2O_3	Fe_2O_3	CaO	MnO	TiO_2	MgO	K_2O	Na_2O	SO_3	P_2O_5
Cashew shell	5,5	26,537	57,4	11,72	6,9	2,55	0,02	1,04	0,96	2,91	0,16	1,56	14,45
Raw coal	8,2	23,500	61,8	25,6	3,4	2,1	0,01	1,2	0,8	1,9	0,1	1	1,3

product, are carried out of the classifier the mill is driven by a squirrel-cage induction motor. The torque of grinding table is driven by the vertical bevel-planetary gearbox which also stands the horizontal and vertical load caused by upper parts weight and grinding load.

2.3 ILC-Calciner

The In-line calciner (ILC) is usually designed as a big vessel located above kiln inlet area. The designs show extended riser duct, so called tube calciners.

Over the years of calciner development and necessity of burning the alternative fuels resulted in increasing of the internal volume of the calciner. It consists of the vessel located in front or behind the preheater tower that ends-up with so called loop (goose) duct. This duct makes a turn very often at the top of preheater tower and lowers onto the bottom cyclones.

The ILC calciners are usually contain the following plants: Loop (goose) duct, Calciner vessel, It's easy to operate, reliable, and have the lowest NOX emission, thanks to our unique high-temperature reduction zone. CO emission is low too, because our mid-calciner notch and specially designed exit duct provide superior mixing and combustion.

The In-Line calciner is suitable for firing both traditional and alternate fuels. And operation is easy, with single entry locations for both the calciner fuel and tertiary air.

The ratio of firing in the calciner is 60% of total which represented 2,1 tph of CNS, while achieving a calcination rate at the kiln inlet of 95% as objective. For buildup and chemistry there is an allowable bypass of kiln gas of 10%. This helps to deal with pre-calciner blockage in case. But it was not an issue of this co-grinding test.

Preheater optimization as well as calciner optimization, the core of the pyro-process of each modern cement kiln is the preheater. Most substantial thermo-chemical reactions of the cement process occur in the riser ducts and in the calciner and certainly in the sintering zone of the kiln. The kiln has five stages with Inline (ILC) calciner this provide a substantial degree of calcination and let us to consider it a core equipment of modern kilns.

2.4 VCM Operation Procedure

This Procedure is intended to provide the are several guidance's and instructions for carrying out the Coal Mill Start-up this is call operation procedure.

The working principle of the vertical coal grinding mill is like that of mill, coal mill normal starting procedure can be summarized as.

Grinding mill motor start: The start working of vertical coal mill is driven by an electric motor.

At this state the following parameter are controlled:

Mill inlet shutoff damper 30% open before feeding
Mill vent fan outlet damper 50% open before feeding
Raw coal transport system fully running without feeding

The raw coal enters the vertical mill bin: The reducer drives the millstone to rotate, and the mixed raw coal and CNS enters the mill through the feed port, scattering on the millstone. The controlling of this parameters is required:

Mill feed 30 tph (Wet basis) for the startup,
Mill DP: 45 mbar,
Mill grinding pressure is arrow 110 bar.

Grinding the raw coal: Because of hydraulic pressure, the extrusion between the grinding rollers and the millstone causes the big pieces of mixture of the raw coal and CNS crushed. However, the small pieces of powder (raw coal with CNS) will be crushed to coal powder by the sliding friction caused by the relative motion between the grinding rollers and the millstone. To succeed this level of action the following parameter are:

Fine coal residue: 12% residue on 90 μm
Mill inlet shutoff damper 100% open
Mill vent fan outlet damper 100% open

Pulverized coal drying: Raw coal is grinded in the vertical coal mill and dried by hot air (usually below 300°) into the vertical roller coal mill, which meets the drying standard of finished pulverized coal. For the safety operation of the following parameter are controlled to avoid the explosion of the coal mill bag filter

Mill outlet Temperature: 80 °C
Mill inlet oxygen level <10% (wet basis)
Bag filter outlet oxygen level <12% (wet basis)

Separation of finished coal powder: The pulverized coal powder is sent to the powder concentrator through the rising hot air stream for separation. The coal powder meeting the fineness requirement is transported to the pulverized coal bin, and the unqualified coarse coal is sent back to the vertical coal mill for further grinding. Actions took,

Adjusting the separator 75%
Increasing the grinding pressure 150 mbar.

2.5 Process Measurement

For the better reading of the pressure and the temperature of the VCM system, some conter-check measurement was done on side. Based on these parameters, the optimized concept and layout for the modification of the VCM is then elaborated, utilizing as much existing equipment as possible and existing fans to keep the production costs low.

3 Results and Discussion

3.1 Fuel Mixing Impact

The raw coal LHV is 23,5Gj/t from the data from analyses Table 4, it can be read 22Gj/t this is showing a heat lost during the grinding process, the main reason of this lost is the presence of the kiln dust in hot gas from the kiln.

The average of co-grinding coal LHV is arrow 20 Gj/t this value normally supposed to be high than raw coal LHV. The fine powder moisture is on target which should be less than 1%.

3.2 VCM Operation and Optimization

During the normal operation the mill parameter was:

Mill feed 40TPH, with the reject load of 0,25TPH or 0.62% of reject, the specific power consumption was 13,78 Kwh/t.

Cashew nutshell Co-grinding parameter was.

Mill feeding 41TPH, the reject load of 0,11TPH or 0.27% of reject, the specific power consumption was 11,8Kwh/t. To reach this parameter some adjustment had been during the operation.

The control of mill feed with Mill DP by adjusting the grinding pressure with feeding load, the mill outlet temperature helps to increase the feeding by following the fine coal moisture and finesse in target. Mill nozzle ring velocity was improved by increasing the recirculation gas to 100%. From these data we can conclude the CNS can contribute to the electrical power saving about 0,98kwt/t.

Table 4 Fine coal analysis after the VCM

Coal and CNS analyses								
Fuels	W(a) [%]	W(t,r) [%]	V(r) [%]	A(r) [%]	S(a,r) [%]	Q(i,d) [kJ/kg]	Q(i,r) [kJ/kg]	P2O5 [%]
Coal	0.90	0.90	20.01	27.71	1.01	22,006	23,990	0,56
	0.80	0.80	19.32	29.17	1.10	23,237	23,223	0,98
	0.90	0.90	18.38	31.68	1.16	22,632	22,617	0,45
	1.00	1.00	18.41	31.59	1.18	22,519	22,502	0,68
	0.80	0.80	18.90	31.31	1.23	22,590	22,576	0,15
	0.90	0.90	18.87	30.82	1.28	22,821	22,806	0,85
	0.90	0.90	17.82	31.26	1.27	22,490	22,474	0,75
Coal + CNS-Charcoal	0.90	0.90	18.58	29.97	1.20	20,039	20,023	4,56
	0.80	0.80	20.01	28.42	1.17	20,384	20,370	5,76
	1.50	1.48	20.03	28.76	1.16	20,039	20,013	8,54
	0.90	0.89	15	19.33	2.61	20,805	20,790	7,45
	0.60	0.60	20.58	28.39	1.14	20,606	20,596	8,25
	0.40	0.40	18.47	27.25	0.89	20,495	20,488	9,15
	0.50	0.50	18.51	27.28	0.98	20,482	20,474	8,95
	1.20	1.20	17.48	27.47	0.95	20,393	20,372	8,78
	0.30	0.30	18.70	27.44	0.98	20,266	20,261	6,56
	0.80	0.80	17.80	28.06	1.04	20,064	20,050	7,85
	0.60	0.60	18.07	28.07	1.03	20,066	20,056	8,15
	0.80	0.80	17.45	23.53	2.09	20,350	20,335	8,54
	0.60	0.60	18.50	28.30	1.00	20,135	20,125	6,45

3.3 *Clinker Quality Impact*

Clinker quality depends on raw material composition and the fuel used for the pyro processing. The direct impact of the fuel on the clinker quality is a high lime saturation factor of raw meal and a low ash content of coal fines will make the f-CaO content of the clinker hard to control, from Table 5, data we can read that lime saturation was in control in both sides.

The phosphate ratio gets small change 0.54% high, but it still in target, the alite, belite, ferrite… all the main component of the clinker are in target also.

Table 5 Clincker quality analysis

Clinker quality																						
FUEL USED	SiO_2	Al_2O_3	Fe_2O_3	CaO	MnO	MgO	SO_3	K_2O	TiO_2	Na_2O	P_2O_5	Sum (%)	fCaO	LW	LSF	SR	AR	C3S	C2S	C3A	C4AF	LQ
Coal	21	5,16	3,83	65,2	0,06	0,13	0,17	0,64	0,43	0,25	0,39	####	1.6	1350	100.56	2.33	1.59	69.00	8.00	9.00	10.00	26.28
	21,1	5,46	3,33	67,2	0,06	0,23	0,7	0,34	0,23	0,15	0,49	99,3	1.4	1365	99.42	2.40	1.64	66.00	10.00	9.00	10.00	24.59
	20,1	4,46	3,13	66,2	0,06	0,28	0,72	0,39	0,28	0,11	0,29	####	1.4	1260	99.01	2.40	1.52	66.00	11.00	8.00	11.00	25.73
	21,8	5,64	3,4	66,3	0,07	0,43	0,63	0,33	0,27	0,15	0,52	99,5	1.2	1300	98.16	2.28	1.57	61.00	13.00	9.00	11.00	26.97
	20,9	5,49	3,45	67,5	0,06	0,61	0,61	0,29	0,29	0,15	0,50	99,9	1.3	1275	100.56	2.33	1.59	69.00	8.00	9.00	10.00	26.28
Coal + CNS-Charcoal	21	5,29	3,35	66,2	0,07	1,46	0,54	0,33	0,29	0,15	1,32	100	1.9	1350	98.16	2.32	1.55	63.00	13.00	9.00	11.00	26.60
	20,8	5,59	3,55	65	0,07	1,68	0,6	0,38	0,29	0,15	1,26	99,4	1.5	1345	99.24	2.32	1.56	67.00	9.00	9.00	11.00	26.56
	20,3	5,35	3,43	66,7	0,07	1,59	0,52	0,29	0,26	0,15	0,98	99,6	1.7	1255	101.21	2.37	1.56	69.00	7.00	8.00	10.00	25.80
	21	5,46	3,33	66,5	0,07	1,65	0,44	0,34	0,27	0,15	0,56	99,8	1.8	1295	98.84	2.39	1.64	62.00	13.00	9.00	10.00	26.01
	21,1	5,29	3,49	65,9	0,06	1,57	0,49	0,29	0,28	0,15	0,95	99,6	1.3	1245	99.01	2.40	1.52	66.00	11.00	8.00	11.00	25.73
	21,8	5,64	3,4	65,3	0,07	0,43	0,63	0,33	0,27	0,15	1,52	99,5	1.2	1300	98.16	2.28	1.57	61.00	13.00	9.00	11.00	26.97
	20,9	5,49	3,45	66,5	0,06	0,61	0,61	0,29	0,29	0,15	0,95	99,3	1.3	1275	100.56	2.33	1.59	69.00	8.00	9.00	10.00	26.28
	21,2	5,44	3,5	65,3	0,07	1,63	0,63	0,33	0,27	0,15	1,22	99,7	1.5	1255	97.28	2.37	1.55	60.00	15.00	9.00	11.00	26.30
	20,4	5,49	3,55	66,2	0,07	1,66	0,54	0,33	0,29	0,15	0,85	99,5	1.3	1300	98.16	2.32	1.55	63.00	13.00	9.00	11.00	26.60

4 Conclusion

Global benefits from the implementation of cashew nutshell co-grinding usage instead of totally fossil fuel would be the reduction of greenhouse gas (GHG) emissions. Provided the cashew nutshell or cashew nut transformation residues are sourced sustainably, biomass is a zero-emission fuel.

Use of nationally sourced biomass would help to retain foreign exchange that would otherwise have been spent on imported fossil fuels.

Our government is promoting biofuel as an alternative fuel source to help reducing the country's dependence on costly imported fossil fuel. Increasing the local transformation of cashew nut will require a processing of more cashew plants resulting in a corresponding increase in the volume of waste that can be tapped for biofuel.

Cashew nutshell availability is subject to seasonal fluctuations due to the vagaries of nature. Hence, although the local population are facing the scarcity of the biofuel and the high cost.

Continuous supply of sufficient quantities of biomass fuels is not assured and risk mitigation measures must be put in place.

References

MAEP. 2013. 4ème recensement national de l'agriculture 2011–2014 : volume 2 : module de base II : résultats chiffrés détaillés sur l'agriculture togolaise, Juin 2013, 575 p.

Sawadogo,M., Tchini Tanoh,S., Sidibé,S., Kpai,N., Tankoano. I., 2018. Cleaner production in Burkina Faso: Case study of fuel briquettes made from cashew industry waste. Journal of Cleaner Production Volume 195, 10 September 2018, Pages 1047–1056

Sivagurunathan, P., Sivasankari, S., Muthukkaruppam, S.M., 2010. Characterisation of cashew apple (Anacardium occidentale L.) fruits collected from Ariyalur District. J. Biosci.Res., Vol.1(2): 101–107

Municipal Solid Waste Energies Substitution Improvement with Kiln By-Pass Systems Operational Optimization

Essossinam Beguedou, Satyanarayana Narra, Mona-Maria Narra, Komi Agboka, Damgou Mani Kongnine, and Ekua Afrakoma Armoo

Abstract The cement industry ranks first among energy-consuming industries in terms of absolute consumption (Sousa et al. Sousa: V., Bogas, J. A., Real, S., Meireles, I., & Carriço, A., 2023. Recycled cement production energy consumption optimization., Sustainable Chemistry and Pharmacy, pp. 32, 101010.). The overall cost of energy (thermal and electrical) accounts for 30 to 40 percent of the total cost of cement manufacture.

1 Introduction

The cement industry ranks first among energy-consuming industries in terms of absolute consumption (Sousa et al. 2023). The overall cost of energy (thermal and electrical) accounts for 30 to 40 percent of the total cost of cement manufacture. Because of this, the cement industry has traditionally placed a high focus on efficient energy utilization (Afkhami et al. 2015). To start the reactions and phase changes

E. Beguedou (✉) · S. Narra · M.-M. Narra · E. A. Armoo
Professorship of Material and Energy Valorisation of Biogenous Residues, Professorship of Waste and Resource Management, University of Rostock, Justus-von-Liebig Weg 6, 18059 Rostock, Germany
e-mail: essossinam.beguedou@uni-rostock.de

S. Narra
German Biomass Research Centre gGmbH, Torgauer Straße 116, 04347 Leipzig, Germany

K. Agboka
West African Service Center for Climate Change and Adaptive Land Use (WASCAL), Lomé, Togo

D. M. Kongnine
Faculty of Science, Department of Physique, Centre d'Excellence Regional Pour La Maitrise de L'Electriciré (CERME), University of Lomé, Lomé, Togo

M. Narra and S. Narra (eds.), *Innovations in Circular Economy and Renewable Energy in Africa*, World Sustainability Series, https://doi.org/10.1007/978-3-031-68330-5_15

required to create the intricate mineral compounds that give cement its special qualities, very high temperatures are required during cement manufacturing (Worrell 2008).

The operational stage that supplies the energy and ambient conditions required for the reaction and phase transition is pyro-processing in massive rotary kilns. This process accounts for most of the cement manufacturing process's energy use and environmental effects. Enhancing energy efficiency lowers carbon dioxide emissions from fuel and electricity use and could potentially lower cement production costs. Energy management, the use of more energy-efficient process equipment, replacing outdated installations with new ones, switching to new types of cement production processes, cement kiln optimization processes, carrying out research and development required to prepare and burn alternative fuels in cement kilns, and developing new cement manufacturers are all ways to improve these processes (Proaño et al. 2020).

In the past few decades, numerous studies have been carried out to determine how to employ alternative fuels and raw materials to increase energy efficiency while also conserving the environment. The use of alternative fuels in cement production has recently drawn a lot of interest in developing nations like Togo due to its success in reducing the requirement for producing thermal energy from fossil fuels and a subsequent reduction of emissions from fossil fuels. Such alternative fuel sources may otherwise have been an environmental annoyance with significant potential carbon emissions due to the inadequate waste management methods in such countries.

Due to their alkaline environment, high temperature, and prolonged residence time, rotating kilns have the potential to burn a wide range of waste and hazardous materials. This article analyses and discusses various relevant research publications as well as recent developments in the use of alternative fuels in the cement industry. Studies on how bypass dust effect environmental emissions are also included in this article. It also describes the situation of kiln bypass now, how they are used in the cement business, and the benefits they have for the environment.

Manufacturing cement is a business with high CO_2 emissions. Fuel burning and the breakdown of carbonates, which typically make up most of the clinker raw mix, both produce CO_2. Around 8% of the total CO_2 emissions from all human activities are attributable to the manufacturing of cement (Alsalman et al. 2021).

These emissions are created because of the secondary constituents' partial volatilization or dissociation at the high temperatures of the kiln and their varying degrees of condensation at lower temperatures in the pre-heater-pre-calciner (Gerassimidou et al. 2021). Intense cycles result in serious issues with functioning, clinker quality, and emissions. A bypass is primarily constructed near the kiln inlet to allow a predetermined portion of the kiln gas to exit the kiln at high temperature to lessen the intensity of circulation phenomena (1000–1200 °C) (Choi et al. 2021). A significant portion of the process' heat is lost through the escape of sensible and latent heat from by-pass gases and dust.

To increase the use of alternative fuels is a major task for cement plants to minimize the specific fuel cost nowadays (Stančin et al. 2020). To achieve "zero" or "negative" specific fuel cost, it is quite common that alternative fuels with high Cl contents will be used. There is a limitation of Cl input based on kiln system; in general Cl input should not exceed 300 g/t clinker (Mokhtar and Nasooti 2020). It is a crucial requirement for bypass systems to be installed in cement plants to ensure that kilns can be operated under stable operation without any significant problems such as or frequent stops due to blockage and/or heavy build-up formation (Chatterjee and Sui 2019).

The chlorine bypass system must be designed properly to minimize a negative impact on kiln operation, especially on kiln specific heat consumption. The bypass rate must be sufficient to allow the kiln to run under stable conditions (Li et al. 2022; Han and Lee 2020). Keeping the bypass rate higher than necessary will have a negative impact on the specific heat consumption. The bypass rate must be closely monitored and controlled in an operating procedure in a cement factory on a daily routine. The basic requirement of designing a chlorine bypass system to achieve maximum efficiency is one of the key results of this article.

A recent study looked at the energetic and exegetic assessments of an Egyptian cement kiln facility, where the raw meal has such high chloride and sulfate concentrations that the entire kiln gas must be vented through a kiln bypass (Papanikola et al. 2019). The process has been examined without considering any kiln gas diversion through bypass for comparative purposes.

There are some studies focused on the kinetics of early hydration in high-calcium aluminate cement (CAC 70) and by-pass cement kiln dusts (BPCKD). Microcalorimeters were used to follow the hydration process and rheological properties, chemical shrinkage, phase assemblage characterization, and microscopic observations were done. Results were compared to those obtained for CAC with ordinary Portland cement additive (Chatterjee and Sui 2019).

Some research papers present a novel one-dimensional matching method of an asymmetric twin-scroll turbine (ATST) with a small scroll bypass wastegate for energy improvement. Using two-times flow parameters and turbine efficiency prediction curves, the effectiveness of the small and large scroll turbines was evaluated. At moderate and high engine speeds, the tiny ATST has higher engine performance and can reduce fuel consumption by 0.5–1.5% (Jin et al. 2020). To increase alkali removal efficiency, Cl-Cu_2O was doped into the Cu_2O nanocrystals' crystal structure (Sun et al. 2021).

2 Technical Considerations for Alternative Fuel Substitution with By-Pass System in Cement Plants

Depending mostly on Cl input through the alternative fuel (AF) substitutions, the required bypass rate varies from one plant to another. It is not advised to estimate the necessary bypass rate using the "rule of thumb: 1% bypass rate for 100 g/t clinker of chlorine input" to reduce the investment costs and energy consumption of the bypass project (Ghenai et al. 2019).

The clinker production process needs to be optimized to improve output on all fronts, including the environment and the energy economy. Energy analysis, which is based on the first law of thermodynamics, is used to decrease heat losses, or improve heat recovery, but it offers no insight into the energy degradation that takes place during the process (Chatterjee and Sui 2019).

Exergy measures both the quantity and quality of energy sources, as opposed to energy, which just considers the quantity. Exergy can be defined as the highest amount of work that a system can produce when it reaches equilibrium with a reference environment (Abdelzaher 2021).

Exergy analysis based on the second law of thermodynamics results in a better comprehension of how thermodynamic phenomena affect the efficiency of a process, a comparison (Ebrahimi-Moghadam et al. 2021) of the weighting of various thermodynamic factors, and the identification of the most efficient ways to enhance the process under consideration (Medes et al. 2022).

To achieve the best sustainable scenario for such an important strategic business, several studies have recently focused on energetic analyses of the cement industry (Stančin et al. 2020). The process's energy and exergy performance has been assessed.

Cement manufacture demands advanced equipment, a full knowledge of the operational principles of the plant, comprehensive approach, and other factors. This section, which is based on review of previous studies, will describe key technical considerations to make when establishing alternative fuels (AF) substitution from municipal solid waste within cement plants based on a by-pass system. This includes most steps used for by-pass system, by-pass dust and gas handling or optimization in AF co-processing.

2.1 *Main Types of Kiln Bypasses in Cement Plants*

(a) Combined by-pass dedusting system to cooler

Kiln plants typically consist of two strings, a rotary kiln, and a rotary cooler. Each string consists of 4 stages cyclone preheater and pre-calciner. Two air ducts conduct hot tertiary air from the cooler to the pre-calciners. Each pre-calciner and the kiln are separately fired with heavy alkali fuel from municipal solid waste (Ahmadi and Bahiraei 2021).

According to some studies, the amount of air required for combustion is typically between 335 and 345 kg per GJ of energy, or about 380 kg, if one allows for a 10% surplus (Abdelzaher 2012). This indicates that an effective kiln using 3.2 GJ of energy per ton of clinker will only require 1,216 kg of air, or 1.216 times as much air as clinker. The additional air required to achieve a tolerable temperature is surplus to requirements because using this air alone in the cooler would result in clinker temperatures of 399 °C.

Yet, a typical Thames-side wet process kiln from the 1950s would require 3.04 kg of air per kg of clinker and consume 8.0 GJ/t, allowing it to achieve very low clinker temperatures without (theoretically) producing waste air. The problem of waste air was not a severe one at the time because grate coolers were created in the historical context of these lesser effective kilns. New bypass systems are designed connected to cooler fans at recuperation zone and should allow to: (1) Improve bypass removal efficiency, (2) Reduce bypass dust amounts, (3) Allow a better emission control, (4) Recover heat partially from bypass gases. The crossflow principle is closely followed in the design of the Taiheiyo Bypass. Temperature profile and vertical particle segregation define the by-pass dust in a Taiheiyo Bypass. The temperature and alkalis concentration difference between the gas and the solid are greatest in a profiled by-pass gas.

With the Taiheiyo Bypass system, the bypass exhaust gas is connected to the pre-heater. The chlorine bypass system, which Taiheiyo Cement Corporation and Taiheiyo Engineering Corporation jointly developed, is an innovative technology that effectively removes chlorine, which is a major contributor to coating issues and is present in raw materials and fuels, from the kiln preheater system. As a result, this technology allows kilns to run continuously while utilizing enormous amounts of waste materials that are chlorinated (Corporation Taiheiyo Engineering 2022).

Common installation concepts in cement plants as presented in previous research include:

Efficient chlorine removal with less bypass rate.

- Best bypass rate is set based on unique simulation applying raw material chlorine content in preheater system as an index.
- Maximum chlorine removal with minimum heat loss.
- In actual operation, 80–90% of chlorine is eliminated with 5 to 10% bypass.
- Bypass rate up to 10% is ready to provide wherever requested.

Less collected dust quantity

- By introducing cyclone, chlorine rich and fine collected dust disposed of from the system is reduced substantially.
- Coarse particles are returned to the preheater system.

Returning bypass gas to preheater system

- By introducing gas cooler, bypass gas volume is reduced and returned to system again.
- As dust adhesion measures, plate type heat exchanger is introduced for gas cooler.
- Small-sized equipment is realized.
- Taiheiyo Bypass system design with hot cyclone for dust separation.
- A TEC Bypass system design with bag house filter for dust separation.

The crossflow principle is closely followed in the design of the A TEC Bypass. Temperature profile and vertical particle segregation define the by-pass dust in a "A TEC Bypass". The temperature and alkalis concentration difference between the gas and the solid are greatest in a profiled by-pass gas (A TEC, 2021).

With A TEC Bypass the bypass exhaust gas is not connected to the cooler neither to pre-heater. An infinite number of perforations in the grate surface are simulated by horizontal air jets. Cooling air sweeps the fluidizable particles to the bed surface by pressurizing the gaps or interstices in the clinker bed. The pressure drop of the bed is only at its lowest when the voids have been totally evacuated.

Common strengths and shortfalls presented in previous research are (Global Cement 2022):

Advantages:

- Low initial investment
- Easy handling of bypass/clinker dust mix

Problems:

- Issues with process stability
- Difficult control of emissions
- Necessary disposal of high amounts from bypass/clinker dust
- Possible limitation of bypass rate

Gritty clinker dust is a common component of the waste air from grate coolers. Early coolers created a lot of fall-out nearby because, like the kilns, little effort was made to contain the dust. Additionally, because of its capacity to indefinitely coat surfaces with hard hydration products, clinker dust is significantly more disagreeable than kiln dust (which is a basically unmodified raw material). The installation costs significantly increased as more complex dust filters had to be fitted to grate cooled exhausts due to mounting environmental demands.

2.2 *Kiln Bypass Operational Process Control*

Hot gas bypass system is commonly used in existing or new pyro lines to properly control the levels of Cl in hot meal and in clinker. It would be fair to mention that "hot gas bypass system" is a proven system. Due to this reason, hot gas bypass system will be the focus of this paper (Schneider 2019).

Based on experience and research articles on by-pass system design, the efficiency of chlorine bypass system mainly depends on:

Take off duct and tapping location.

- It is very crucial to identify a proper tapping location where chlorine concentration in the kiln gas is the highest with minimum concentration of kiln dust.
- Tapping location need be far away from meal flowing from cyclones and kiln bed charge.
- Take off duct can be made of normal steel with refractory lining inside alternatively it can be made without any refractory inside by using stainless steel type 1.4878. Boiler plate 1.0425 is certified to 450 °C, 1.5415 is certified to 550 °C and 1.4878 is certified to 850 °C. For health and safety reasons, take off duct which is made from steel without refractory lining is not recommended because it cannot handle gas temperature during upset condition.

- In case of using take off duct which is made by stainless steel without refractory lining, the distance between tapping location and quenching chamber must be short enough so that hot kiln bypass gas can immediately be quenched down to 400 °C (in case of double stage quenching) and to 200 °C (in case of one stage quenching).
- Take off duct can be of "round" or "square" shape, but it must be mentioned that square (or rectangular) shapes may cause some refractory problems due to thermal stress on corners. On the other side, square and rectangular shapes can be kept clean easier with air cannons installed at corners.
- Low dust concentration is expected to place opposite to the material charge in the kiln (-Side)
- For calciner kiln, on the top of the kiln and front side wall of the kiln riser duct, below the kiln restriction (- Top)
- In case of double string preheater, tapping location is also foreseen in the mid of the kiln (- Central)

Dust concentration at tapping location (Gerassimidou 2021).

- Dust concentration at tapping location is influenced by kiln inlet design and position of 2nd stage meal splash box above the tapping location.
- Minimum dust concentration with maximum Cl and Alkalis concentration can be expected by having a good and proper kiln inlet slope.

A typical dust concentration need be in the range of 250–500 g/Nm^3.

Dust concentration at 200 g/Nm^3 is possible to achieve with Chloride content in bypass dust above 8%.

Dust concentration can be even higher in case of

- Short circuiting of hot meal from 2nd stage cyclone.
- Extremely fast gas speed inside the kiln (>> 10 m/s)
- Extremely fast gas speed at the narrowest section at kiln inlet (>> 21 m/s).

Fierro et al. (2020).

Average gas speed at the narrowest section at kiln inlet and take off velocity for frontal and clean side extraction with dust concentration are as follows:

- Frontal extraction concept, at 15 m/s of gas speed at the narrowest section at kiln inlet and 4 m/s of take-off velocity, dust concentration should be below 180 g/Nm^3.
- Side extraction concept, at 19 m/s of gas speed at the narrowest section at kiln inlet and 3.5 m/s of take-off velocity, dust concentration should be below 200 g/Nm^3.

Dust concentration can be estimated based on gas speed at the narrowest section at kiln inlet, take off velocity, and location of take-off duct.

In kiln system, in which the hot meal pipes are introduced from the side and are not coming from the back of kiln inlet tray, the kiln inlet is usually significantly dusty and dust concentration up to 1.000 g/Nm^3 are frequently reported.

Bypass dust transport system needs to be designed for 1000 g/Nm3 or 50% greater than the nominal capacity. Additional safety margin to design bypass dust transport system should not be considered again when using dust concentration at 1000 g/Nm^3 (Shah et al. 2020).

2.3 Splash Box Design and Location

The design and location of splash box must be verified and if necessary, the location must be modified far from the tapping location (>1.5–2.0m). It is also recommended to check gas speed at kiln riser duct to ensure that gas speed at that location is sufficient to lift meal from 2nd stage cyclone (Fan et al. 2021).

2.4 Take off Velocity

The size of the take-off area needs be big enough to obtain take off velocity at 4–12 m/s. Due to low velocity, high chlorine concentration with minimum dust concentration is expected from the take off point. The quenching chamber must be carefully designed (Computational Fluid Dynamics (CFD) simulation is strongly recommended) especially when take-off velocity is lower than 6 m/s to minimize the risk of build-up formation in take-off gas duct (Long et al. 2022).

It is obvious that a lower take off velocity is better (less negative impact on kiln performance especially on kiln specific heat consumption), but it is sometimes required to bring additional meal (not only volatiles in vapor) to reduce a stickiness of bypass dust. Additional meal is required in some cases as well to ensure that the Cl concentration in the bypass dust can be obtained below limit and can be used later in the plant (Yörük et al. 2020).

2.5 Distance from Tapping Location to Quenching Chamber

The distance between the tapping location and quenching chamber must be as short as possible to avoid any unwanted clogging due to high temperatures.

Quenching chamber and quench air fan: The optimum dimension of the quenching chamber must be checked and confirmed together with the supplier. There are some cases in which the bypass rate cannot reach the designed point due to an underestimation on the dimension of the quenching chamber (Sani et al. 2020).

The quenching chamber can be designed to cool down the hot kiln gas directly from 1100 °C to 200 °C in a single step. However, the first step quenching must be adjusted directly to 450 °C to avoid the formation of dioxin and furan.

It is important to avoid the contact of water droplets with the quenching chamber lining as build-up formation may occur. Small amount of water injection (500 L/h) to the quenching chamber needs to be considered. There is some positive experience from some plants in respect to SO_2 emission reduction by injecting water in the quenching chamber.

Quench air needs to have two connections with high air velocity at 30–40 m/s for better mixing. It needs to be angled, and horizontal areas must be avoided. Regulated damper should be installed after quench air fan (to avoid tripping on motor). Quenching fans must have 50% of flow reserve (Berisha et al. 2022).

Proper sizing of fresh air damper at appropriate location shows significant energy savings when compared with forced draft fan.

The amount of quench air must be calculated based on temperature of mixed air-gas lower than 180 °C (Gas temperature at 180 °C is advised as the preferred system is taking the bypass gas back to the clinker cooler).

An efficient quenching chamber is based on volume of quenching chamber, velocity& momentum of inlet velocity of bypass gas to quenching chamber, velocity and momentum of inlet velocity of quench air to quenching chamber, velocity and momentum ratio between quench air and bypass gas, retention time of gas inside quenching chamber, number of inlets for quench air, etc. If the quenching chamber could not properly mix, the bypass gas can be quenched below the acid dew point which is the root cause of corrosion due to over quenching (Cheng et al. 2019).

2.6 Cut off Gate

A cut off gate must be installed to avoid a large amount of false air coming to the kiln line while the bypass unit is stopped. This cut off gate needs to be designed to be either fully opened or fully closed. The cut off gate should be preferably installed after the quenching chamber (due to higher reliability and lower maintenance costs on cut off gate), but refractory lining inside quenching chamber is mandatory (Rathoure 2019).

Cut off gas construction material must be stainless steel such as type 1.4878; if operated at gas temperature above 450°C. In any case, stainless steel type 1.5415 is still workable at emergency high temperatures. This is also valid for the frame of the cut off gate.

Quenching chambers without refractory must be made from stainless steel type 1.4841, 1.4878 or 1.4828. The cut off gate must be placed before the quenching chamber. A dedicated cooling fan is also recommended to protect the quenching chamber in case of any interruption of kiln operation (Malacarne et al. 2021).

Gas conditioning tower must be used where there is no crisis of water at the plant level, the water quality needs to be free from solids, and wherever requires low investment. Gas cooling tower is not recommended, especially for new bypass project (Sani et al. 2021).

Gas conditioning tower usually works with high inlet gas temperature (approx. 350 °C) and may create additional harmful substances in the bypass gas. Moreover, it was observed within some plants that cooling towers were highly corroded in relatively short operating time.

Modern gas conditioning systems are equipped with the twin fluid nozzle for better control of the droplet size. The investment cost is low due to a requirement of low evaporation distance (smaller gas conditioning tower), but the operational cost is increased due to high compressor air consumption. On the other hand, mono fluid system demands high investment cost due to long evaporation distance (bigger gas conditioning tower) but with less energy consumption.

The bypass gas after the quenching chamber must be continued to cool down between 150 to 200 °C either by a secondary air quenching system or by a water-cooling system in the gas conditioning tower or by a heat exchanger.

In case that bypass gas is taken to the main stack and is scrubbed from SO_2, SO_3 and HCl by the support of the gas conditioning tower and injection of lime hydrate, the design and operation temperature must be 150 °C with a residual dust moisture below 1.7%.

To keep the gas volumes at this low temperature within reasonable limits, the cooling is affected in the heat exchanger. According to the experience from the COUVROT plant in France, there are not any serious problems encountered so far.

In any case, by having a very limited experience with this kind of heat exchanger in the bypass gas circuit, it would be rather difficult to conclude the performance of gas cooling by heat exchanger. However, heat exchanger could be a solution when a bypass gas is recirculated back to the clinker cooler, especially when bypass rate is higher than 10% (Bersiha et al. 2022).

Electric precipitator (EP) or Bag filter (BF).

Old bypass de-dusting installations are equipped with electric precipitators whereas new bypass systems must be equipped with a bag filter (Ebrahimi et al. 2020).

- Hot gas electric precipitator needs to be avoided as this work sufficiently well only between 280 and 400 °C where PCDD/PCDF develops is most likely.
- Warm gas electric precipitators operate well at 150 °C. In this sense, it tends to corrode if the gas is not impinged with lime hydrate to absorb SO_3 and HCl. Warm gas electric precipitators shall be constructed in duplex stainless steel such as 1.4362 or 1.4462 or a combination of both.
- While treating the kiln bypass gas in the bag filter, the filtration velocity must be maintained below 0,8 $m^3/m^2/min$ at nominal bypass rate plus 10% reserve due to fine and sticky dust. One chamber filter is preferred to minimize the dead low temperature corners and to reduce corrosion possibilities. Glass fiber bags with PTFE coating must be used. One or two emergency air flaps on the gas ducting must be equipped. The main process bag filter shall not be used for bypass gas, it is recommended to use separate bag filters between the kiln exhaust gas and bypass gas.

2.7 Segregation of Coarse and Fine Bypass Dust

Installing a cyclone between the quenching chamber and the bypass filter can separate a coarse bypass dust which has very low Cl content and can be returned to the kiln process. A space for such cyclone must be left at least. This option can be found in the bypass system in the Ketton and Couvrot plant, in South-West Europe region (Singh 2021).

The cyclone efficiency shall not be designed to be higher than 40 to 50% due to high bypass dust fineness. This way, it is possible to reduce the total amount of bypass dust by approx. 50 to 60%. The bypass rate must have a sufficient margin because some portion of chloride is returned to the kiln system through the coarse bypass dust (Pope et al. 2009).

By installing the cyclone after the quenching chamber, the dust concentration at the inlet of the bag filter can be reduced; and filter size thereby can be optimized to a certain extent. However, electric precipitators are not qualified as these work particular poor above alkali-salt contents <20%.

In general, coarse bypass dust (>10 micron) contains less Cl (<20%) and represents more than 70% of total amount of bypass dust.

Fine bypass dust (<10 micron) contains more Cl (>80%) and represents less than 30% of total amount of bypass dust.

Fine bypass dust must be discarded and stored in a dedicated storage system. Depending upon the Cl concentration, a small portion of the fine bypass dust can also be used in the cement mills (up to 0.1% Cl max in the cement). The chloride concentrations in the fine bypass dust may vary from 2 to 25% and total alkali salt concentrations are higher than 30%. It becomes very difficult to transport this material up to storage system pneumatically.

When the Cl in the collected dust is too high, then it limits the total Cl input to the cement mill. Due to recycling of certain amount of Cl dust, bypass systems are operated at higher rate than expected (Abdel Maksoud et al. 2022).

2.8 Volatile Matters

Thresholds (Cl and SO_3 content in hot meal, Cl in clinker, etc.) are very often used, but the author's opinion is that they are not corresponding to current experiences with cement plant operations, especially when burning higher quantities of alternative fuels. From graphs showing relations between chloride and sulfur contents, one can see areas of "difficult" or "very frequent blockages" that are not matching with experiences from day-to-day operation in cement plants. However standard/conservative thresholds (Cl and SO_3 content in hot meal, Cl in clinker, $SO_3 + 2Cl$ in hot meal, etc.) can be used in the first step to evaluate a need of a bypass installation when chloride inputs would increase due to whatever reason (Hannoun et al. 2019).

Typical evaporation factor, Cl content in hot meal can be estimated by Eq. 15.1.

$$Hotmeal\ (lossfree) = \frac{imput\ (lossfree)}{(1 - evaporationfactor)} \quad (15.1)$$

Evaporation factor for Cl is approx. 99% and for SO_3 is approx. 60–70%. There is an example of a plant where 80% of alternative fuels are going to be used. Chloride input will increase from 0.194 g/kg clinker to 1.46 g/kg clinker, so definitively the gas bypass is required. It needs to be noted that hot meal chloride content with 99% evaporation factor would be 12.4% (15%LOI). Such level of chloride would certainly not allow kiln operation at all. The next question comes from which level of chloride in clinker should be used to estimate required chloride bypass. This depends on many factors: It can be 0.2–0.3 g/kg clinker (0.02–0.03%) in one plant and can be 0.5 g/kg clinker (0.05%) in other plant. It is always a combination of local process conditions, burnability of the raw mix, sulfur cycle, and others.

2.9 *Bypass Sizing*

It is possible to rapidly estimate the amount of the kiln bypass rate by using the rule of thumb (removal chloride (loss free basis) multiplied by 100 for calciner kiln and by 50 for suspension preheater kiln). For example, to remove 0.05% (500 g/t clinker) of chloride via bypass dust, then bypass rate can be 5% for calciner kiln. However, this is a rule of thumb, and each plant condition must be estimated separately (Xu et al. 2015, Dissanayake and Karunawardena 2018).

The overdesigned bypass system will negatively impact on the mixing efficiency inside the quenching chamber due to much lower fresh air velocity and lower inlet gas velocity to the quenching chamber. There is a risk that some of quench air can reverse to take off duct which can create build-ups inside ducting to the quenching chamber. The overdesigned bypass system must be avoided to maximize efficiency and minimize the investment cost. In general, the bypass system must be able to handle from the nominal maximum bypass rate down to 60% (Aguilera and Marchetti 1998).

Conservative figures, like 0.8% Cl in hot meal, are normally used as target to design a bypass system in a new pyro line (Ramsay and Khator 1986). This figure must be checked from existing operations and can be changed to minimize the investment cost. There are some plants which handle smooth kiln operation at 3.5% Cl and 4.0% SO_3 in hot meal, with a need of sufficient & efficient cleaning tool, that are well maintained and controlled by air cannons, cardoxes and eventually water cleaning. The target of Cl and SO_3 in hot meal will be different from one plant to the other plant. Optimum targets of Cl and SO_3 in hot meal can be observed under stable/continuous operation without any significant stops due to preheater tower, feed shelf cleaning, cyclone build-ups, etc. An average of chloride and sulfur in hot meal in the suspension preheater kiln is 2.0% Cl and 4.5% SO_3, and average of chloride and sulfur in hot meal for a calciner kiln is 1.5% Cl and 2.0% SO_3 (Singh 2021).

- An average of chloride input to suspension preheater kiln is approx. 2100 g/ t cli.
- An average of chloride input to calciner kiln is approx. 1200 g/ t cli.
- Maximum chloride in fuel mix is 1.0%.
- Maximum chloride in kiln feed is 0.1%.
- Maximum chloride in clinker is 0.05%.
- Required bypass rate can be calculated by

$$Hotmeal(lossfree) = \frac{imput(lossfree)}{(1 - evaporationfactor)(1 - \frac{bypass\%}{100})} \quad (15.2)$$

Equation 15.2: Evaporation factor for alkaline in hot meal.

The evaporation factor (Eq. 15.2) for Cl is approx. 99% and for SO_3 is approx. 60–70%. For kiln bypass sizing, the choice of the equipment design shall also be taken into consideration depending on the type of the kiln that the bypass installation shall be placed in. In the suspension preheater kiln (without calciner and tertiary air duct), it is usually that most of the total fuel input (approx. 85%-100%) comes from the main burner, thus the specific amount of kiln inlet gas volume is often more than double of the specific amount of kiln inlet gas volume from a kiln line equipped with the calciner. Naturally in the suspension preheater kiln, the chlorides vapor concentration in the kiln gas is lower than in the calciner kiln gas (referring to kiln inlet/riser duct location). Due to this reason, the trapping efficiency of the bypass solution for suspension preheater kiln is on the one hand lower. On the other hand, due to lower kiln inlet temperatures (compared with the calciner kiln), the suspension preheater kiln can accept higher input of alternative fuels (means chlorides) than the calciner kiln, thus bypass solution can be installed later in the time of development of alternative fuel burning (Mohamedzein et al. 2022).

Basis for gas volume at kiln inlet: Each plant has a different fuel mix, heat demand via main burner, oxygen content at kiln inlet, calcination degree, etc. Due to this reason, specific amount of kiln inlet gas volume must be calculated by considering actual conditions on each plant (Ashteyat et al. 2022).

2.10 Environmental Considerations and Gas Handling

To ensure that emission can be controlled within acceptable limits after installing the bypass system, understanding of bypass gas characteristics as well as bypass gas handling options are listed below.

2.11 Reducing Emissions and Environmental Impact

Bypass gas taken out from kiln inlets are considered as kiln gas with all potential risks and as a source of gaseous and solid emissions. Bypass gas could have some influence on emission levels (especially on SO_2 and HCl). Risk from PCDD/PCDF formation is not high. Bypass gas after bag filter shall not have its own stack but shall connect to main kiln line stack at least. These emissions are related to 10% O_2 by the stack monitoring; therefore, the gas will be able to mix into the main kiln line stack only with low chlorine bypass rate (Darweesh 2020).

Directing the bypass gas to the grate cooler dedusting is not recommended. It would at least require installing the continue gas emission measurements unit on such stack and worse could require using costly additional emission reduction solutions. Visible plume at the main stack could be formed by aerosols if the ammonia from the kiln, SNCR, or high NH3 background level reacts with Cl or SO_2 coming from bypass gas and by this increasing the stack SO_2 and HCl emissions. Reduction of emissions (SO_2 and HCl) at stack can be achieved after re-routing the bypass gas to the clinker cooler in the recuperation zone. It is recommended to have lime hydrate injection provisions for all cases of returning the gas back to the clinker cooler to avoid corrosion caused by SO_3 and HCl. The injection of lime hydrate shall be at gas temperature less than 250 °C (Brouwers et al. 1987).

There are solutions which can reduce or even eliminate the impact of bypass on emissions like controllable water injection into the quenching chamber, increase of dust entering from the kiln, injection of high surface lime hydrate, maximizing of water injection for bypass GCT, bypass gas scrubbing in raw mill system, etc. (Mohamedzein et al. 2022).

Low dust concentration in bypass take off is favorable for increase of SO_2 emission. In some cases, the increase of dust concentration in bypass take off can be one of the solutions to reduce SO_2 emission. Sulfur is collected better when increasing relative moisture of bypass gas streams (water injection to quenching chamber). Sulfur then reacts with free lime content in bypass dust and enhanced humidity and is partially scrubbed too solid. However, if the gas is taken back to the clinker cooler, the application of water injection in the quenching chamber is advised only with small amounts and not by a GCT at all. In general, low temperature operating baghouse filters are more efficient in the collection of sulfur than the EP. It is essential to have a very fast cooling system of bypass gas which enables to lower the temperature below 250 °C to prevent PCDD/PCDF formation. Higher temperature exceeding 240–260 °C elevates the risk of PCDD/PCDF formation. The risk of dioxins formation is high within a temperature window of 200 to 350 °C and therefore it is required to maintain the bag filter/EP inlet temperature below this critical temperature range. Special attention should be considered for systems with higher VOC which is in combination with chlorine content and some types of metallic catalyst (Cu) good precursors for synthesis of PCDD/PCDF (Vojvodíková et al. 2021).

2.12 Bypass Gas Handling

There are 3 main options of bypass gas handling which are listed below.

- Connection to the kiln main stack: the least "washing/scrubbing" effect is with connection of bypass gas directly to the kiln main stack (Biege et al. 2012).
- The highest "washing/scrubbing efficiency can be obtained after maximizing bypass gas to raw mill circuit, the "washing/scrubbing" effect could be up to 100% (Rathoure 2019).

The amount of gaseous SO_2 depends on the sulfur input to the system, alkalis and sulfur ratio, and excess air level. Sulfur at the kiln inlet can be above 1000 mg/Nm^3, and in some case can be reached to 10,000 mg/Nm^3 as well. These levels of sulfur at the kiln inlet are not only related to sulfur input from fuels as a large portion due to liberation of sulfur from i.e., K_2SO_4 during reducing conditions.

- Connection to the cooler baghouse: This is a cheap solution to use the spare capacity of cooler vent air filter to separate the bypass dust (Dembla and Mersman 2021).

A blend of bypass dust and clinker dust is reasonable transportable. Bypass dust segregation occurs very often and controlling of Cl in cement produced is not as accurate and precise as separate dosing of bypass dust. For higher bypass rates and Cl inputs, it is necessary to dispose a part of such a blend out of the process regularly. Due to this reason, it is highly recommended to have a separate bypass dust storage and bypass dust dosing system (Ahmed and Ragei 2021).

With collection of bypass gas to the cooler stack, it needs to be considered that bypass gas, a part of kiln gas, includes pollutants as CO, NOx, SO_x, HCl, HF, VOC, etc. It is generally required by states or environmental departments to measure all pollutants (in some cases continuously), control all pollutants, report all pollutants, and pay for all emissions released to the atmosphere. CEMS (continuous emission monitoring) shall be applied very often in case that we exceed certain specific mass flow (Nocuń-Wczelik and Stolarska 2019).

Connection to the cooler´s recuperating zone: It is highly recommended to connect the bypass gas to the front part of cooling fans of a grate cooler. Those fans supply the secondary and tertiary air for combustion, thus bypass gas containing some emission substances could be utilized in the kiln system (Altun 2022).

Connection of bypass gas to the cooler´s recuperating zone is limited with approx. 10% of bypass rate (without an installation of HEX). Based on available data, there is no objection to connect bypass gas to cooler fixed inlet, but connecting of the bypass gas to fixed inlet fans is not recommended because there are additional issues and limitations (Abdel-Gawwad et al. 2019).

Generally, the mixed gas underneath the grate shall not be colder than 120 °C. The best cooling efficiency is obtained if ambient air is used for cooling. Bypass gas can have a temperature of approx. 180 °C once reaching the manifold, where the bypass gases are mixing with the further clinker cooler intake air, prior being conveyed with the cooler fans to the under grate. Cooler fans shall be mechanically designed for such parameters and shall be bigger to deliver similar specific air volumes (Miller and Hansen 2004).

The clinker temperature at the outlet of the clinker cooler can be slightly higher; therefore, additional cooling air will be required. It is necessary to clean the deposits at the grates and at the under-grate departments periodically (during major kiln stops). Ducting from bypass fan up to the cooler fans need to be insulated to prevent condensation as much as possible.

Some types of new coolers (lane coolers) are designed with internal hydraulics which is located under the lane system. These hydraulic units are sensitive for higher air temperatures and sometimes may prevent the bypass gas cooler recirculation. In such specific case, HEX can be installed after bypass bag filter (Vojvodíková et al. 2021).

Outlet temperature from HEX shall be at 120 °C (must be above the acid dew point, which can be only sustained by injecting lime hydrate to stem the developed SO_3 and HCl). Surface temperature of the HEX must be higher than the acid dew points as well. This can only be obtained by a co-current system. Ideally the HEX shall be a dual string design: 1st stage from i.e., 450 °C to 250 °C in cross counter current and 2nd stage in cross co-current design) (Gosh et al. 2022).

Due to this reason, total investment of HEX plus new generation coolers (lane coolers) is significantly higher than using grate cooler (IKN) without HEX. HEX should be installed when designing to recirculate 15% bypass rate to recuperation zone of clinker cooler (Sharma et al. 2021).

Kiln control system must have additional control procedure to prevent an operation of bypass unit while cooler fans are stopped or even partly stopped. Under that situation, it can create serious process impact on kiln operation. Local authorities may request clear procedure to assure that bypass gas will not leave the kiln system in uncontrolled manner. This solution is considered as "no emission" (Rathoure 2019).

2.13 Dust Handling

Maximizing the amount of bypass dust and minimizing a variation of cement quality can be achieved by having a proper transport and storage system. Bypass dust is a corrosive material. Options to use bypass dust out of the plant to minimize disposal costs is an important topic to minimize the production cost of clinker and cement.

The transportability of bypass dust depends mainly on concentration of salts. This is generally expressed by content of chloride (Cl) which can vary in most cases between 5 to 25% (Al- Bakri et al. 2022).

It is desired on one side to reduce the amount of bypass dust produced to a minimum, but on the other side this results in an increase in the concentration of bypass salts included in dust blend to a maximum. This sticky, low-density blend is very difficult to transport pneumatically over large distances if not pre-treated. Mechanical conveyors are more beneficial in case that higher concentration of salts is expected (compressing screw conveyors, chain conveyors) (Vojvodíková et al. 2021).

In case that pneumatic transport is used, a dilution of bypass dust with other available materials is needed. Additional raw meal or CKD is a common additive, targeting chloride concentrations of ca. 10% max. This solution is acceptable if the bypass dust is not disposed or transported over greater distances to i.e., grinding plants (Zawrah et al. 2022).

Typically, the collected bypass dust from the bypass filter is dosed either to cement mills by using a dedicated weigh feeder (where good control over cement quality is expected) or to a clinker transport system (where the bypass dust is combined with clinker product). The first case involves higher investment cost, whereas another case shows that segregation of the bypass dust from the clinker and eventually leads to quality variation in cement (Sengupta and Sengupta 2020).

To keep the investment cost at lower side, sometimes bypass dust can be fed directly to cement mills. In this case, there are some chances of having dosing fluctuations which may lead to quality variations in cement. Further, bypass dust is possible only when the cement mills are in operation. On the other hand, no bypass dust is available to the cement mills when the kiln is shut down. Bypass dust should be fed into the separator and should not fed directly to mill itself (Ray 2022, Zawrah et al. 2022).

To overcome the issues related to direct dosing system, bypass dust can be stored in a dedicated silo and later fed controlled to the cement mills with a proper dosing system. In this case, additional equipment and high investment cost will be involved. Due to the sticky nature of the bypass dust, some handling problems are to be expected. A flat bottom silo with a centered discharge screw is a recommended option for bypass dust storage (Darweesh 2022).

The addition of bypass dust into cements has an impact on properties of final products (setting time, early cement strengths, etc.), so, it is not recommended to change the percentage of bypass dust in cement.

Disposal of bypass dust is very problematic and costly. Due to this reason, it is very common and favorable to reuse the maximum amount of bypass dust in the process directly. The most common way to utilize bypass dust is to feed into cement with consideration to maximal content of Cl which is generally 0.1%. To achieve maximum limit of Cl in cement, it is necessary to have a precise feeding system to the cement mills. Mechanic discharge out of a silo provides often better and more precise results which enable to control Cl in Singh (2021).

2.14 Safety Related to Bypass Dust Handling

Bypass dust is a corrosive material with high concentration of salts and heavy metals such as lead, cadmium, zinc, copper, chromium, and manganese. There are limitations in contents of soluble heavy metal compounds which form restriction related to landfilling of wastes on standard dump sites (for instance content of soluble Lead

(Pb) compounds must be <5 mg/l in most cases). On the other side, there is a risk that bypass dust could be considered a hazardous waste for all costs related to its landfilling and handling.

Bypass dust is very often considered as skin corrosive, with possible damage to eyes and toxic to repeated exposure. It can cause severe skin burns, eye damage, allergic skin reactions, and it is potentially suspect to causing lung cancer.

It is necessary to use/apply all cautions to safe the health of the workers. MSDS must be prepared in case that the bypass dust is sold as a product to any potential customers. Operating procedures related to bypass dust handling need to be set up for plant workers who can get in direct contact with bypass dust.

Personal protective tools must be worn such as safety glasses with side shields (non-vented goggles in case of higher dust concentration in the air), long sleeve shirts and trousers avoiding direct contact with skin, respirators or facemasks approved by MSHA/NIOSH, protective gloves with wrist cuffs, etc. (Dissanayake and Karunawardena 2019).

2.15 Use of Bypass Dust Out of the Plant

To dispatch or dispose "diluted" bypass dust out of the process can nevertheless be very expensive. This shall also be evaluated during bypass designing and the installation of separating cyclone prior to bypass filter shall be considered as an option to reduce the overall amount of bypass dust produced.

Chlorine washing: This type of bypass dust utilization can be considered as an environmentally friendly solution. On the other hand, it is the most difficult and expensive way. The whole washing plant requires a rather big space (in ROHOZNIK plant—CRH, it is installed in a 900 m^2 hall). The main product is KCl with a purity >95% which can be used for production of fertilizers. Another product is a blend of salts which is usable for instance as road de-icing tools during winter season. The only problematic part is a heavy metal "cake" which is considered as hazardous, but the total amount of this residue is not very high. The overall operating cost of this equipment mainly depends on the purity of the final products (Darweesh 2020). Regarding their marketing and selling price, it is estimated that the final cost for bypass dust disposal can be between 30–150 EUR/t. Nevertheless, project costs for the installation of equipment and de-bottlenecking of the operation are very high (Ashteyat et al. 2018).

Landfilling is an option for bypass dust, but this need be the last option to be considered. It is necessary to take local laws, surrounding conditions, limitations, and real composition and properties of bypass dust which is produced into account when deciding in landfilling.

2.16 Instrumentation and Controls (Loops and Interlocking)

Automatization must be in place to ensure a smooth operation of bypass system.

The instrumentation and gas temperature sensors need be considered during design phase and proper installation of temperature sensor to avoid build-ups on wear protection of temperature sensors must be considered during construction phase. Response time on temperature sensors must be fast enough to ensure "proper" adjustment of the chlorine bypass system (Czapik et al. 2020).

A pressure sensor which is installed at take-off gas duct from kiln riser duct to chlorine bypass system can be used to monitor a build-up condition of take-off gas duct. It is sufficient to install one pressure sensor on each location except at the take-off gas duct which is recommended to install additional sensors (at least 2 sensors in operation at the same time) (Stevulova et al. 2022).

An online dust measuring device is recommended to be installed when the chlorine bypass system has a separated stack or when introducing bypass gas to the clinker cooler. Since bypass gas can be introduced to the recuperation zone of the clinker cooler, it is mandatory to install an online dust measuring device to avoid clogging on grate plate of the clinker cooler to avoid the wear out of the fan impeller. Good and reliable "broken bag" detectors are a real need (optical type with analog output is more advantageous than the standard triboelectric one giving a binary signal) (Abdelgader et al. 2022).

Gas flow measuring device of quenched air and water flow measuring device for GCT: The amount of extracted gas from kiln riser duct and percentage of bypass rate can be calculated based on the amount of quenched air and water injection in GCT. Both of instruments are recommended to be installed to closely monitor and optimize the bypass rate all the time to obtain optimum specific heat consumption (Saleh et al. 2020).

A piezo-metric ring is mounted on the fan suction cannot be applied when connecting bypass gas to the cooler´s recuperating zone. There need to be industrial "PRANDTL" tubes or "VENTURI" flowmeters ("TORRBAR", "DEBIMO" blades). They need to be operated with higher gas temperatures (>200 °C) and shall be stainless steel quality. Along the gas velocity measurement, it is also necessary to determine the intake air temperature (Geng et al. 2019).

Here, it is simple to reach the final temperature of 80 °C for cooled clinkers. This is made feasible by fan batteries positioned below the grate, send air cross currents when hot clinkers pass over the grate. The cooling air usage in these coolers ranges from 1.4 to 1.8 Nm^3/kg of clinker. These numbers provide an explanation for the higher thermal efficiency of grate cooler-equipped kilns. At the furnace entrance, the secondary air temperature is easily capable of reaching 1000 °C.

Bypass installation, once in operation, acts as an additional kiln ID-fan. It is recommended to prepare the control system for any stops or fluctuations in its operation (Singh 2021).

Main interlocking likes opening emergency fresh air damper in front of dedusting unit when gas temperature in front of dedusting reaches H limit, stopping on chlorine bypass system when gas temperature in front of dedusting reaches HH limit, stopping on chlorine bypass system when observing wet bottom problem on GCT, stopping on chlorine bypass system when bypass dust temperature reaches HH limit must be in place.

It is recommended to use a bypass control concept (Abdel Maksoud et al. 2022) in order to determine the total input of volatiles into the kiln system (g/t cli) as well as the hot meal concentrations.

This information will be deduced from the bypass rate. Based on the bypass rate the amount of quench air necessary to reach a certain temperature (e.g., 200 °C) can be calculated or taken from tables.

Adjustment of the quench air fan settings to supply this amount of air needed can occur.

It is crucial to identify a proper tapping location of take-off duct in the kiln riser duct where high enrichment of volatiles can occur and to extract partial amount of these hot kiln gas with a minimum amount of kiln dust. The best location would be frontal on the top of the kiln and front side wall of the kiln riser duct (Soliman and Moustafa 2020).

Modified bypass system to have lower dust concentration will allow for an increase bypass rate like in the Radotin, Lengfurt, and Devnya plant. Normally, bypass performance can be observed by monitoring dust concentration (g/Nm^3), chloride content in bypass dust, etc. Dust concentration below 300 g/Nm^3 can be easily achieved for new kiln line project since gas speed inside the kiln, gas speed at the narrowest section at kiln inlet, and take off velocity are well below design criteria (El-Sayed Seleman et al. 2020).

Removal rate of bypass system can be calculated based on chlorine removal in bypass dust versus average of total chlorine input and output. Usually, removal rate for chloride is 25–75% and for sulfate is 1–20%. These parameters need to be considered for performance guarantee for the bypass system.

Concerning the shape of the take-off duct, square (or rectangular) shapes may cause some refractory problems due to thermal stress on corners. On the other side, square and rectangular shapes can be kept clean easier with air cannons installed at corners.

A good and proper kiln inlet slope ensures that the hot meal enters to the kiln in a smooth way and all volatiles suspend in a gaseous state which can immediately be trapped through the take point with minimum dust concentration.

The distance between the take off point and quenching chamber shall be as short as possible to avoid any unwanted clogging due to high temperatures.

A cut off gate shall be preferably installed after the quenching chamber (higher reliability + lower maintenance costs), but it is crucial to choose the correct construction material (Abdel Gawwad et al. 2020).

The limit of chloride and sulfur is different from plant to plant and each case needs be evaluated separately. Normally, a suspension preheater kiln can handle higher SO_3 and Cl in hot meal when comparing with a calciner kiln. An overdesigned bypass system must be avoided to maximize efficiency and minimize the investment costs.

However, the nominal operation range shall be designed for 60–100% of the capacity (Al-Tahan et al. 2021).

It is recommended to connect bypass gas to coolers fans in recuperation zone of clinker cooler. Based on available data, there is no objection to connect bypass gas to a cooler fixed inlet, but if possible "connecting of the bypass gas to fixed inlet fans shall be carefully considered" because there are additional issues and limitations. Connecting bypass gas to the cooler´s recuperating zone is considered as "no emission" (Salamanova et al. 2022).

Electric precipitators need to be avoided. They usually have poor dedusting efficiency of bypass dust compared to bag filter, cost usually more and tend to corrode in similar speed like gas cooling towers if operated with a gas conditioning tower. Hot gas electric precipitators may also cause harmful substances generation such as PCDD/PCDF.

Bypass gas highly diluted by quench air still contains quite a lot of NOx, SO_x, HCl, and HF that are naturally coming from the kiln. Due to this reason, bypass gas after bag filter shall not have its own stack but shall connect to main kiln line stack at least (Zawrah et al. 2022).

Transportability of bypass dust depends mainly on concentration of its de-sublimated alkali-salts. Mechanical conveyors are much more beneficial in case that higher concentration of salts is expected. This is also valid for storage of bypass dust in silos where the compressed air aeration used for dust discharge is not as easy as mechanic systems (propellers, flat bottom silos with screw discharge, etc.).

Cleaning frequency and cleaning methods will impact on optimum bypass rate. Inefficient cleaning will require higher bypass rate than necessary which will impact negatively on kiln specific heat consumption. Increasing 1% of bypass rate for suspension preheater kiln will increase specific heat consumption approx.15–25 kJ/kg clinker and for calciner kiln will increase specific heat consumption approx.10 kJ/kg clinker. Number and location of air blaster, cleaning port, and cardox must be sufficient at proper location to have an efficient cleaning. High pressure water must be used by well-trained staff (Athira et al. 2021). Too high sulfur content in hot meal due to insufficient excess air level and improper combustion condition inside the kiln could be a cause of cyclone blockage even by keeping bypass in operation. Due to this reason, sulfur volatility is normally used as combustion indicator for stable kiln operation especially when burning high amount of solid alternative fuels at main burner (with/without satellite burner). Sulfur volatility shall be below 0.7. sulfur volatility can be calculated by (Eq. 15.3) (Kaliyavaradhan et al. 2020).

$$Sulphur\ volatility = 1 - \frac{\frac{SO_3\ clinker}{(100 - LOI\ clinker)}}{\frac{SO_3\ hotmeal}{(100 - LOI\ hotmeal)}} \tag{15.3}$$

Equation 15.3*: Sulfur volatility estimation.*

It is recommended to clean the extraction orifice and quenching chamber once per shift (or more frequently if needed) to avoid a clogging of the system (intelligent cleaning is a preferable solution). It is recommended to close cleaning slots tightly all the time after finishing the cleaning; otherwise, false air and further clogging/malfunction of the bypass will occur. It is recommended to regularly check refractories for damages and detect cold traps to eliminate them. It is essential to consume maximal possible and allowable amount of generated bypass dust into produced cement. To ensure this, a precise feeding equipment is a must (keeping chloride level in cements as close to limits defined in cement norms). To keep this, it is also important to evaluate a proper size of the intermediate bypass storage capacity considering needed flexibility but also project costs/budget. Homogeneity of bypass dust is important and can be improved by recirculating bypass dust to bypass silo for example like in Vác plant in Hungary (Kaliyavaradhan et al. 2021).

The cost of bypass dust disposal can be up to 150 EUR/t. Standard value of bypass dust is normally approx. 0.002 kg/kg clinker for each 1% bypass rate, so total amount of bypass dust will be approx. 20,000 ton based on 1 million ton of clinker at 10% bypass rate. With this information, maximizing bypass dust can bring a significant saving to Cement plant (for example, saving by 300,000 EUR for 2,000 bypass dust). Experience from Mokra plant already show us that only 35% of bypass dust can be utilized in cement grinding system which can result in 65% discarding if it cannot be sold as soil stabilizing product (Putri et al. 2022).

It is a must to evaluate overall benefit after increasing alternative fuel by considering the right amount and the correct price of "discarded" bypass dust.

3 Methodology

Based on the extensive literature review carried out in section "Technical considerations for alternative fuel substitution with by-pass system in cement plants", a case study was carried out. The methodology for the case study is like that described in Murray and Price (2008). The case study was categorized into two major activities:

Performance investigation of the plant under various substitution ratios of coal to alternative fuels (AF)

An assessment of the quality of the clinker (final product) produced using XRF in the laboratory.

3.1 Operation and Technical Performance

The grate clinker coolers (GCC) are employed in the cement industry to recover heat from hot clinker produced in rotary kilns (RK) and bypass exhaust heat during kiln operation (Fig. 15.3). This is based on the measurement made on the Specific Energy Consumption (SEC) of the RK, as well as the first and second law efficiencies of both cooler units (Fig. 15.2). Recovering energy from the grate cooler or kiln bypass is crucial for cutting production costs and addressing the environmental issues that the cement industry is facing. In this study, the grate clinker cooler was modeled and connected to a bypass gas duct for recuperation. First, a thermodynamic efficiency study and process flow model were run at 85% of AF substitution (the gas from bypass was not connected, see Fig. 15.1).

The following steps elaborate the various activities carried out as part of the case study:

Step 1: Raw meal, raw coal, and alternative fuel chemical composition design to fit with clinker quality.

Step 2: Fuel injection and data record, the alternative fuel stated increased in grade of 10 tons per hours to 85% of substitution of the standard fossil fuel.

Step 3: By-pass exhaust gas management as exergies for the which will booster kiln second and 3rd air in oxygen.

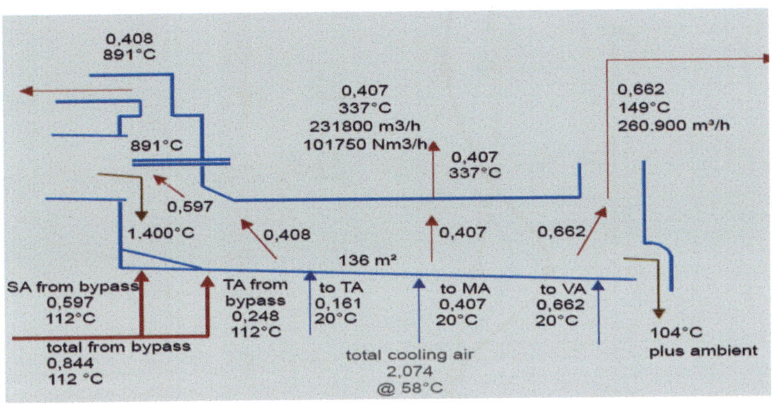

Fig. 15.1 Recirculation to air cooling from by-pass (AF 85%)

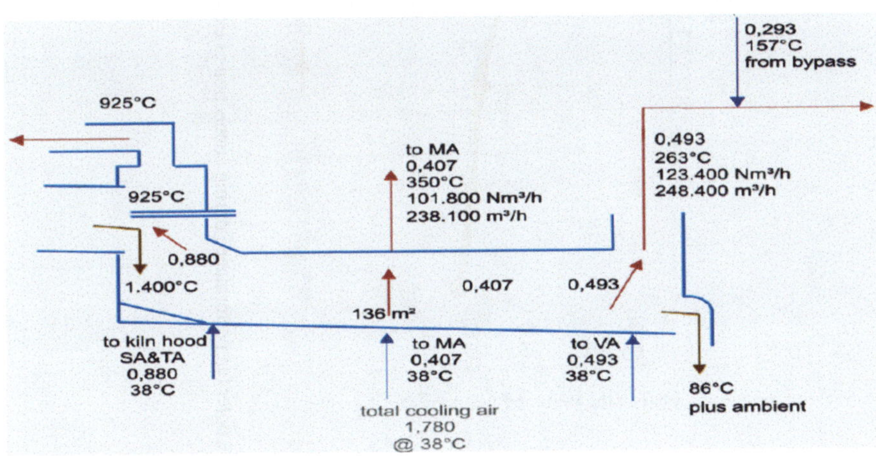

Fig. 15.2 No recirculation to cooler exhaust (AF 55%)

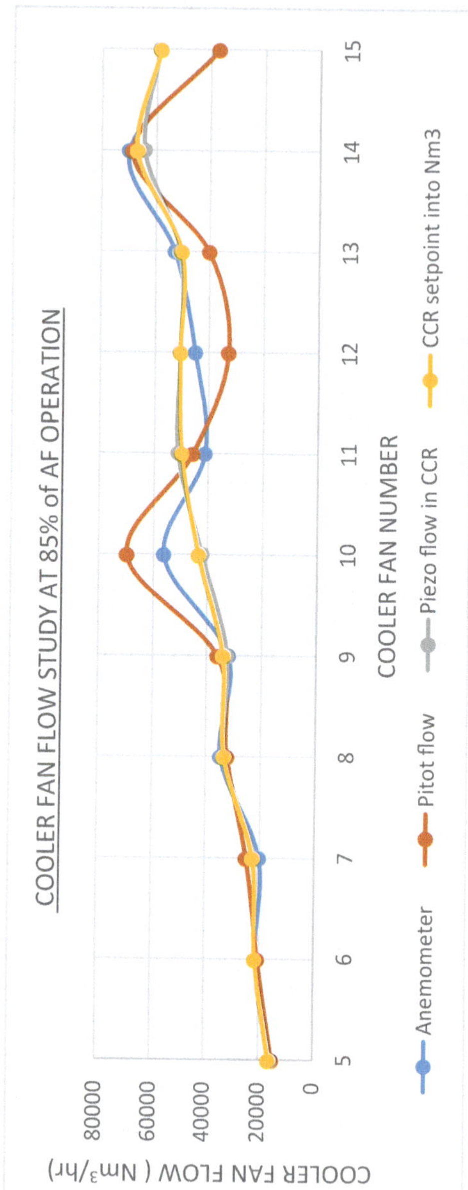

Fig. 15.3 Cooler fan performance during connection of exergy gas

Step 4: Cooler exergies optimalisation by maintaining the clinker temperature below 110 °C.

The air flow side countermeasure was taken by using pitot tube and anemometer while increasing AF-Fuel substitution up to 85%. This was done by recuperating by-pass exhaust gas as exergies and their effects on green industrial cooler operation management (Mossi et al. 2021).

3.2 *Laboratory Analysis of the Final Product (Clinker)*

The clinker was analyzed by using x-ray spectrometer (XRF) in the laboratory.

30g of clinker were taken from the cooler outlet (on pan conveyer) at a rate of one sample each hour and taken to the laboratory for investigation.

4 Results and Discussion

The effects of cooling air temperature and AF substitution rate on the functionality of the grate clinker coolers (GCC) and rotary kiln (RK) are investigated. It is discovered that raising the AF substitution rate to 85% can raise the first to second law efficiencies of the RK by 35.62 to 35.29%. The SEC for second law efficiency was 3,221.8 kJ/kg clinker compared to first one which was 5,035.4 kJ/kg clinker.

Table 15.1 shows the average operation data of the co-process and exergy of cooler. The under-grate pressure of the cooler (in mbar) was measured simultaneously with the air flow for the cooling clinker from the central control room (CCR).

According to some previous studies, fuel data sheets indicate that, the amount of air required for combustion is typically between 335 and 345 kg per GJ of energy, or about 380 kg if one allows for a 10% surplus (El-Nahas 2016). This indicates that an effective kiln using 3.2 GJ of energy per ton of clinker will only require 1216 kg of air, or 1.216 times as much air as clinker. The additional air required to achieve a tolerable temperature is surplus to requirements because using this air alone in the cooler would result in clinker temperatures of 399°C (Suntivarakorn and Treedet 2003).

Nonetheless, a typical Thames-side wet process kiln from the 1950s would need 3.04 kg of air for every kilogram of clinker and consume 8.0 GJ/t clinker, enabling it to achieve very low clinker temperatures without (theoretically) creating waste air (Nadziakiewicz and Kozioł 2012). Because grate coolers were developed in the historical context of these, the waste air issue was not a serious one at the time.

The basic ingredients in this sort of kiln are heat, air, and fossil or alternative fuels (primary and secondary). The final tier of the preheater tower houses the major flour input for the rotary kiln circuit (Radwan 1995). It enters the rotary section through a derivative route right below the final cyclone, which is independent of the exit gases in the kiln (first stage). A feed circuit is used to introduce the fossil fuel (coal, heavy

Table 15.1 Cooler operation data 85% AF substitution with connection of by-pass exhaust gas

Under-grate pressure(mbar)			Air Flow (Km^3/h)			Air flow		
Fans	Measured	CCR	Measured Air flow	CCR	Deviation (%)	Nm^3/kg cli	Grate Area	Km^3/h
Fan 5	75,3	−2	16,87	15,43	9	0,07	2,78	16,654
Fan 6	98	−2,1	23,86	23,03	4	0,11	3,52	19,674
Fan 7	−2,6	102,07	24,39	23,93	2	0,11	3,52	19,674
Fan 8	−1,5	69	37,34	34,93	7	0,16	9,05	35,956
Fan 9	−1,8	68,6	39,10	35,09	11	0,16	9,05	35,956
Fan 10	−0,9	107,8	63,73	45,15	41	0,20	13,58	53,934
Fan 11	−1,1	100,63	57,63	52,31	10	0,23	13,58	53,934
Fan 12	−1,1	84,12	47,84	33,34	43	0,15	13,58	53,934
Fan 13	−1	71,15	55,60	52,02	7	0,23	13,58	53,934
Fan 14	−1,6	130,8	77,98	70,76	10	0,32	18,11	71,912
Fan 15	−1,5	106,15	61,19	61,38	0	0,27	16,6	63,859
Total			506	447		2,01	117	479
Volume Air flow (Nm^3/kg cli)			2.108	2.075				2.088

fuel oil, or natural gas) to the main burner. When it comes to alternative fuels, they are either added simultaneously with fossil fuels in the main burner or in the same input as the flour (below the last cyclone) (Birch 1990).

The basic ingredients in this sort of kiln are heat, air, and fossil or alternative fuels (primary and secondary). The final tier of the preheater tower houses the major flour input for the rotary kiln circuit (Radwan 1995). It enters the rotary section through a derivative route right below the final cyclone, which is independent of the exit gases in the kiln (first stage). A feed circuit is used to introduce the fossil fuel (coal, heavy fuel oil, or natural gas) to the main burner. When it comes to alternative fuels, they are either added simultaneously with fossil fuels in the main burner or in the same input as the flour (below the last cyclone) (Birch 1990).

Figure 15.2 depicts a cooler heat and mass balance in the use of alternative fuels, which is at 55%. The total cooling air ratio is 1.7 at 38 °C, the clinker temperature is 1400 °C, and the bypass exhaust gas is not linked to the clinker cooler. The tertiary air temperature is around 925 °C, whereas the secondary air temperature is 925 °C. The raw material drying process can utilize exhaust gas at a temperature of 263 °C. 86 °C is the cooling clinker temperature.

The comparison of site manual measurements (Anemometer and Pitot tube) with automatic readout and central control room (CCR) is shown in Fig. 15.3. According to the graph, the pitot tube exhibits more deviations than the Anemometer from cooler fans 10 to 15.

Figure 15.3 shows a cooler heat and mass balance when alternative fuels are used, with an efficiency of 85%. The clinker temperature coming out of the kiln is 1400 °C, the total cooling air is 2.1 at 68 °C, and the bypass exhaust gas is connected to the clinker cooler at a temperature of 112 °C. While the secondary air temperature is 891 °C, the tertiary air temperature dropped to about 891 °C. Exhaust gas can be used in the raw material drying process at 149 °C. The cooling clinker temperature is 104 °C.

Clinker and exhaust gas are the primary byproducts. A few meters after the main burner's head (the end of the clinkerization zone); at the cooler outlet, the clinker exits the rotary cylinder (Afkhami et al. 2015). The riser pipe that connects the rotating cylinder (rotary kiln) to the final cyclone of the preheater tower is where the gases leave the rotary part of the kiln. On the final level of the preheater tower, the gases come out of the rotary kiln circuit (Ghalandari et al. 2019).

New strategies to lower energy usage per unit weight of product given the volume of cement production need to be considered. The effectiveness of the cooler is essential to the efficiency of this operation as it pre-heats the air required for combustion in the calciner and rotary kiln and recovers heat from the hot clinkers (Vaccaro 2006).

Heat energy is heavily consumed by the clinker plant. Therefore, it is crucial from an economic standpoint to produce clinker with the least amount of heat required and to fully utilize the fuels. Table 15.2 illustrates the biggest range of quality variation among the major fuel types, with SO_3 content less than 1%, for coal and alternative fuel combined. The finesse of the raw mix has improved, which was done to maximize the kiln system's efficiency. Table 15.3 summarizes the data regarding the changes.

The system proved very stable when coal and SRF were used as the primary fuel. For 100% coal usage, the raw mix fineness (Fins %) was maintained at 12% on 90 μm.

The quality of the clinker is constant with no significant changes in the four oxides (Silica oxide (SiO_2—22%), alumina oxide (Al_2O_3—5%), carbon oxide (CaO—63 to 66%), and iron oxide (Fe_2O_3 −3,3%), and the loss on ignition (LOI) was in the range of 36%.

Consequently, it is necessary to identify the ideal values of operating parameters for the cooler to reduce the energy consumption in a cement factory. The energy needed to drive the clinker bed and the heat losses from the cooler to the surroundings determine how much energy is used in the cooler. As a result, the effectiveness of the cooler's heat recovery from the clinkers and the losses from the cooler surface (via conduction, convection, and radiation) to the environment are its key determinants of efficiency (Pedersen et al. 2017).

Table 15.2 The industry analyses of fuels

Fuel	Mad	Aad	Stad	Fcad	Vad	SO_3	Qnet.ad/(kJ/kg)
	%	%	%	%	%	%	
Coal	10	14,1	1,38	45,7	28,72	3,39	22,595
Coal + 45%SRF	6,67	0,35	1,12	83,25	9,73	0,86	22,895
Coal + 85%SRF	5,2	0,24	0,86	82,33	8,45	0,57	24,690

Table 15.3 Raw mix and clinker quality analyses during alternative fuel substitution

Items	Fins	LOI	SiO_2	Al_2O_3	Fe_2O_3	CaO	MgO	K_2O
	%	%	%	%	%	%	%	%
Raw mix for 100% Coal usage	12	36,21	13,7	3,17	2,16	42,8	1,02	0,124
Raw mix for 55% Coal usage	10	36,98	14,8	2,17	2,16	41,8	1,02	0,324
Raw mix for 15% Coal usage	9	36,61	13,9	3,07	2,06	42,47	1,012	0,144
		36,11	13,77	3,17	2,16	42,8	1,03	0,134
Clinker quality for 100% coal usage			21,94	5,4	3,38	63,07	1,602	0,202
			20,94	5,8	2,48	62,07	1,502	0,302
			22,01	5,45	3,38	61,07	1,602	0,202
Clinker quality for 55% coal usage			21,94	5,4	3,38	63,01	1,602	0,202
			21,94	5,4	3,38	62,03	1,602	0,202
			21,94	5,4	3,38	63,2	1,602	0,202
Clinker quality for 15% coal usage			21,94	5,4	3,38	64,17	1,602	0,202
			20,94	5,8	3,48	65,04	1,502	0,302
			22,01	5,45	3,38	64,07	1,602	0,202
			21,94	5,4	3,38	66,07	1,602	0,202
			21,94	5,4	3,38	66,07	1,602	0,202
			21,94	5,4	3,38	66,07	1,602	0,202

The cooler's clinker temperature change has a significant impact on the quality of the cement the plant produces. An effective cooler will have a minimum clinker outflow temperature and acceptable characteristics. The ideal cooling air rate, clinker input rate, clinker inlet temperature, cooler length, number of air apertures, and grate speed to increase the efficiency of the cooler used in cement businesses therefore need to be selected.

By altering the operating parameters in GCC, energy consumption is reduced by 34%, saving 1813.6 kJ for every kilogram of clinker produced. The applications can potentially reduce CO_2 emissions into the atmosphere by 6740.66 tons each year.

5 Conclusions

Cement production requires sophisticated equipment and an in-depth knowledge of the plant's operating principles. Co-processing municipal solid waste with coal is eco-friendly and should be granted eco-friendly status. Import-substitute status should be granted for re-claimed alternative fuel. The results of this study suggest that cement businesses can adapt and include the usage of alternative fuels, such as municipal solid waste in this case, at high ratios up 85% with bypass exhaust gas recuperation. The thermographic investigation of alternate fuel behavior in kiln pre-heater needs more study.

As one of the most energies and pollution intensive industrial processes, cement production also contributes to clinker emissions of CO_2, NOX, and SO_2 as well as some heavy metal discharge from the pre-calciner kiln system. As demonstrated in this study report, this can get worse if the kiln bypass operation technique is poorly planned.

Further investigations on standardizing kiln bypass operations and bypass dust utilization/recirculation still needs to be carried out to optimize recirculation options and minimize landfilling.

References

A Tec, 2021: Available: https://www.atec-ltd.com/en/.

Abdel Maksoud, M. I. A., Sallam, O. I., Kassem, S. M., Fahim, R. A., & Awed, A. S., 2022. Novel strategy for hazardous cement bypass dust removal: structural, optical and nuclear radiation shielding properties of CBD-bismuth borate glass., Journal of Inorganic and Organometallic Polymers and Materials, pp. 32(9), 3533–3545.

Abdelgader, H. S., Amran, M., Kurpińska, M., Mosaberpanah, M. A., Murali, G., & Fediuk, R., 2022. Cement kiln dust. In Sustainable Concrete Made with Ashes and Dust from Different Sources, Woodhead Publishing., pp. 451–479.

Abdel-Gawwad, H. A., Heikal, M., Mohammed, M. S., Abd El-Aleem, S., Hassan, H. S., García, S. V., & Alomayri, T., 2019. Sustainable disposal of cement kiln dust in the production of cementitious materials., Journal of Cleaner Production, pp. 232, 1218–1229.

Abdel-Gawwad, H. A., Sanad, S. A., & Mohammed, M. S., 2020. A clean approach through sustainable utilization of cement kiln dust, hazardous lead-bearing, and sewage sludges in the production of lightweight bricks., Journal of Cleaner Production, pp. 273, 123129.

Abdelzaher, M. A., 2021. Experiential investigation on the effect of heavy fuel oil substitution by high sulfur petcoke on the physico-mechanical features and microstructure of white cement composites., Engineering Research Express, pp. 3(1), 015028.

Afkhami, B., Akbarian, B., Beheshti, N., Kakaee, A. H., & Shabani, B., 2015. Energy consumption assessment in a cement production plant., Sustainable Energy Technologies and Assessments, pp. 10, 84–89.

Aguilera, N., & Marchetti, J. L., 1998. Optimizing and controlling the operation of heat-exchanger networks., AIChE Journal,, pp. 44(5), 1090–1104.

Ahmadi, A. A., Bahiraei, M., 2021. Thermohydraulic performance optimization of cooling system of an electric arc furnace operated with nanofluid: A CFD study.,» Journal of Cleaner Production,, pp. 310, 127451.

Ahmed, D. A., & Ragei, S. M, 2021. An investigation of the mechanical properties of ternary blend cement pasts containing a large amount of cement kiln dust., Journal of Scientific Research in Science, pp. 38(part 1 (Basic Sciences)), 71–90.

Al-Bakri, A. Y., Ahmed, H. M., & Hefni, M. A., 2022. Cement kiln dust (CKD): potential beneficial applications and eco-sustainable solutions., Sustainability, pp. 14(12), 7022.

Alsalman, A., Assi, L. N., Kareem, R. S., Carter, K., & Ziehl, P., 2021. Energy and CO2 emission assessments of alkali-activated concrete and Ordinary Portland Cement concrete: A comparative analysis of different grades of concrete., Cleaner Environmental Systems, pp. 3, 100047.

Al-Tahan, M. A., Shatat, M. R., Abdallah, M. H., & Taher, M. A., 2021. Influence of the thermally and hot water treatment of cement kiln dust on the Physico-chemical features of Portland cement., Journal of Building Engineering, pp. 38, 102170.

Altun, O, 2022. Air classification performances of the components within the varied feed blends.,» Powder Technology, pp. 399, 117092.

Ashteyat, A. M., Haddad, R. H., & Obaidat, Y. T., 2018. Case study on production of self compacting concrete using white cement by pass dust., Case Studies in Construction Materials, pp. 9, 190.

Ashteyat, A. M., Al Rjoub, Y. S., Murad, Y., & Asaad, S., 2022. Mechanical and durability behaviour of roller-compacted concrete containing white cement by pass dust and polypropylene fibre., European Journal of Environmental and Civil Engineering, pp. 26(1), 166–183

Athira, V. S., Bahurudeen, A., Saljas, M., & Jayachandran, K., 2021. Influence of different curing methods on mechanical and durability properties of alkali activated binders., Construction and Building Materials, pp. 299, 123963.

Berisha, S., Živković, I., Kotnik, J., Mlakar, T. L., & Horvat, M., 2022. Temperature fractionation of mercury in the cement production process using quadrupole mass spectrometry., Cement and Concrete Research, pp. 162, 106970.

Birch, E. R. I. K.., 1990. Energy Savings in Cement Kiln Systems., Energy Efficiency in the Cement Industry, pp. 12756, 112.

Brouwers, J. J. H., Verbeek, P. H. J., & Thomson, W. R, 1987. Analytical system availability techniques., Reliability Engineering,, pp. 17(1), 9–22.

Global Cement, 2022. A TEC to install ReduChlor chlorine bypass system at Holcim Schweiz's Eclépens cement plant. On line available: https://www.globalcement.com/news/item/13519-a-tec-to-install-reduchlor-chlorine-bypass-system-at-holcim-schweiz-s-eclepens-cement-plant.

Chatterjee, A., Sui, T., 2019. Alternative fuels–effects on clinker process and properties., Cement and Concrete Research, pp. 123, 105777.

Cheng, Z., Guo, Z., Tan, Z., Yang, J., & Wang, Q., 2019. Waste heat recovery from high-temperature solid granular materials: Energy challenges and opportunities., Renewable and Sustainable Energy Reviews,, pp. 116, 109428.

Choi, H., Kim, J., Choi, S., & Kim, S., 2021. KCl Extracted from Chlorine Bypass Dust as Activator for Plain Concrete., Materials 2021, pp. 14, 6091.

Corporation, TAIHEIYO ENGINEERING, «Taiheiyo Chlorine Bypass System,» TAIHEIYO ENGINEERING, 2012–2022. 2022. Online available: https://www.taiheiyo-eng.co.jp/en/engineering/taiheiyo-chlorine-bypass-system.html#:~:text=The%20chlorine%20bypass%20system%20which,from%20kiln%20preheater%20system%20efficiently. Last accessed: 06 03 2023.

Czapik, P., Zapała-Sławeta, J., Owsiak, Z., & Stępień, P, 2020. Hydration of cement by-pass dust.,» Construction and Building Materials, pp. 231, 117139.

Darweesh, H., 2020. Hazardous-Effects-of-Electrostatic-Precipitator-Cement-Kiln-Bypass-Dust-Waste-on-both-Environment-and-Public-Human-Health. Malik Badshah, editor. Prime Archives in Biosciences., Hyderabad, India: Vide Leaf., p. np.

Dembla, A., & Mersmann, M., 2021. Data-Driven Thermal Energy Management Including Alternative Fuels and Raw Materials Use for Sustainable Cement Manufacturing., In Intelligent and Sustainable Cement Production CRC Press., pp. 141–197.

Dissanayake, M. C. W., Karunawardena, A., 2019. Embankment Construction on Power Blender Cement Mixed Soft Soil Treated Ground for CKE Bypass. In ICSBE 2018: Proceedings of the 9th International Conference on Sustainable Built Environment.

Ebrahimi-Moghadam, A., Deymi-Dashtebayaz, M., Jafari, H., & Niazmand, A., 2020. Energetic, exergetic, environmental and economic assessment of a novel control system for indirect heaters in natural gas city gate stations., Journal of Thermal Analysis and Calorimetry,, pp. 141, 2573–2588.

El-Nahas, S., Mohamed, A. E., Ahmed, R. R., & Abd El-Sadek, M. S., 2016. Improvement of boiler's efficiency using heat recovery and automatic combustion control system., Energy Procedia, pp. 100, 193–197.

El-Sayed Seleman, M. M., El-Kheshen, A. A., Kharbish, S., & Ebrahim, W. R., 2020. Utilization of Cement Kiln Dust for the Preparation of Borosilicate Glass., Interceram-International Ceramic Review, pp. 69, 26–33.

Fan, S., Wang, Y., Cao, S., Sun, T., & Liu, P., 2021. A novel method for analyzing the effect of dust accumulation on energy efficiency loss in photovoltaic (PV) system., Energy, pp. 234, 121112.

Fierro, J. J., Escudero-Atehortua, A., Nieto-Londoño, C., Giraldo, M., Jouhara, H., & Wrobel, L. C., 2020. Evaluation of waste heat recovery technologies for the cement industry., International Journal of Thermofluids, pp. 7, 100040.

Geng, Y., Wang, Z., Shen, L., & Zhao, J., (2019). Calculating of CO2 emission factors for Chinese cement production based on inorganic carbon and organic carbon.,» Journal of Cleaner Production, pp. 217, 503–509.

Gerassimidou, S., Velis, C. A., Williams, P. T., Castaldi, M. J., Black, L., & Komilis, D., 2021. Chlorine in waste-derived solid recovered fuel (SRF), co-combusted in cement kilns: A systematic review of sources, reactions, fate and implications., Critical Reviews in Environmental Science and Technology, pp. 51(2), 140–186.

Ghalandari, V., Majd, M. M., & Golestanian, A., 2019. Energy audit for pyro-processing unit of a new generation cement plant and feasibility study for recovering waste heat: A case study., Energy, pp. 173, 833–843.

Ghenai, C., Inayat, A., Shanableh, A., Al-Sarairah, E., & Janajreh, I., 2019. Combustion and emissions analysis of Spent Pot lining (SPL) as alternative fuel in cement industry., Science of The Total Environment, pp. 684, 519–526.

Ghosh, S. K., Parlikar, U. V., Karstensen, K. H., Ghosh, S. K., Parlikar, U. V., & Karstensen, K. H., 2022. Cement Manufacturing—Technology, Practice, and Development. Sustainable Management of Wastes Through Co-processing,, pp. 73–90.

Han, M. C., Lee, D. J., 2020. Engineering Characteristics of Blast Furnace Slag Cement Mortar Using Chlorine Bypass System-Dust as Alkali Activator., Journal of the Korean Recycled Construction Resources Institute, pp. 8(2), 235–244.

Hannoun, G., Jaouad, A., Schebek, L., Belkziz, J., & Ouazzani, N., 2019. Energetic potential and environmental assessment of solid wastes as alternative fuel for cement plants., Appl. Ecol. Environ. Res, pp. 17, 15151–15168.

Jin, J., Pan, J., Lu, Z., Wu, Q., Xu, L., Fan, B., & Quaye, E. K., 2020. An investigation on performance of an asymmetric twin-scroll turbine with a small scroll bypass wastegate for a heavy-duty diesel engine., Journal of Engineering for Gas Turbines and Power, pp. 142(6), 061006.

Kaliyavaradhan, S. K., Ling, T. C., & Mo, K. H., 2020. Valorization of waste powders from cement-concrete life cycle: A pathway to circular future., Journal of Cleaner Production, pp. 268, 122358.

Kaliyavaradhan, S. K., Ling, T. C., & Mo, K. H., 2021. Bowtie Method: Study of Occupational Health and Safety Risks in Cement Production Process., Media Kesehatan Masyarakat Indonesia, pp. 17(4), 124–133.

Li, Y. M., Wei, W. M., Hao, J., Gao, P. F., Liu, P. F., Li, G. N., ... & Wang, Z. J., 2022. Demonstration application of Chinese-made chlorine bypass system for stable cement kiln using calcium carbide residue as main raw material., Journal of Material Cycles and Waste Management, pp. 24(2), 618–632.

Long, H., Liao, Y., Cui, C., Liu, M., Liu, Z., Li, L., ... & Yan, D., 2022. Assessment of popular techniques for co-processing municipal solid waste in Chinese cement kilns.,» Frontiers of Environmental Science & Engineering, pp. 16, 1–13.

Malacarne, C. S., da Silva, M. R. C., Danieli, S., Maciel, V. G., & Kirchheim, A. P., 2021. Environmental and technical assessment to support sustainable strategies for limestone calcined clay cement production in Brazil. Construction and Building Materials, pp. 310, 125261.

Mendes, B. C., Pedroti, L. G., Vieira, C. M. F., Carvalho, J. M. F., Ribeiro, J. C. L., Albuini-Oliveira, N. M., & Andrade, I. K. R., 2022. Evaluation of eco-efficient geopolymer using chamotte and waste glass-based alkaline solutions. Case Studies in Construction Materials, pp. 16, e00847.

Miller, S. W., & Hansen, J. P, 2004. Methods for reducing SO/sub 2/emissions., in Cement Industry Technical Conference, In IEEE-IAS/PCA.

Mohamedzein, Y., Hassan, H., & Al-Hashmi, K., 2022. Effect of cement-by-pass dust stabilized-expansive soil subgrade on pavement rutting., Innovative Infrastructure Solutions, pp. 7, 1–11.

Mokhtar, A., Nasooti, M., 2020. A decision support tool for cement industry to select energy efficiency measures., Energy Strategy Reviews, pp. 28, 100458.
Mossie, A. T., Wolde, M. G., Beyene, G. B., Palm, B., & Khatiwada, D. A., 2021. Comparative Study of the Energy and Environmental Performance of Cement Industries in Ethiopia and Sweden., in 2021 International Conference on Electrical, Computer, Communications and Mechatronics Engineering (ICECCME), pp. 1–5. IEEE.
Murray, A., & Price, L., 2008. Use of alternative fuels in cement manufacture: analysis of fuel characteristics and feasibility for use in the Chinese cement sector.
Nadziakiewicz, J., & Kozioł, M., 2012. Different possible ways for saving energy in the cement production., Advances in Applied Science Research, pp. 3(2), 1162–1174.
Nocuń-Wczelik, W., & Stolarska, K., 2019. Calorimetry in the studies of by-pass cement kiln dust as an additive to the calcium aluminate cement., Journal of Thermal Analysis and Calorimetry, pp. 138, 4561–4569.
Papanikola, K., Papadopoulou, K., Tsiliyannis, C., Fotinopoulou, I., Katsiampoulas, A., Chalarakis, E., ... & Lyberatos, G., 2019. Food residue biomass product as an alternative fuel for the cement industry., Environmental Science and Pollution Research, pp. 26, 35555–35564.
Pedersen, M. N., Nielsen, M., Clausen, S., Jensen, P. A., Skaarup Jensen, L., & Dam-Johansen, K., 2017. Imaging of flames in cement kilns to study the influence of different fuel types., Energy & Fuels, pp. 31(10), 11424–11438.
Pope, M. A., Lee, J. I., Hejzlar, P., & Driscoll, M. J., 2009. Thermal hydraulic challenges of gas cooled fast reactors with passive safety features., Nuclear Engineering and Design, pp. 239(5), 840–854.
Proaño, L., Sarmiento, A. T., Figueredo, M., & Cobo, M., 2020. Techno-economic evaluation of indirect carbonation for CO2 emissions capture in cement industry: A system dynamics approach. Journal of Cleaner Production, pp. 263, 121457.
Putri, E., Lestari, M., Novrikasari, N., Andarini, D., Camelia, A., Fujianti, P., & Nurhaliza, T., 2022. Conversions of Cement bypass waste to Nano, hydroxyapatite exploited in water purification., p. sp.
Radwan, A. M., 1995. Energy efficiency of China's cement industry., Energy, pp. 20(7), 669–681.
Rathoure, A. K., 2019. Fugitive dust control in cement industries., in Zero Waste, pp. 317–327.
Ray, S, 2022. Effect of control parameters on erosion wear performance of glass-, epoxy composites filled with waste marble powder., pp. 23–31.
Salamanova, M., Murtazaev, S. A., Saidumov, M., Alaskhanov, A., Murtazaeva, T., & Fediuk, R., 2022. Recycling of Cement Industry Waste for Alkali-Activated Materials Production., Materials, pp. 15(19), 6660.
Saleh, H. M., Salman, A. A., Faheim, A. A., & El-Sayed, A. M, 2020. Sustainable composite of improved lightweight concrete from cement kiln dust with grated poly (styrene)., Journal of Cleaner Production, pp. 277, 123491.
Sani, M. M., Noorpoor, A., & Motlagh, M. S., 2020. Multi objective optimization of waste heat recovery in cement industry (a case study)., Journal of Thermal Engineering, pp. 6(4), 604–618.
Schneider, M, 2019. The cement industry on the way to a low-carbon future. Cement and Concrete Research, pp. 124, 105792.
Sengupta, P., & Sengupta, P., «Cement manufacturing and process control., 2020. Refractories for the Cement Industry, pp. 61–76.
Shah, K., An, N., Ma, W., Ara, G., Ali, K., Kamanova, S., ... & Xing, L., 2020. Chronic cement dust load induce novel damages in foliage and buds of Malus domestica., Scientific reports, pp. 10(1), 12186.
Singh, T., 2021. Utilization of cement bypass dust in the development of sustainable automotive brake friction composite materials. Arabian Journal of Chemistry, pp. 14(9), 103324.
Soliman, N. K., & Moustafa, A. F., 2020. Industrial solid waste for heavy metals adsorption features and challenges; a review., Journal of Materials Research and Technology, pp. 9(5), 10235–10253.
Sousa: V., Bogas, J. A., Real, S., Meireles, I., & Carriço, A., 2023. Recycled cement production energy consumption optimization., Sustainable Chemistry and Pharmacy, pp. 32, 101010.

Stančin, H., Mikulčić, H., Wang, X., & Duić, N., 2020. A review on alternative fuels in future energy system., in Renewable and sustainable energy reviews, pp. 128, 109927.
Stevulova, N., Junak, J., Strigac, J., & Junakova, N, 2022. Recycling cement bypass dust and granulated blast furnace slag into hydraulic road binder., In IOP Conference Series: Materials Science and Engineering, pp. Vol. 1252, No. 1, p. 012010.
Sun, D., Li, Z., Huang, S., Chi, J., & Zhao, S., 2021. Efficient mercury removal in chlorine-free flue gas by doping Cl into Cu2O nanocrystals., Journal of Hazardous Materials, pp. 419, 126423.
Suntivarakorn, R., & Treedet, W., 2003. Co-combustion of sludge with coal. Applied Energy, pp. 75(3–4), 239–248.
Vaccaro, M., 2006. Burning alternative fuels in rotary cement kilns.,» in IEEE Cement Industry Technical Conference, 2006. Conference Record, In IEEE Cement Industry Technical Conference.
Vojvodíková, B., Procházka, L., & Boháčová, J., 2021. X-ray diffraction of alkali-activated materials with cement by-pass dust., Crystals, pp. 11(7), 782.
Worrell, E., 2008. Energy efficiency improvement opportunities for the cement industry, p. np.
Xu, T., Jin, L., Fan, X., Zhou, Y., Pasupathy, S., & Talwadker, R., 2015. Hey, you have given me too many knobs!: Understanding and dealing with over-designed configuration in system software., in Proceedings of the 2015 10th Joint Meeting on Foundations of Software Engineering.
Yörük, C. R., Uibu, M., Usta, M. C., Kaljuvee, T., & Trikkel, A, 2020. CO 2 mineralization by burnt oil shale and cement bypass dust: effect of operating temperature and pre-treatment., Journal of Thermal Analysis and Calorimetry, pp. 142, 991–999.
Zawrah, M. F., Shehata, N., Saddek, A. B., & Ismail, A. R. G., 2022. Utilization of white-cement kiln dust in presence of raw/waste materials for production of alkali-activated products, International Journal of Environmental Science and Technology, pp. 1–14.

Palm Kernel Shell and Raw Coal Co-processing for Sustainability Goals in Clinkerization Pyro-Processes

E. Beguedou, S. Narra, K. Agboka, E. A. Armoo, and M. K. Damgou

1 Introduction

Humanity has been using the African oil palm (*Elaeis guineensis* Jacq.) as a source of oil and other goods for thousands of years. The cultivation of oil palms has grown dramatically over the past 50 years, to the point that palm oil is now a significant commodity in international trade and the oil palm is a key source of vegetable oil (Owoyele 2014). The African oil palm and its cousins in terms of their origins, evolution, distribution, and use; discusses the development of the palm oil business. The manufacture of biodiesel and different oleochemicals is among the growing list of non-food uses for this resource, which is used for food in around 77% of cases (Henson 2018). The oil palm business produces a lot of leftovers and garbage, including empty fruit bunches, palm kernel shells, plant trunks, fiber, leaves, and others. When palm oil is harvested and processed, substantial levels of organic debris, suspended matter, and oil and grease are also produced in the effluents. These wastes

E. Beguedou (✉) · S. Narra · E. A. Armoo
Professorship of Material and Energy Valorisation of Biogenous Residues, Professorship of Waste and Resource Management, University of Rostock, 18059 Rostock, Germany
e-mail: essossinam.beguedou85@gmail.com

S. Narra
German Biomass Research Centre gGmbH, Leipzig, Germany

K. Agboka
West African Service Center for Climate Change and Adaptive Land Use (WASCAL), University of Lomé, P.O. Box 1515, TogoLomé
e-mail: kagboka@univ-lome.tg

M. K. Damgou
Department of Physique, Centre d'Excellence Regional pour la Maîtrise de l'Electricité (CERME, University of Lomé, P.O. Box 1515, TogoLomé
e-mail: kmani@univ-lome.tg

M. Narra and S. Narra (eds.), *Innovations in Circular Economy and Renewable Energy in Africa*, World Sustainability Series, https://doi.org/10.1007/978-3-031-68330-5_16

harm the ecology and have an adverse impact on community health (Sridhar 2009a, b).

The mapping of palm grove areas shows a very good geographical representation of oil palm cultivation in the agricultural landscape of the study area. Indeed, the plant is grown all over the area, making it the first cash crop of the Maritime region and the municipalities and along the Mono in Togo in sum, 3 major palm groves stand out in the study area.

These are the palm grove area of the Zio and Haho valleys (comprising the prefectures of Agou, Zio and Avé), that of the lower Mono valley (comprising all the municipalities of Benin and the prefectures of Yoto, Bas-Mono, Vo, and Lakes), and that of the high valley of Mono (prefectures of Haho, Ogou and Middle Mono).

Oil palm tree is the first cash crop in the study area in Togo. Four types of combustible residues are produced from oil palm. These are leaves, diet stalks, fruit pulp, and nut shells. The latter two (fruit pulp, and nut shells) have so great an energy potential that they are subject to sustained marketing despite a considerable degree of self-consumption. Indeed, the hulls are sought among others by blacksmiths and cement production plants to feed their kilns and incidentally in the building to replace the gravel in the concrete (Pamungkas 2019).

Pulp (called "kpelebs") is generally sold to households as a fuel or accelerator for the combustion of other fuels. In recent years it is noticed that almonds are less and less processed into palm oil. They are sold directly to traders who export them to Nigeria and Ghana.

The total number of palm groves in Benin is estimated at approximately 19,050 hectares based on the data available in the territorial economic diagnoses of Mono-Couffo in 2014. For a total area of 183,372 ha of natural palm grove recorded in Togo as part of the national census of agriculture (MAEP 2013), about 77,145 hectares are found in the Maritime region (42.1%) and 93,804 hectares in the Plateau region (51.2%). Most oil palm fuel deposits are found in the prefectures of Zio, Yoto, Ogou and Haho. With an average level of own consumption of 46.13% for the hulls, the quantity of fuel available in the study area amounts to 73,153 tons of hulls, including 15,995 tons in Benin and 57,158 tons in Togo.

Palm Kernel Shell Availability through middlemen collecting from farming community at different locations. Farther location will pose a challenge in terms of cost transportation. Currently this is the biggest availability. Material is received ready to feed (Sridhar 2009a, b).

To achieve these objectives, Togolese Greenfield Clinker Plant wishes to recycle ecological waste, especially palm kernel shell, by gradually replacing it with fossil fuels. Plant managers are challenged to the matter of finding an appropriate balance between pre-processing and co-processing.

There is a simple play-off between the two: The more effort spent on pre-processing by washing and drying PKS, the less complicated the co-processing process needs to be in pyro-system. This is intuitive, as finely ground and intensively dried AF particles will burn easily, for example in the In-line calciner (ILC), without the need for complicated combustion technologies. In contrast, large, wet, and only slightly pre-processed lumpy AF particles may need a special reactor to help them fully combust. Once this pre-processing of PKS has been achieved, the co-processing in the pyro-system was easily carried out.

The lake of pre-processing, and the increasing urge to use a higher amount of mostly wet and hard to ignite biomass such as palm kernel shell, leads to the trend that clinker producers are increasingly looking for technologies that can handle any type or quality of alternatives fuel.

The pre-processing provides flexibility to their Alternative Fuels procurement process and, at the same time, saves on pre-processing cost, regardless of whether the plant does that itself or has it delivered by an external supplier.

To guarantee the quality of the finished goods, the palm kernel shell carbonization machine needs to be fitted with pretreatment tools. To fulfill EU emission regulations, the raw material is crushed, dried, transported to a carbonizing furnace, where it produces combustible gas, which is then cleaned using spray dedusting and condenser systems (Favre 2016; Holling 2013).

Inactive coal mines have negative effects on the environment and public safety. Due to climate change and global warming, the period of decarbonization is accelerating, and the region of former coal mines has many environmental issues. Agriculture, animal husbandry, and forestry can all be done to increase the cement plant substitution rate. However, the fundamental concern is how to co-processing agricultural waste smoothly in cement plant (Holling 2013).

An efficient way to restore damaged soils is with biochar. It removes CO_2 from the environment and can be exchanged for carbon credits on the carbon market. Economic considerations are also significant, thus the alternative that offers the greatest economic advantages will be chosen.

For their reactivity to be maximized, the pks co-processing must be extracted and processed with great care. This can help lower the carbon footprint of the cement industry. Calcined PKS can be produced on an industrial scale using currently available methods.

2 Material and Methodes

2.1 ILC-Calciner

The In-line calciner (ILC), This type of calciner is usually designed as a big vessel located above kiln inlet area. The designs show extended riser duct, so called tube calciners (Wurman 2002).

Over the years of calciner development and necessity of burning the alternative fuels resulted in increasing of the internal volume of the calciner. It consists of the vessel located in front or behind the preheater tower that ends-up with so called loop (goose) duct. This duct makes a turn very often at the top of preheater tower and lowers onto the bottom cyclones (Beguedou 2023).

The ILC calciners are usually contain the following plants: Loop (goose) duct, Calciner vessel,

It's easy to operate, reliable, and have the lowest NOX emission, thanks to the unique high-temperature reduction zone. CO emission is low too, because of the mid-calciner notch and specially designed exit duct provide superior mixing and combustion.

The In-Line Calciner is suitable for firing both traditional and alternate fuels. And operation is easy, with single entry locations for both the calciner fuel and tertiary air.

The ratio of firing in the calciner is 56,4% of fine coal and PKS was about 3,6%, which represented 1,71 tph of Palm Kernel Shell, while achieving a calcination rate at the kiln inlet of 95% as objective. For buildup and chemistry there is an allowable bypass of kiln gas of 10%. This help to deal with pre-calciner blockage in case. But it was not an issue of this co-grinding test.

Preheater optimization as well as calciner optimization. The core of the pyro process of each modern cement kiln is the preheater. Most substantial thermo-chemical reactions of the cement process occur in the riser ducts and in the calciner and certainly in the sintering zone of the kiln. The used kiln for this study has 5 stages with Inline (ILC) calciner, this provides a substantial degree of calcination and lets the researchers to consider it a core equipment of modern kilns.

2.2 *Physical Aspects of Palm Kernel Shell*

The quality of the Palm kernel shell (PKS), received at the study plant, was not good. It had unwanted material. There was a lot of mud and moisture. The company that delivered the PKS had transported 30% of unwanted material such as mud, sand, and water with the PKS.

2.3 *Laboratory Analyses of Palm Kernel Shell Before Pre-processing*

The Table 16.1 is showing the laboratory result where quality issues with the Net Calorific Value, which is 13,10Gj/t, with the minimum of 10,95Gj/t the moisture is 18,02 in average with 23.86 as maximum, occurred. The sand is 12,45% in average with the maximum point of 13,04. The density of palm kernel shell gets high 1,25 in average with maximum point of 1,3.

Table 16.1 Analysis of the palm kernel shells

BIOMASS	Moisture	Density	Sand	V(r)	A(r)	S (a.r)	Q (i.r)
		T/m^3	[%]	[%]	[%]	[%]	[KJ/kg]
PKS01	20,07	1,05	9,57	61,23	20,01	0,03	12,198
PKS02	19,86	1,25	10,04	62,02	21,01	0,02	12,105
PKS03	22,02	1,3	12,58	60,02	18,5	0,01	10,958
PKS04	20,07	1,05	9,57	61,23	20,01	0,03	12,198
PKS05	23,86	1,25	13,04	62,02	21,01	0,02	13,105
PKS06	18,02	1,3	12,45	60,02	18,5	0,01	11,958
SAMPLE AVREGE	20,65	1,20	11,21	61,09	19,84	0,02	12,087,00

Table 16.2 Palm kernel shell costs

Biomass	Moisture	Density	Q (i.r)	Purchase price	Energy cost as landed	Energy cost at pre-calciner
	%	T/m^3	[GJ/t]	Euro/t	Euro/GJ/m^3	Euro/GJ/m^3
Palm k. Shell	20,65	0,62	12,087	27,10	4,56	5,77

2.4 *PKS Cost Analysis*

In Table 16.2 the purchasing price of PKS was 27,10 Euro/t with Table 16.2 data it can be estimated that the energy cost as landed in euro per giga-joule per meter cube. The energy cost at pre-calciner is the main power fee added to energy cost as landed which represented the reel expenses of the plant by using palm kernel shell.

2.5 *Penalties*

The Table 16.3 is showing the penalty decided for the supplier of palm kernel shell, the physical aspect of the palm kernel shell need to be new, the contamination percentage need to be less than 1% to be pass, the moisture or water content need to be less or egal to 5% to be ok, the net caloric value need to be up to 16Gj/t.

By applying this principle, the company have deducted some tones of delivered material.

Moisture = 20,65%, Volume to be deducted = (20,65–5) % of 100t = 100*15,65% = 15,65 tons,

Contamination = 11,21%, Volume to be deducted = (11,21–1) % of 100t = 100*10,21% = 10,21 tons,

Table 16.3 Penalties data

Parameters	Target	Conformity
Physical aspect	Old	NOK
	New	OK
Contamination (Sand, mud, waste)	<1% of volume	OK
	>1% of volume	NOK
Moisture	≤5%	OK
	>5%	NOK
Net calorific Value (NCV)	≥16GJ/T	OK
	<16GJ/T	NOK
Penalties on the suppliers		
Parameters	Target	Penalities
Moisture	Moisture > 5%	We deduct (M - 5) % from the total volume delivered
Contamination (Sand, mud, waste)	Contamination > 1% of volume	We deduct (C - 1) % from the total volume delivered
Net calorific value	<16GJ/T	We deduct (16 GJ/T – X GJ/T) = 16-X of 100t, = (16-X)/100 = (16-X) %

Net Caloric Value = 12,08GJ/T, Volume to be deducted = (16GJ/T - 12GJ/T) = 3,92 of 100t = 3,92 tons.

2.6 Palm Kernel Shell Pre-processing

A dry process to separate palm kernel shell and unwanted material was developed using a series of winnowing columns, each powered by a single blower unit.

The blower was installed at the bottom of the column to create a forced-draught system with positive pressure along the column.

Additionally, the blower system was incorporated with a damper flap to enable airflow rate adjustment to achieve good separation of palm kernel shell and sand.

2.7 Laboratory Results After Pre-processing

Table 16.4 is showing the laboratory result after the pre-processing of PKS, the Net Calorific Value is 19,42Gj/t, with the minimum of 19,95Gj/t; The moisture is 3,79% in average with 4,52% as maximum. The sand is 0,62% in average with the maximum point of 0,9%. The density of palm kernel shell gets low 0,62 in average with maximum point of 0,68.

Table 16.4 Analysis of palm kernel shells after pre-processing

BIOMASS	Moisture	Density	Sand	V(r)	A(r)	S (a.r)	Q (i.r)
		T/m3	[%]	[%]	[%]	[%]	[KJ/kg]
PKS01	3,56	0,65	0,9	51,23	20,01	0,03	19,198
PKS02	5,28	0,59	0,7	61,12	21,01	0,02	19,105
PKS03	2,45	0,62	0,6	61,02	18,5	0,01	19,958
PKS04	4,52	0,61	0,6	61,23	20,01	0,03	19,198
PKS05	2,96	0,68	0,5	63,02	21,01	0,02	19,105
PKS06	3,99	0,59	0,4	63,02	18,5	0,01	19,958
SAMPLE AVREGE	3,79	0,62	0,62	60,11	19,84	0,02	19,420,33

Table 16.5 Palm kernel shell pre-processing costs

Biomass	Moisture	Density	Q (i.r)	Purchase price	Energy cost as landed	Energy cost at pre-calciner
	%	T/m^3	[GJ/t]	€/T	€/Gj/m^3	€/Gj/m^3
PKS-Prepo	3,79	0,62	19,42	41,22	3,56	4,77
PKS-Charcoal	1,9	0,92	17,92	31,72	2,56	3,24

2.8 *PKS Cost Analysis*

The Table 16.5 is showing the purchasing price of palm kernel shell that have been pas to the pre-processing statement.

The purchasing price from pre-processing company is 41,22 €/Gj with Table 4 data we estimate the energy cost as landed 3,56 Euro/Gj/m^3 and the energy cost at pre-calciner is 4,77 €/Gj/m^3.

3 Results and Discussion

The conventional separation system for the recovery of palm kernel shell from its sand contamination or palm kernel shell mixture and using water as process media generates a considerable amount of waste effluent that harms the environment.

The aim of using this dry separation processing for the recovery of palm kernel shell by using winnowing columns. A commercial system consisting of a series winnowing columns was developed and installed at a local palm oil mill.

The performance of the separation process was evaluated in terms of its palm kernel shell losses, sand content, palm kernel shell recovery rate operation cost minimization almost 4,12 Euro/t of PKS.

The average moisture losses based on mud palm kernel shell by this drying processed were found to vary from 15,2 to 21,4%, with most of the values obtained being below the targeted limit of 5%. The sand content was in the range 0,59–0,63%, which was mostly below the targeted limit of 1%.

The palm kernel shell recovery rate was in the range 92–95%, with most of the values achieving the minimum targeted limit of 85%. The system operates under completely dry conditions and, therefore, produces few wastes arrow 8%.

The capital expenditure costs of installing equipment for this highly processed PKS do not differ too much from the pre-processed company. However, the cost analyses show a benefit of using a highly-quality pre-processing palm kernel shell and this represents a remarkably lower operating expenditure of clinker production, this increases the kiln performance index to 96% instead of 88% as operation coefficient of the kiln.

This, and the increasing urge to use a higher amount of mostly wet and mud palm kernel shell to substitute the fossil fuel, leads to the trend that clinker production lost was high about 700tclk/day, to face this issue the managements are increasingly looking for technologies that can handle any type or quality of palm kernel shell.

This solution of pre-processing (traditional charcoal) provides flexibility to their pks procurement process and, at the same time, saves on pre-processing cost, regardless of whether our plant does that itself or has it delivered by our external supplier, the fuel cost saving can be estimated to 1,9 €/Gj/m^3.

In many companies like ours, the preferred approach tended towards is putting the penalty with no effort to produce a clean PKS that can easily burn in the pyro-system.

However, as the main motivation for PKS carbonizations' use is reducing cost, Togolese green clinker plant seeks out every opportunity to optimize the balance between our pre-processing and co-processing costs on our own terms. comparing this operation fees impact, we realize that our company make more saving by financing the pre-processing project.

4 Conclusions

The Sustainable Development Goals (SDGs), also known as the Global Goals, were adopted by the United Nations in 2015 as a universal call to action to end poverty, protect the planet, and ensure that by 2030 all people enjoy peace and prosperity this was the aim of Togolese green clinker plant.

The penalty appalment on procurement agreement is limiting the goals of sustainability this let us on another approach of institute or legal framework that can handle this statement, the EU Taxonomy.

The pre-processing and co-processing are related in sustainable goals and on also the plant cost saving with a good energy recovery, putting the penalty in agreement to solve the quality issue is unpowerful.

References

Beguedou, E. N. (2023). Alternative Fuel Substitution Improvements in Low NOx In-Line Calciners. Clean Technologies, 5(2), 713-743.

Favre, C. M. 2016. Emissions control technologies to meet current and future European vehicle emissions legislation. (p. 16.06. 16). URL: www. aecc. eu (дата обращения:).

Henson, K. E.-B. 2018. Cohort profile: prescriptions dispensed in the community linked to the national cancer registry in England. BMJ open, 8(7), e020980.

Holling, B. K. 2013. CO2 recovery from industrial hydrogen facilities and steel production to comply with future European Emission regulations. Energy Procedia, 37, 7221-7230.

MAEP. 2013. 4ème recensement national de l'agriculture 2011–2014 : volume 2 : module de base II : résultats chiffrés détaillés sur l'agriculture togolaise, Juin 2013, 575 p.

Owoyele, B. V. 2014. Traditional oil palm (Elaeis guineensis jacq.) and its medicinal uses: A review. CELLMED, (vol. 4, no 3, p. 15–22.).

Pamungkas, B. C. 2019. Potential of biomass utilization in rotary kiln of nickel processing plant. In IOP Conference Series. Materials Science and Engineering IOP Publishing., Vol. 588, No. 1, p. 012006.

Sridhar, H. B. (2009a). Why do birds participate in mixed-species foraging flocks? A large-scale synthesis. Animal Behaviour, 78(2), 337-347.

Sridhar, M. K. (2009b). Palm oil industry residues. In Biotechnology for agro-industrial residues utilisationSpringer, Dordrecht., 341–355.

Wurman, M. A. (2002, May). Commissioning a 2.2 mt/y cement plant in Midlothian, Conference Record. In IEEE-IAS/PCS Cement Industry Technical Conference (pp. 125–137). Texas.: IEEE.

Uncited References

Biege, N. W., Smith, D. L., & Shenk, R. E, 2001. Strategies for reducing cement kiln baghouse corrosion problems., in Cement Industry Technical Conference. IAS/PCA.

Dembla, A. (2007). Experiences with the design of large size cement plants-Process and layout considerations in pyroprocess systems. 2007 IEEE Cement Industry Technical Conference Record, 114-122.

Ebrahimi, M., Ghomi, S. M. T. F., & Karimi, B., 2020. Application of the preventive maintenance scheduling to increase the equipment reliability: Case study-bag filters in cement factory., Journal of industrial and management optimization, pp. 16(1), 189–205.

Henson, I. 2015. A Brief History of the Oil Palm. Production, Processing, Characterization, and Uses, 1–29.

Sawadogo, M., Tchini Tanoh, S., Sidibé, S., Kpai, N., Tankoano. I., 2018. Cleaner production in Burkina Faso: Case study of fuel briquettes made from cashew industry waste. Journal of Cleaner Production Volume 195, 10 September 2018, Pages 1047–1056

Sharma, P., Sheth, P. N., & Mohapatra, B. N, 2019. Waste-to-Energy: Issues, Challenges, and Opportunities for RDF Utilization in Indian Cement Industry., In Proceedings of the 7th International Conference on Advances in Energy Research, Singapore.

SPRINGER NATURE

GPSR Compliance

The European Union's (EU) General Product Safety Regulation (GPSR) is a set of rules that requires consumer products to be safe and our obligations to ensure this.

If you have any concerns about our products, you can contact us on ProductSafety@springernature.com

In case Publisher is established outside the EU, the EU authorized representative is:

Springer Nature Customer Service Center GmbH
Europaplatz 3
69115 Heidelberg, Germany

The manufacturer's authorised representative in the EU is Springer Nature Customer Service Centre GmbH, Europaplatz 3, 69115 Heidelberg, Germany. If you have any concerns regarding our products, please contact ProductSafety@springernature.com

Printed and bound by CPI Group (UK) Ltd, Croydon, CR0 4YY
23/09/2026
02221546-0009